TOPOLOGY
PRODUCT & QUOTIENT SPACE
AND
CONVERGENCE

TOPOLOGY
PRODUCT & QUOTIENT SPACE AND CONVERGENCE

By

S.C. Sharma

DISCOVERY PUBLISHING HOUSE
NEW DELHI-110002

Reprinted - 2019

First Published - 2006

ISBN: 978-81-8356-120-4

Topology

Product & Quotient Spaces and Convergence

Published by:

DISCOVERY PUBLISHING HOUSE PVT. LTD.

4383/4B, Ansari Road, Darya Ganj

New Delhi-110 002 (India)

Phone: +91-11-23279245, 23253475; 43596065

E-mail: discoverybooksindia@gmail.com

discoverypublishinghouse@gmail.com

web: www.discoverypublishinggroup.com

Printed at:

Infinity Imaging Systems

Delhi

PREFACE

This book "Topology" has been written for the students of various universities. In the earlier chapters, proof are given in considerable detail, as our subject unfolds through the successive chapters and the reader acquires experience in following abstract mathematical arguments, the proof become briefer and minor details are more and more left for the reader to fill in for himself. It is a basic principle in the study of mathematics, and one too seldom emphasised that a proof is not really understood until the stage is reached at which one can grasp it is a whole and see it as a single idea. In achieving this end much more is necessary than merely following the individual steps in the reasoning.

We hope this book will be found useful by the students and teacher in the various universities of India.

Author

Contents

1

PRODUCT AND QUOTIENT SPACE

LATTICE

Definitions: (i) *A partially ordered set A is said to be a* lattice *iff every pair iff of elements in A has infimum and supremum.*

We shall denote the infimum and the supermum of any pair of element x, y of A and x A y and $x \vee y$ respectively.

COMPLETE LATTICE

(ii) *A lattice A is said to be a complete lattice iff every non-empty subset of A has an infimum and a supermum.*

(iii) *A non-empty subset B of a lattice A is said to be sub-lattice iff $x \wedge y$ and $x \vee y$ are members of B for every pair of members, y of B.*

Theorem 1:

Let X be any non-empty set and let $\{X_\lambda, X_\lambda : \in L\}$ be an arbitrary non-empty collection of topological spaces. spaces. For each $l \in L$ let f_λ be mapping of X into X_λ and let $\mathfrak{I}$ be the week topologies v on X determined by $\{f_\lambda : \lambda \in \mathfrak{I}\}$. Than $\mathfrak{I}$ is the intersection of all topologies V on X such that f_λ is v $\mathfrak{I}$ continuous for each $\lambda \in \mathfrak{I}$, that is, $\mathfrak{I}$ is the coarsest topology on X fir which f_1 is continuous for each $l \in \mathfrak{I}$.

Proof :

We first prove that f_l is $\mathfrak{I}$-$\mathfrak{I}_\lambda$ continuous for each $l \in \mathfrak{I}$. Let *G* be any $\mathfrak{I}_\lambda$-open set. Then by definition of week to be log y we have $f_\lambda^{-1}[G] \in \mathfrak{I}$. Hence f_λ is $\mathfrak{I}$-$\mathfrak{I}_\lambda$ continuous.

Now let v be topology on *X* such that fl is v-$\mathfrak{I}_\lambda$ continuous for each $\lambda \in$ L. Then $f_\lambda^{-1}[G] \in v$ for each $G \in \mathfrak{I}_\lambda$ and for each $\lambda \in \Lambda$. Since *v* is *a* topology on *X*, *v* contains all the unions of finite intersections of member of the collection $\{f_\lambda^{-1}[G] : l \in \Lambda$ and $G \in \mathfrak{I}_\lambda\}$. This means that *v* contains $\mathfrak{I}$, that is $\mathfrak{I}$ is weaker than *v*.

Thus, we have shown that $\Im$ is the coarsest topology on X such that f_λ is $\Im$-$\Im_\lambda$ continuous for each $\Im \in \Lambda$. This implies that $\Im$ is the intersection of all topologies v on X such that f_λ is v-$\Im_\lambda$ continuous for each $\lambda \in \Im$

Theorem 2:

Let C be the class of all topologies on a non-empty set X. There C is a complete lattice with respect to the relations $\leq$ defined by setting $\Im 1 \leq \Im 2$ iff $\Im_1$ is weaker than $\Im_2$ Moreover, c has a least member and a greatest member.

Proof:

It is easy to see that C is *a* partially ordered set with respect to the relation "is weaker that". If $\Im_1$, $\ddot{A}_2$ are any two members of C, then the $\Im_1 \wedge \Im_1 \cap \Im_2$ and $\Im_1 \vee \Im_2$ is the intersection of all topologies $\Im_1$ such that $\Im_1 \leq \Im_1$ and $\Im_2 \leq \Im 1$ [Note that the intersection of arbitrary collection of topologies on X is a topology on X]. Hence C is lattice. It is also a complete lattice. For if C′ is any subclass of C then inf C′ is the intersection of all the members of C′ is the intersection of all the members of C′ and sup C′ is the intersection of all those members of C which are stronger than every members of C′.

Further, the indiscrete topology I is the weakest topology on X and the discrete topology D is the strongest typology on X. Thus, I is the least and D is the greatest member of C.

Definition : *Let X be any non-empty set and let $A = \{(X_\lambda, \Im_\lambda) : l \in \Lambda\}$ be any arbitrary non-empty collection of topological spaces. For each $\lambda \in \Lambda$, let f_λ be a mapping of X into X_λ. Then the topology L on X generated by the collection $B = \{f_\lambda^{-1}[G] : l \in \Lambda$ and $G \in \Im_\lambda\}$ is called the weak topology (or the initial topology) on X determined by the collection all finite intersections of members of A is called the defining subbase of $\Im$ and the collection B of all finite intersections of members of A is called the defining base of $\Im$.*

Theorem 3:

Let $(X, \Im)$ and (Y, v) be two topological spaces. Then the collection $E = \{G \times H : G \in \Im$ and $H \in v\}$ *is a* base for some topology for $X \times Y$.

Proof:

We shall show that E satisfies the conditions [B*1] and [B*2].

Since $X \times Y \in E$,

we have $X \times X = \cup \{G \times H : G \in \Im$ and $H \in v\}$

Thus [B*1] is satisfied. Now let $G_1 \times H_1$ and $G_2 \times H_2$ be any two members of E. We then have

$$(G_1 \times H_2) \cap (G_2 \times H_2) = (G_1 \cap G_2) \times (H_1 \cap H_2). \qquad ...(1)$$

Since $\mathfrak{I}$ and v are topologies, we have

$G1 \in$ Ä. $G_2 \in \mathfrak{I}$

$\Rightarrow$ $G_1 \cup G_2 \in \mathfrak{I}$

and $H_1 \in v, H_2 \in v$

$\Rightarrow$ $H_1 \cup H_2 \in v.$

Hence, it follows from (1) that $(G_1 \times H_1) \cup (G_2 \times H_2) \in E$. Thus, we have shown that the intersection of any two members of E us again *a* member of E and so $[B*2]$ is also satisfied. Therefore is a base for some topology for $C \times Y$.

Example 1:

Consider the set N of all natural numbers. This set is partially ordered with respect to the relation ≤ defined by setting a ≤ by iff a divided b. If a, be are any two natural numbers, then a Λ b is the highest common factor (H.C.F >) of a, and b a ∨ b is the least common multiple (L.C.M.) of a and b. Hence (N, ≤) is a lattice. It is however not complete lattice since {a ∈ N : a ≥ 1} has no supermum.

Example 2:

Consider the set R of all real numbers. This set is partially ordered by ≤ in the usual sense. If x, y be any two real numbers, then x ∧ y = min {x, y} and x ∨ y = max {x, y}. Hence R is a lattice. But it is not a complete lattice since {x ∈ R : x ≥} is a subset of R which has no supermum. In fact it has no upper bound.

The set N of all natural number and the set Q of all rational numbers are sublattices of R since

x ∈ N, y ∈ N

$\Rightarrow$ x Λ y ∈ N

and x ∨ y ∈ N

and x ∈ Q, y ∈ Q

$\Rightarrow$ x Λ y ∈ Q

and x ∨ y ∈ Q.

Example 3:

Let P (X) be the power set of a non-empty set X so that P (X) is the collection of all subsets of X. Then P (x) is partially ordered with respect *to the relation two members of F, then $f \Lambda g$ is the real valued function on X defined by $(f \notin g)$ = min {f (x), g(x)} and $f \vee g$ is that defined by $(f \vee g)(x)$ = max {f (x), g(x)}. Hence F is a lattice here.*

PRODUCT OF TWO SPACES

Definition : *Let $(X, \mathfrak{I})$ and (Y,v) be two topological spaces. Then the topology w whose base is $E = \{G \times H : G \in \mathfrak{I}$ and $H\ \mathfrak{I}\ v\}$ is called the Product topology for $X \times Y$ and $(X \times Y, w)$ is called the* product space *of X and Y.*

We know that *E* is actually *a* base for sine tonology *w* of the above definition.

Theorem 1:

The product topology on a non-empty set $X \times Y$ is the weak topology for $X \times Y$ determined by the projection maps p_x and p_y from the topology on X and Y

Proof:

The weak topology has for a subbase

$$\{C_\lambda : G_\lambda = p_x^{-1} [U_\lambda\} \text{ or } G_\lambda$$
$$= p_y^{-1} [V_\lambda]. \text{ for some } U_\lambda \text{ open in } X \text{ or } V_\lambda \text{ open in } Y\}.$$

Then intersection

$$p_x^{-1} [U_l] \cap \ldots \cap p_x^{-1} [U_m] \cap p_y^{-1} [V_l] \cap \ldots \cap p_y^{-1} [V_n]$$
$$= (U_l \times Y) \cap \ldots \cap (U_m \times Y) \cap (X \times V_n) \ldots \cap (X \times V_n) \text{ of a}$$

finite number of such sets has the form

$$(U_l \cap U_2 \cap \ldots \cap U_m) \times (V_l \cap V_2 \cap \ldots \cap Vn)$$
$$= U \times V$$

we have already prove that where *U* is open in *X* and *V* is open in *Y*. Hence weak topology has the same base as the product topology, and so the two topologies are the same

Note that $\pi_x^{-1} [U_1] = U_1 \times Y, \pi_y^{-1} [V_1] = X \times V_1$ etc.].

Theorem 2:

Let $(X. \mathfrak{I})$ and (Y, V) be two topological spaces and B a base for $\mathfrak{I}$ and c a base for v. Then,

$$E = \{B \times C : B \in B \text{ and } C \in c\}$$

is *a* base for the product topology w for $X \times Y$.

Proof:

Let (x, y) be any point of $X \times Y$ and let N be a w-nhd of (x, y). Since $E = \{G \times H : G \in \mathfrak{I}$ and $H \in v)$ is *a* base for w, there exist *a* member $G \subseteq H$ of E such that

$$(x, y) \in G \subseteq H \subset N,$$

Since G is $\mathfrak{I}$-open and B is *a* base for $\mathfrak{I}$, there exists some $B \in B$ such that $x \in B \subset G$. Similarly there exist some $C \subset c$ such that $y \in C \subset H$:. It follows that

$$(x, y) \in B \subseteq C \subset G \times H.$$

Hence, from (1) and (2), we get $(x, y) \in B \times C \subset N$. This implies that E is a base for w.

Example:

Consider the topology

$\mathfrak{I} = \{\varnothing, \{a\}, X\}$ *for* $X = \{a, b, c\}$ *and the topology*

$v = \{\varnothing, \{p\}, \{q\}, \{p, q\}, \{r, s\}, \{p, r, s\}. \{q. r, s\}, Y\}$

for $Y = \{p, q, r, s\}$ and find *a* base for the product topology of $X \times Y$.

Solution :

We know that

$B = \{\{a\}, X\}$ is *a* base for $\mathfrak{I}$ and

$C = \{\{p\}, \{q\}, \{r, s\}\}$ is *a* base for v. Hence *a* base for the product topology is given by

$$E = \{\{a\} \times \{p\}, \{a\} \times \{q\}, \{a\} \times \{r, s\},$$
$$X \times \{p\}, X \times \{q\}, X \times \{r, s\}\},$$

$$\{\{(a, p)\}, \{(a, q)\}, \{a, r\}, (a, s), \{(a, p), (b, p), (c, p)\}\},$$
$$\{(a, p), (b, q), (c, q), (c, q)\}, \{(a, r), (a, s), (b, r), (b, s), (c, r), (c, s)\}\}$$

PRODUCT OF TWO METRIC SPACES

Theorem:

Let (X, d) and (Y, e) be two metric spaces. Define δ by setting $\delta((x_1, y_1), (x_2, y_2)) = \sqrt{\{d^2(x_1, x_2) + e^2(y_1, y_2)}$. Then δ is a metric for $X \times Y$ and the metric topology is the same as the product topology.

Proof:

First we prove that d is a metric for $X \times Y$. Evidently $\delta((x_1, y_1). (x_2, y_2)) \geq 0$ for all $(x_1, y_1), (x_2, y_2) \in X \times Y$

Also $\delta((x_1, y_1), ((x_2, y_2)) = 0$

$\Leftrightarrow$ $d^2(x_1, x_2) + \varepsilon^2(y_1, y_2) = 0$

$\Leftrightarrow$ $d(x_1, x_2) = 0$

and $e(y_1, y_2) = 0$

$\Leftrightarrow$ $x_1 = x_2$ and $y_1 = y_2$

$\Leftrightarrow$ $(x_1, y_1) = (x_2, y_2)$.

Thus [$m1$] and [$m2$] are satisfied

Now $\delta((x_1, y_1), (x_2, y_2)) = \sqrt{\{d_2(x_1, x_2) + e_2(y_1, y_2))\}}$

$= \sqrt{\{d^2(x_2, x_1) + e^2(y_2, y_1)\}}$

$= \delta((x_2, y_2), (x_1, y_1)\}$.

so that [m_3] is also satisfied. Finally we have

$\delta((x_1, y_1), (\acute{x}_2, y_2)\} = \sqrt{\{d^2(x_1, x_2) + e^2(y_1, y_2)\}}$

$\leq [\{d(x_1, x_3) + d(x_3, x_2)\}^2 + \{e(y_1, y_3) + e(y_3, s_2)\}^2]^{1/2}$

[using triangle inequality for d and e]

$\leq [d^2(x_1, x_3) + e^2(y_1, y_3)]^{1/2} + [d^2(x_3, x_2) + e^2(y_3, y_2)]^{1/2}$

[using Minkowski inequality]

$= \delta((x_1, y_1), (x_3, y_3)) + \delta((x_3, y_3), (x_2, y_2))$.

Hence, triangle inequality is also satisfied for δ. Accordingly is a metric for $X \times Y$. We now show that the topology induced by δ is the same as the product topology for $X \times Y$. Let w denote the product topology for $X \times Y$. We must show that every d-open set is w-open and conversely. Let G be any δ open subset of $X \times Y$ and let $(x_0, y_0) \in G$. Then there exists a d-open sphere $S((x_0, y_0), \varepsilon$ with centre at (x_0, y_0) and radius $\varepsilon > 0$ such that $S((x_0, y_0), \varepsilon) \subset G$. Consider

$A = S(x_0, \varepsilon/\sqrt{2})$

and $B = S(y_0, \varepsilon/\sqrt{2})$.

Then A is *a* d-open set containing x_0 and B is an e-open set containing y_0. It follows by the definition of product topology that $A \times B$ is *a* w-open set containing (x_0, y_0). It will now be shown that $A \times B \subset G$. Let (x, y) be any arbitrary member of $A \times B$. Then $x \in A$ and $y \in B$ so that $d(x_0, x) < \varepsilon/\sqrt{2}$ and $\varepsilon(y_0, y) < \varepsilon/\sqrt{}_2$. Hence $\delta[(x_0, y_0), (x, y)]$

$= [d^2(x_0, x) + e^2(y_0, y)]^{1/2} < [(\varepsilon/\sqrt{2})^2 + (\varepsilon/\sqrt{2})^2]^{1/2} = \varepsilon$

It follows that $(x, y) \varepsilon S((x_0, y_0), \varepsilon) \subset G$. Thus we have shown that

$(x, y) \varepsilon A \times B \Rightarrow (x, y) \varepsilon G$.

Thus $(x, y) \varepsilon A \times B \subset G$ we know tnat is a w-open subset of $X \times Y$.

Conversely, let G be any w-open subset of $X \times Y$ and let $(x_0, y_0) \varepsilon G$. By the definition of product topology, there must exist *a* d-open set D and an e-open set E each that

$$(x_0, y_0) \varepsilon D \times E \subset G.$$

Then $x_0 \in D$ and $y_0 \in E$ and so there exist a d-open sphere S

(x_0, ε_1) such that $S(x_0. \varepsilon_1) \subset D$ and e-open sphere $S(y_0, \varepsilon_0)$ such that $S(y_0, \varepsilon_2) \subset E$. Let $\varepsilon = \min\{\varepsilon_1. \varepsilon_2\}$. We claim that

$$S((x_0. y_0), \subset S(x_0, \varepsilon_1) \times S(y_0, \varepsilon_2).$$

Let (x, y) be any arbitrary element of $S((x_0, y_0). \varepsilon)$.

Then, since $\delta((s_0, y_0), (x, y))$

$$= [d^2(x_0, x) + e^2(y_0, y)]^{1/2}, \text{ we have}$$

$$d(x_0, x) \leq \delta((x_0, y_0)) < \varepsilon \leq \varepsilon_1$$

and $\quad e(y_0, y) \leq \delta((x_0, y_0), (s, s)) < \varepsilon \leq \varepsilon_2.$

It follows that

$$(x, y) \in S(x_0, \varepsilon_1) \times S(y_0, \varepsilon_2).$$

Hence, $\quad (x_0, y_0) \in S((x_0, y_0), \varepsilon_1) \times S(y_0. \varepsilon_2).$

But $\quad S(x_0, \varepsilon_1) \times S(y_0, \varepsilon_2) \subset D \times S \subset G$

Therefore $\quad (x_0, y_0) \in S((x_0, y_0), \varepsilon) \subset G$

Thus, we have shown that every point of G is contained in a d-open sphere which is contained in G and consequently G is an open set with respect to the metric δ.

PROJECTION MAPPINGS

Definition: *The mappings*

$$\pi_x : X \times Y \to X : \pi_x((x, y)) = x \not\vee (x, y) \in X \times Y$$

and $\pi_y : X \times Y \to Y : \pi_y((x, y)) = \not\vee \vee (x, y) \in X \times Y$

are called the projection of the product $X \times Y$ on X and Y respectively.

Theorem 1:

Let $(x, \mathfrak{J})$ and $(Y. v)$ be two topological spaces and let $(X \times Y, W)$ be the product space of two spaces. Then the projection px are continuous and open mappings. Further the product topology W is the coarsest topology for which the projections are continuous.

Proof:

Recall that π_x is a mapping of $X \times Y$ into X defined by $\pi_x((x, y)) = x$ for every $(x, y) \in X \times Y$. Let G be any $\mathfrak{I}$-open set. Then it is evident from the definition of π_x that $\pi_x^{-1}[G] = G \times Y$ which is a basic w-open subset of $X \times Y$. [$\because G \in \mathfrak{I}, Y \in v \Rightarrow G \times Y \in$ where E is the base for W]. Hence π_x is a W-$\mathfrak{I}$ continuous mapping. Similarly π_y is W-v continuous mapping. Now let U be any w-open subset of $X \times Y$. Then by the definition of the base E for W, we have

$$U = \cup \{G \times H : G \in \mathfrak{I}, H \in v \text{ and } G \times H \in E \subset\}$$

$$\text{Hence } \pi [U] = \pi_x [\cup \{G \times H : G \in \mathfrak{I}, H \in v \text{ and } G \times G\ E\}$$

$$= \cup \{\pi_x \{G \times H\} : G \in \mathfrak{I}, H \in v \text{ and } G \times G \in E\}$$

$$= \cup \{G : G \in \mathfrak{I} \text{ and } G \times H \in E\}$$

[By the definition of π_x] $\in \mathfrak{I}$.

It follows that π_x is an open mapping. Finally let W be any topology for $X \times Y$ which the projection are continuous and let U be any W-open set. Then

$$U = \cup \{G \times H : G \in \mathfrak{I}, H \in v \text{ and } G \times H \in E'\} \text{ where } E' \subset E,$$

$$= \cup \{(G \cap X) \times (Y \cap H) : G \in \mathfrak{I}, H \in S \text{ and } G \times H \in E'\}$$

$$= \cup \{(G \times Y) \cap (X \times H) : G \in \mathfrak{I}, H \in v \text{ and } G \times H \in E'\}$$

$$= \cup \{\pi_x^{-1}[G] \cap \pi_y^{-1}[H] : G \in \mathfrak{I}, H \in v \text{ and } G \subseteq H \in E'\}$$

$\because$ π_x is W-v continuous $\Rightarrow \pi_y^{-1}[G] \in W$ and

π_y is W-$\mathfrak{I}$ continuous $\Rightarrow \pi_y^{-1}[H] \in W$ etc.

Thus every W-open set is W open and so W is coarser than W. Since W is any topology for $X \times Y$ for which the projections are continuous, it follows that W is the coarsest topology for which projections are continuous.

Theorem 2:

Let y_0 be a fixed element of Y and let $A = X \times \{y_0\}$. Then the restriction of π_x to A is a homeomorphism of the subspace A of $X \times Y$ onto X. Similarly the restriction of π_y to $B = \{\pi_0\} \times Y$, $\pi_0 \in X$ is a homeomorphism.

Proof:

Let g_x denote the restriction of π_x to A, that is, let

$$g_x : \to X : g_x((x, y_0) = x \vee (x, y_0) \in A. \text{ Then}$$

$$g_x((x_1, y_0)) = g_x((x_2, y_0) \Rightarrow x_1 = x_2$$

$$\Rightarrow (x_1, y_0) = (x_2, y_0)$$

$\Rightarrow g_x$ is one-one.

g_x is evidently onto. Since by the preceding theorem π_x is continuous, it follows that g_x, is also continuous. We now show that g_x is open. Let V be any open subset of A. Then $V = A \cap U$ for some open subset U of $X \times Y$. But

$$U = \cup \{G \times H : G \in \mathfrak{I}, H \in v \text{ and } G \times H \in E'\} \text{ where } E' \in E$$

We then have

$$g_x [V] = g_x [A \cap U]$$
$$= g_x [A \cap [\cup\{G \times H : G \in \mathfrak{I}, H \in v \text{ and } G \times H \in E'\}]$$
$$= g_x \{\cup\{A \cup (G \times H) : G \in \mathfrak{I}, H \in v \text{ and } G \times H \in E'\}]$$

[Distributive law]

$$= \cup \{g_x [(X \times \{y_0\}) \cap (G \times H)] : G \in \mathfrak{I}, H \in v \text{ and } G \times H \in E'\}$$
$$= \cup \{g_x [X \cap G) \times (\{y_0\} \cap H)\} : G \in \mathfrak{I}, H \in v$$

and $G \times H \in E'\}$

$$= \cup \{g_x[\times (\{y_0\} \cap H] : G \in \mathfrak{I}, H \in v \text{ and } G \times H \in E'\} \quad ...(1)$$

If $y_0 \notin H$, then it is easy to see from (1) that $g_x [V] = \varnothing$. If $y_0 \in H$, then (1) gives

$$g_x [V] = \cup \{g_x [G \times \{y_0\}] : G \in \mathfrak{I}, H \in v$$

and $G \times H = \in E'\}$

$$= \cup \{G : G \in \mathfrak{I} \text{ and } G \times H \in E'\} \in \mathfrak{I}.$$

This implies that g_x is an open mapping as well. Thus we have shown that g_x is one-one, onto, continuous and an open mapping and consequently it is a homeomorphism.

PRODUCT INVARIANT PROPERTY FOR FINITE PRODUCTS

Theorem 1:

The product spaces $X \times Y$ is connected if and only if X and Y are connected.

Proof:

Assume the $X \times Y$ is connected. Since the projections π_x and π_y are continuous and open mappings, that X and Y are also connected spaces.

Conversely, let X and Y be connected spaces. To show that $X \times Y$ is also connected. Let (x_1, y_1) and (x_2, y_2) be any two points of $X \times Y$. The $\{x_1\} \times Y$ as homeomorphic to Y and $X \times \{y_2\}$ is homeomorphic to X. Hence $\{x_1\} \times Y$ and $X \times \{y_2\}$. They intersect in the point (x_1, x_2) and hence their union is a connected set by theorem. Since this union contains (x_1y_1) and (x_2, y_2), it follows that $X \times Y$ is connected.

Theorem 2:

The product space $X \times Y$ is compact if and only if each of the spaces X and Y is compact.

Proof:

Let $X \times Y$ be compact. Since the projection maps $\pi_x : X \times Y \to X$ and π_y $X \times Y \to Y$ are continuous and onto, we hence already prove it

Conversely, let X and Y be compact spaces. We want to show that $X \times Y$ is compact, it suffices to show that every basic open cover of $X - Y$ has a finite sub-cover. Since a basic open set in $X - Y$ is of the form $G \times H$ where G is open in X and H is open in Y, we may denote a basic open cover by $c = \{G_\lambda \times H_\lambda : 1 \in \Lambda\}$ where G_λ is open in X and H_λ is open in Y. For a given point $x \in X$. the set $\{x\} \times Y$ is homeomorphic to Y is therefore, compact. Since $\{x\} \times Y$,, being a subset of $X \times Y$ is covered by c and $\{x\} \times Y$ is compact, there exist a finite sub-family of c, say

$\{G_\lambda \times H_\lambda : i = 1, 2,..., n\}$, which covers $\{x\} \times X$. Let $\bigcap_{i=0}^{n} G_\lambda = G(x)$. Then $G(x)$ is open in X and contains x since each G_i contains x. Hence $\{G(x) \times H_\lambda,: 1, 2, ..., n\}$ is still a finite open cover of $\{x\} \times Y$.

Proceeding in this manner of each $x \in X$, we contract the collection $\{G(x) : x \in X\}$ of open sets in X which covers X. By compactness of X, there exists a finite sub-cover $\{G(x,) : j = 1, 2..., m\}$ of this cover for X. Since each $G(x_j)$ is an intersection of open sets in X which were used to from c. We may select on open set $G_{\lambda_j} \in c$ such that $G(x)_{\lambda j_j}) \subset G_{\lambda_j}$ for $j = 1, 2,..., m$. Therefore $\{Gx_{\lambda_j} : j = 1, 2, ..., ... m\}$ is a finite open cover of X, and for each j, $1 \leq j \leq m$. $\{G_{\lambda_x} \times H_\lambda : i = 1, 2,...n\}$ covers the subset $G(x_j) \times Y$ of $X \times Y$. By its construction the collection $\{G_{\lambda_{xj}} \times H_\lambda : i = 1, 2, ..., n = 1.2, ... m\}$ is then a finite sub-cover of c for $X \times Y$ and therefore $X \times Y$ is compact.

Theorem 3:

The product space of Hausdorff spaces is Hausdorff.

Proof:

Let X and Y by two Hausdorff spaces. To show that $X \times Y$ is also a Hausdorff space, Let (x_1, y_1) and (x_2, y_2) be any two distinct points of $X \times Y$. Then either $x_1 \circ x_2$ or $y_1 \circ y_2$. Take x_1 x_2. Since X is Hausdorff, there exist open sets G_1 and $G_2 \times X$ such that $x_1 \in G_1$ and $x_2 \in G_2$ and $G_1 \cup G_2 = \emptyset$. Then $G_1 \times Y$ and $G_2 \times X$ are open subsets of $X \times Y$ such that

$(x_1, y_1) \in G_1 \times Y$, $(x_2, y_2) \in G_2 \times Y$ and

$(G_1 \times Y) \cup (G_2 \times Y)$

$= (G_1 \cup G_2) \times Y = \varnothing \times Y = \varnothing$

It follows that X × Y is Hausdorff.

Theorem 4:

The product space of two second countable spaces is second countable.

Proof:

Let X and Y be two second countable spaces. To show that X × Y is also countable. Let B and C be countable bases for X and Y respectively. Consider the collection

$$D = \{B \times C : B \in B, C \in c\}.$$

Then D is surely a countable collection we already know that D is countable base for X × Y.

GENERAL PRODUCT SPACES (TECHONOFF TOPOLOGY)

Recall that if $\{X_\lambda : \lambda \in \Lambda\}$ is an arbitrary collection of sets indexed by L then the **Cartesian product** *of this collection is the set of all mapping x defined by $x : \Lambda \rightarrow \cap \{X_\lambda : \lambda \in \Lambda\}$ such that $x(\lambda) \in X_\lambda$ for all $\lambda \in \Lambda$ and is denoted by $\pi \{x_\lambda : \in \Lambda\}$ or by $\times \{X_\lambda : \lambda \in \Lambda\}$. The set X_λ is called the λth* **coordinate** *set of the product.*

it is customary to use the symbol x_λ for the image $x(\lambda)$ of λ; under the mapping x. Thus if $X = \cap \{X_\lambda : \in \Lambda\}$.

Then $\times \{X_\lambda : \lambda \in \Lambda\}$

$= \{x : x : \Lambda \rightarrow X,$

where $x(\lambda) \in X_\lambda \ \forall \lambda \in \Lambda\}$.

If one of the sets is empty then $\times \{X_\lambda : \lambda \in \Lambda\}$ is empty. Also very often we shall denote the mapping x by $\{x_\lambda : \lambda \in \Lambda\}$ where $x_\lambda \in X_\lambda$, x_λ is called the λth coordinate of x. If $X_\lambda = X$ for all $\lambda \in \Lambda$, then for $\times \{X_\lambda : \lambda \in \Lambda\}$, we write X^Λ Thus X^Λ is the set of all mappings on Λ into X and is called the Λth power of X. For any set X. The set of all mapping on the empty set ∅ into X consists of ∅ alone. Thus $X^\varnothing = \{\varnothing\}$.

Definition: *Let $X \rightarrow \{X_\lambda : \lambda \in \mathfrak{J}\}$. Then the mapping $x_\lambda : X \rightarrow X_\lambda$ defined by $\pi_\lambda(x) = x_\lambda$ for all $x \in X$ is called the λ projection.*

Thus π_λ is the mapping on X which carries each point x of the product onto its λth coordinate x_λ.

Example:

Describe the set π_λ^{-1} $[A_\lambda]$ where $A_\lambda \subset X_\lambda$ in words.

Solution:

π_λ^{-1} $[A_\lambda]$ is the set of all x in $\times$ $[X_\lambda : \lambda \in \Lambda]$ whose λ^{th} coordinates are members of A_λ. Equivalently. π_λ^{-1} $[A_\lambda]$ is the set of all those members of $\times$ $\{X_\lambda : \lambda \in \Lambda\}$ whose λ^{th} coordinates are restricted to lie in A_λ and all other coordinates are unrestricted in the respective coordinate sets. Symbolically, π_λ^{-1} $[A_\lambda] = \{x : x \in X$ and $x_\lambda = \in A_\lambda\}$ where $X = \times$ $[X_\lambda : \lambda \in \Lambda\}$.

Note that π_α^{-1} $[A_\alpha] = \times$ $\{Y_\lambda : \lambda \in \Lambda\}$ where $Y_\alpha = A_\alpha$ and $Y_\lambda = X_\lambda$ whenever $\lambda \neq \alpha$.

EVALUATION MAP

Definition : *For each λ in L let f_λ be a mapping of a set X into a set Y_λ. Then the mapping $e : X \rightarrow \times$ $\{Y_\lambda : \lambda \in \Lambda\}$ defined in such a manner that for every $x \in X$ (e, (x)) (λ) $= f_\lambda$ (x) for all $\lambda \in \Lambda$ is called the evaluation up of X into the cartesian product $\times$ $\{Y_\lambda : \lambda \in \Lambda\}$ determined by the family $\{f_\lambda : \lambda \in \Lambda\}$.*

Observe that e (x) is an element of the cartesian product $\times$ $\{X_\lambda : \lambda \in \Lambda\}$ and so by the definition e (x) is itself a mapping of L into $\cup$ $\{\Psi\lambda : \Lambda\}$ such that (e (x) (l)) $\in Y_\lambda$ for all $\lambda \in \Lambda$. Also f_λ is a pre-assigned mapping of X into Y_λ so that f_λ (x) $\in Y_\lambda$. It is this pre-assigned element f_λ (x) of Y_λ which we take to be equal to (e, (x)) (l). In short, the lth coordinate of e (x) is taken to be f_λ (x). That is why the map e is said to be determined by the family $\{f_\lambda : \lambda \in \Lambda\}$.

Thus $\qquad$ e (x) $= \{f_\lambda$ (x) :. $\lambda \in \Lambda\}$ for all $x \in X$ $\qquad$ ...(1)

and (π_λ o e) (x) $= \pi_\lambda$ (e (x) = (e (x)) (λ) $= f_\lambda$ (x) so that

π_λ o e $= f_\lambda$ fol all $\lambda \in \Lambda$. $\qquad$ ...(2)

Remark : *If e' is a mapping of X into $\times$ $\{Y_\lambda : \lambda \in \Lambda\}$ such that π_λ o e' for all $\lambda \in \Lambda$, then e' = e.*

Proof:

Suppose, if possible, that e $\neq$ e. Then e' (x) $\neq$ e (x) for at least one x $\in$ X. Let. e' (x) $= \{y_\lambda$ $\lambda : \in \Lambda\}$ where $y_\lambda \in Y_\lambda$. Also by definition of e, we have e (x) $= \{(f_\lambda$ (x) : $\lambda \in \Lambda\}$.

Since e' (x) $\neq$ e (x) there is at lest one $\lambda \in \Lambda$; say λ_0, such that $y_{\lambda_0} \neq$ (x). Hence for this particular λ_0, we have

$(\pi_{\lambda_0} \text{ o } e')(x) = \pi_{\lambda_0}(e'(x)) = \pi_{\lambda_0}((y_\lambda : \lambda \in \Im\}\} = \pi_{\lambda_0} \neq (x)$

which is a contradiction since π_λ o $e' = f_\lambda$ for all $\lambda \in \Lambda$.

(b) *f is a mapping X into* $\times$ *{Yl : L ∈ L}, then f is the evaluation of X into* $\times \{Y_\lambda : \lambda \in \Im\}$ *determined by they family* $(\pi_\lambda\ 0\ f : \lambda \in \Lambda\}$.

Proof:

Since f (x) is an element of the product $\times \{Y_\lambda : \lambda \in \Lambda\}$, it follows by definition that f (x) is a mapping of $\Im$ into $\cup \{Y_\lambda : \in \Lambda\}$ such that $(f(x))(\lambda) \in Y_\lambda$ for all $\lambda \in \Lambda$. Let

$$(f(x))(\lambda) = y_\lambda \text{ where } y_\lambda \in Y_\lambda. \quad ...(1)$$

We may then write $f(x) = \{y_\lambda : \lambda \in \Lambda\}$.

Now $\quad (\pi_\lambda \text{ o } f)(x) = \pi_\lambda(f(x)) = (\{y_\lambda : \lambda \in \Lambda\}) = y_\lambda \quad ...(2)$

From (1) and (2), we have $(f(x))(\lambda) = (\pi_\lambda : \text{o } f)(x)$.

Hence f is the evaluation of X into $\times \{Y_\lambda : \lambda \in \Lambda\}$, determined by the family $\{\pi_\lambda \text{ o } f : l \in \Lambda\}$.

Definition (a) : *The family* $F = \{f_\lambda : \lambda \in \Im\}$ *defined, is said to distinguish points or to separate points of X iff, for every pair of distinct elements* x_1 *and* x_2 *of X, there exists a*$\lambda \in L$ *such that* $f_\lambda(x_1) \neq f_\lambda(x_2)$.

(b) *F is said to* **Distinguish Points and Closed Sets** *iff for every closed subset H of X and each point* $x \in X - H$,*, there exists some* f_λ *in F such that* $f_\lambda(x) \notin (f_\lambda[H])$.

Theorem:

The family $\{f_\lambda : \lambda \in \Lambda\}$ *distinguishes points of X iff the evaluation map of X into* $\times \{Y_\lambda : \lambda \in \Lambda\}$ *determined by* $\{f_\lambda : \lambda \in \Lambda\}$ is one-one.

Proof :

If $\{f_\lambda : \lambda \in \Lambda\}$ distinguishes points of X. Then $x_\lambda \neq x_2 \Rightarrow f_\lambda(x_\lambda)\ ^1\ f_\lambda(x_2)$ for some $\lambda \in \Lambda \Rightarrow e(x_1) \neq e(x_2)$.

[$\because$ e (x_1) and e (x_2) differ at least in one coordinate].

Hence e is one- one

Conversely, let e be one-one. Then

$$x_1 \neq x_2 \Rightarrow e(x_1) \neq e(x_2)$$
$$\Rightarrow f_\lambda(x_1) \neq f_\lambda f(x_2) \text{ for some } \lambda \in \Im$$
$$\Rightarrow \{f_\lambda : \lambda \in \Lambda\} \text{ distinguishes points of X.}$$

DIAGONAL MAP

Definition : *Let Λ be any non-empty set. Then the mapping $d : X \to X^{L}$ defined in such a manner that for every $x \in X$, $(d(x))(\lambda) = x$ for every $\lambda \in \Lambda$ is called the diagonal map of a set X into the Λth power X^{Λ} of X.*

Since d (x) is an element of $X^{\wedge}$, it follows from the definition of product that d (x) is a mapping of Λ into X so that $(d(x))(\lambda) \in X$ for all $\lambda \in \Lambda$. In the definition of d, we take this element $(d(x))(\lambda)$ of X equal to the pre-assigned element x of X for all $\lambda \in \Lambda$. Thus d (x) is a constant functions of Λ into X.

Theorem:

The diagonal map d is one-one.

Proof :

$d(x) = d(y) \Rightarrow d$ is one-one.

The subset d [X] of X^{3} is called the diagonal of X^{Λ}. The diagonal of X^{Λ} is the set of all constant functions of Λ into X. Also d is a one-one mapping of X onto d [X] and for each $\lambda \in \Lambda$, the restriction of the projection π_{λ} of X^{Λ} into X to d [X] is the inverse of d.

EMBEDDING

Definition. *By an embedding of a topological space X into another space Y, we mean a map $f : X \to Y$ which defines a homeomorphism of X onto $f[X]$.*

Example:

If X is a subspace of a space Y, then the inclusion map $i : X \to Y$ defined by $i(x) = f$, for every $x \in X$ is an embedding.

PRODUCT TOPOLOGY (OR TYCHONOFF TOPOLOGY)

Definition : *For each λ in an arbitrary index set Λ, let $(X_{\lambda}, \mathfrak{I}_{l})$ be a topological space and let $X = \times \{X_{\lambda} : \lambda \in \Lambda\}$. Then topology $\mathfrak{I}$ for X which has subbase the collection.*

$$B = \{\pi_{\lambda}^{-1}[G_{\lambda}] : \lambda \in \mathfrak{I}, G_{\lambda} \in \mathfrak{I}_{\lambda}\}$$

is called the product topology (or the Tychonoff Topology) for X. and $(X, \mathfrak{I})$ is called the product space of the given spaces.

Note that here π_{λ} denotes as usual the lth projection. The collection B, is called the defining subbase for $\mathfrak{I}$.

The collection B of all finite intersections of elements of B, would then from a base for $\mathfrak{I}$..

In particular, if we have a countable collection of topological spaces $\{X_1, X_2, X_3, \ldots\}$ the product space $X = \times \{X_n : n \in N\}$ consists of all sequences $x = \{x_1, x_2, x_3, \ldots\}$ where $x_n \in X_n$

Moreover, $\pi_a^{-1} [G_a] = X_1 \times X_2 \times \ldots X_{a-1} \times X_a \times X_a + 1 \times \ldots$

Theorem 1:

Let $\{(X_\lambda, \Im_\lambda) : \lambda \in L\}$ be an arbitrary collection of topological space and let $X = \times \{X_\lambda : \lambda \in \Lambda\}$. Let $\Im$ be a topology for X. Then the following statements are equivalent:

(a) *$\Im$ is the product topology for X.*

(b) *$\Im$ is the smallest topology for which the projections are continuous.*

Proof:

(a) ⇒ (b) : Let π_λ be the λth projection maps and let G_λ be any $\Im_\lambda$-open subset of Xl. Then by (a), $\pi\lambda^{-1} [G_\lambda]$ is a member of the subbase for $\Im$ and hence $\pi_\lambda^{-1} [G_\lambda]$ must be $\Im$-open. It follows that $\pi\lambda$ is $\Im$-$\Im$ continuous. Now let v be any topology on X such that $\pi\lambda$ is v-$\Im_\lambda$ continuous for each $\lambda \in \Lambda$. Then $\pi^{-1} [G_\lambda]$ is open for every $G_\lambda \in \Im_\lambda$. Since V is a topology for X, V contains all the unions of finite intersections of members of the collection

$$\{\pi_\lambda^{-1} [G_\lambda] : \lambda \in \Im \text{ and } G_\lambda \in \Im_\lambda\}.$$

This implies that V contains $\Im$, that is, $\Im$ is coarser than v. It follows that $\Im$ is the smallest topology for X such that π_λ is $\Im - \Im_\lambda$ continuous for each $\lambda \in \Im$.

(b) ⇒ (a) : Let B. Be the collection of all sets of the form $\pi^{-1} [G_\lambda]$ where G_λ is an open subset of X_λ for $\lambda \in L$. The topology v for X will make all the projections π_λ continuous iff B, $\subset$ v.

We know that the smallest topology for X which makes all the projections continuous is the topology determined by B, as a subbase. That is it is the product topology for X.

Note : Throughout this chapter, whenever we refer to the product topology, we shall mean the topology generated by the collection of all sets of the from $\pi_\lambda^{-1} [G_\lambda]$ where $\lambda \in \Im$ and $G_\lambda \in \Im_\lambda$.

Note that $\pi_a^{-1} [G_a]$ consists of all points $x = \{x_\lambda : \lambda \in \Lambda\}$ in X such that $\pi_a (x) \in G_a$. In other words $\pi_a^{-1} [G_a] = \times \{Y_\lambda : \lambda \in \Lambda\}$ where $Y_a = G_a$ and $Y_\lambda = X_\lambda$ whenever $\lambda \neq \alpha$. Λ base member for this topology is a set B of the from $\cap \{\pi_\lambda^{-1} [G_\lambda] : \lambda \in \Im\}$ where Λ' is a finite subset of Λ and $G_\lambda \in \Im_\lambda$ for some $\lambda \in \Lambda$. Observe that $B = \times \{Y_\lambda : \lambda \in \Lambda\}$ where for each $\lambda \in \Lambda$, $Y_\lambda \in \Im_\lambda$ and $Y_\lambda = X_\lambda$ for all but a finite number of λ' s, It follows that every open set in the product space X must contain a set (a base member) of in

Λ. Thus all but a finite number of coordinates of points of a non-empty open subset of the product space are unrestruct d in the respective coordinate spaces. Hence if L is infinite then a set H of the form $\times \{G_\lambda : \lambda \in \Lambda\}$ is surely not open in the product space if G_λ is an open product subset of X_λ for every $\lambda \in L$ since all the points H are restricted to lie in the respective G_λ' s. It is easy to see that the collection c of all sets of the from $\times \{G_\lambda : \lambda \in \Lambda\}$, where $G_\lambda \subset \mathfrak{I}_\lambda$, from each $\lambda \in \Lambda$, is a base for some topology $\mathfrak{I}^*$ for the product set $\times \{X_\lambda : \lambda \in \Lambda\}$, but as seen above this topology $\mathfrak{I}^*$ is not in general, the same as the product topology $\mathfrak{I}$. The reader may ask. "why is the definition adopted for the cartersian product of two spaces, is not extended to the product of an arbitrary collection of spaces". The reason simply is that the topology $\mathfrak{I}$ as mentioned above is not so useful to the mathematical as the topology $\mathfrak{I}$ as defined, since $\mathfrak{I}$ does not yield desired theorems and results.

Theorem 2:

Let $\{(X_\lambda, \mathfrak{I}_\lambda) : \lambda \in \Lambda\}$ be an arbitrary collection of topological spaces and let $X = \times \{X_\lambda : \lambda \in \Lambda\}$. Show that the collection c of all sets of the form $\times \{G_\lambda : \lambda \in \Lambda\}$ where $G_\lambda \in \mathfrak{I}_\lambda$ for each $\lambda \in \Lambda$, is a base for some topology for X

Proof:

We know that c satisfies the conditions [B*1] and [B*2]

[B*1] : Let $x \in X$ so that $x = \{x_\lambda : \lambda \in \Lambda\}$ where $x_\lambda \in X_\lambda$. Them the there exists a $\mathfrak{I}_\lambda$ open set G_λ (which may be X_λ) such that $x_\lambda \in G_\lambda$.

Hence, x is an element of a set of the form $\times \{G_\lambda : \lambda \in \Lambda\} = G$ say. Thus to each $x \in X$, there exists a member G of x such that $x \in G$. It follows that

$$X = \cup \{G : G \in c\}$$

[B*2] Let $G \in c$ and $G' \in c$. Then $G = \times \{G_\lambda : \lambda \in \Lambda\}$

and $G' = \times \{G_\lambda : \lambda \in \Lambda\}$

where $G_\lambda \in \mathfrak{I}_\lambda$ and $G_\lambda' \in \mathfrak{I}_\lambda$ for every $\lambda \in \Lambda$. Now

$$G \cap G' = (\times \{G_\lambda : \lambda \in \mathfrak{I}\}) \cap (\times \{G_\lambda' : \lambda \in \Lambda\}) \quad ...(1)$$

$$= \times \{G_\lambda \cup G_\lambda' : \lambda \in \Lambda\}$$

Since $\mathfrak{I}_\lambda$; is a topology for X_λ, we have

$$G_\lambda \in \mathfrak{I}_\lambda' \text{ and } G_\lambda' \in \mathfrak{I}_\lambda \Rightarrow G_\lambda \cap G_\lambda' \in \mathfrak{I}_\lambda.$$

It follows from (1) that $G \cap G' \in c$.

Thus [B*2] is also satisfied.

Example:

Let B be a non-empty member of the base B for the product topology. Show that π_λ [B] = X_λ for all except a finite number of λ' s. Deduce that if G is a non-empty open subset of the product space X, then π_λ [G] = X, for all except a finite number of λ' s.

Solution:

Now B = × {Y_λ : $\lambda \in \Lambda$} where $Y_\lambda \in \Im_\lambda$ and $Y_\lambda = X_\lambda$ for all but to finite number of λ's. Hence by the definition of the projection map π_λ, we have π_λ [B] = Y_1 where $Y_\lambda = X_\lambda$ for all except a finite contains a non empty open subset G of the product space functions a non-empty member B of the base B, we also have π_λ [G] = X_λ for all except finite number of λ' s.

Theorem 3:

Each projection map π_λ is an open map.

Proof:

Let $\alpha \in \Lambda$ and let B be any member of the base B for the $Y_\lambda \in \Im_\lambda$ and $Y_\lambda = X_\lambda$ and $Y_\lambda = X_\lambda$ for all but a finite number of λ' s. So that each Y_λ is an open subset of X_λ. Now by definition of π_α, we have pa {B] = Y_α which is open in X_α. Thus the image under π_α of every basic open subset of X is open. Now let G be any open subset of X. Then G is the union of collection λ of base members of the form B. Hence.

π_α [G] = π_α[$\cup$ {A : A $\in \Lambda$}] = $\cup$ {π_α [A] : A $\in \Lambda$}.

As shown above π_α [A] is an open subset of X_α for each A $\in \Lambda$ and hence their union π_α [G] must also be open in X_α. Thus we have shown that image under π_α of every $\Im$-open set G is $\Im_\alpha$-open. It follows that π_α is an open mapping.

Theorem 4:

Let f be a mapping of a space Y into product space X = × {Xl : l $\in \Im$}. Then f is continuous iff the composition pl o f : Y $\rightarrow X_\lambda$ is continuous.

Proof:

Let f be continuous. Since all projections are continuous. We know that π_λ of is also continuous.

Conversely, let each composition π_1 of be continuous and ant U be any member of the defining subbase B, of the product space X. Then U=π_λ^{-1} [G] for some $\lambda \in \Im$ and some G $\in \Im_1$.

Also $\quad f^{-1}$[U] = f^{-1} {π_λ^{-1}[G]] = (π_λ of –1[G].

Since π_λ of is continuous, it follows that (π_λ) f^{-1} {G} = f^{-1} {U} is open in Y. Thus, we have shown that the inverse image under f of every sub-basic open set in the product space X is open in Y. It follows that f is continuous.

Theorem 5:

Let X be the non-empty product space $X [X_\lambda : \lambda \in \Lambda\}$. Then a non-empty product subset $F = \{F_\lambda : \times \{F_\lambda : \lambda \in \Lambda\}$ is closed in X if and only if each F_λ is closed is X_λ.

Proof:

Let F_λ be closed in X_λ for every $\lambda \in \Im$. Since the projection p_λ is continuous for each $\lambda \in \Im$. p_λ^{-1} $[F_\lambda]$ is closed in X. It is easy to see that

$$F = \cup \{p_\lambda^{-1} [F_\lambda] : \lambda \in \Lambda\}.$$

It follows that F is closed in X, being an intersection of closed sets.

Conversely, let $F = \times \{F \in \Lambda\}$ be closed in X. To show that each F_λ is closed in X_λ. Let $\mu \in \Lambda$ be arbitrary. We shall show that F_μ is closed. Let z_μ be any limit point of F_μ in X_μ. Consider the point z where $\pi_\mu (z) = z_\mu$ and $p_\lambda (z)$ is any element of F_λ for $\lambda \neq \mu$. Let G be any basic open set for the product topology containing z. Then $\pi_\mu [G]$ is open and contains z_μ. Since z_μ is a limit point of F_μ, $F_\lambda [G]$ must contain a point x_μ of F_μ different from z_μ. Therefore, G contains the point x where $\pi_\lambda (x) = \pi_\lambda (z)$ for $\lambda \neq \mu$ and $\pi_\mu (x) = x$. Evidently $x \in F$. Also since and z differ in μ^{th} coordinate, we have $x \neq z$. Thus we have shown that every basic open set containing z contains a point of F different from z. Hence, z is a limit point of F. Since F is closed in X, $z \in F$ which implies that $\pi_\mu (z) \in \pi_\mu [F]$. Since of F. $\pi_\mu (z) = z$ and $\pi_\mu [F] = F$ we have $z_\mu \in F$. Thus F contains all its limit points and so Fm is closed. Since μ was arbitrary, we see that F_λ is closed for every $\lambda \in \Lambda$.

Theorem 6:

Let $\times (X_\lambda : \lambda \in L\}$ be a product space and $A_\lambda \subset X_\lambda$ for every $l \in L$.

Then $\overline{\times \{A_\lambda : \lambda \in \Lambda\}} = \times \{\overline{A_\lambda} : \lambda \in \Lambda\}$

Proof:

By theorem (above), $\times \{A_\lambda : \lambda \in \Lambda\}$ is closed since each $\overline{A_\lambda}$ is closed so that $\overline{\times \{A_\lambda : \lambda \in \Lambda\}} = \times \{\overline{A_\lambda} : \lambda \in \Lambda\}$

Now $\times \{A_\lambda : \lambda \in \Im\} \subset \times \{\overline{A_\lambda} : \lambda \in \Lambda\}$.

$\Rightarrow \times \{\overline{\times \{A_\lambda : \lambda \in \Lambda\}} \subset \overline{\times \{A_\lambda : \lambda \in \Lambda\}} = \times \{\overline{A_\lambda} : \lambda \in \Lambda\}$ ...(1)

If for some λ, $A_\lambda = \varnothing$,

then $\times \overline{\{A_\lambda : \lambda \in \Lambda\}} = \times \{\overline{A_\lambda} : \lambda \in \Lambda\} = \varnothing$.

Now assume that $G_\lambda \neq \varnothing$, for all λ and let $x = \{x_\lambda : \lambda \in \Im\}$ be any element of $\times \{\overline{A} : \lambda \in \Lambda\}$ so that $x_\lambda \in \overline{A_\lambda}$ for every $\lambda \in \Lambda$ and therefore x_λ is an adherent point of A_λ. Let G be any basic open set for the product topology containing x. Then $G = \times \{G_\lambda : \lambda \in L\}$ where G_λ is open in X_λ and $G_\lambda = X_\lambda$ for all our a finite number of λ's.

Let $G_\lambda = X_\lambda$ for $\lambda \neq \lambda_1, \lambda_2 \ldots, \lambda_3$ where n is finite. Since $x_\lambda \in \overline{A}$ for every $\lambda \in \Im$, $x_{\lambda_2} \in G_{\lambda_3}$ implies G_{λ_2} contains a point of AG_{λ_2}, for k = 1, 2, ..., n, by the definition of adherent point. Also for $\lambda \neq \lambda_1 \ldots$, ln, $G_\lambda = X_\lambda$ so that $A_\lambda \subset G_\lambda$ in this case.

It follows that G contains a point of $\times \{A_\lambda : \lambda \in \Im\}$. Thus we have shown that every basic open set containing x contains a point of $\times \{A_\lambda : \lambda \in \Im\}$ and therefore

$$x \in \times \{A_\lambda : \lambda \in \Im\}$$

It follows that,

$$\times \{\overline{A} : \lambda \in \Lambda\} \subset \times \overline{\{A_\lambda : \lambda \in \Lambda\}} \quad \ldots(2)$$

From (1) and (2),

$$\times \overline{\{A_\lambda : \lambda \in \Lambda\}} = \times \{\overline{A_\lambda} : \lambda \in \Lambda\}$$

Theorem 7:

Let B, be a sub-base for a topology on a space Y and let s be sequence in Y. Also let x ∈ Y. If s is eventually in each member of B. Containing x, then s converges to x.

Proof:

Let N be a nhd of x. Then there exists a basic nhd $M_1 \cap M_2 \cap \ldots M_k$ of x, such that

$$x \in M_1 \cap M_2 \cap \ldots \cap M_k \subset N.$$

where $M_i \in B$, for i = 1, 2, ... k.

Since s is eventually in each M_i for each i = 1, 2, ... k, there exists a positive integer m_i such that s (n) ∈ M, for n > m.

If we set m = max $\{m_1 \cdot m_2 \ldots m_1\}$.

We have s (n) $\in M_1 \cap M_2 \cap \ldots \cap M_k \subset N$ for every $n \geq m$.

Thus we have shown that S is eventually in each nhd of x. It follows that s coverages to x.

Theorem 8:

Let s be a sequence in the product space $X = X\{X_2 : \lambda \in A\}$. Then s converges to a point x in X iff π_λ as converges to $\pi_\lambda(x)$ (= x) for each $\lambda \in A$.

Proof:

Let the sequence s converge to $x \in X$. Since each projection π_λ is continuous. It sequence π_λ os converges to π_λ (x) = x_λ.

Conversely, let the sequence π_λ os) *i.e.* the sequence π_λ (s_λ). π_λ ((s_2). (s_3)... converge to π_λ (x) (= x_λ) for every $\lambda \in \Lambda$, nd let π_λ^{-1} [G] be any member of the sub-base B, of the product space X such that $x \in \pi_\alpha^{-1}$ [G], where G $\in \Im_\alpha$. Then $\pi_\alpha(x) \in G$. That is. $x_\alpha \in G$. Also by hypothesis. π_λ os converges to π_λ (ξ) (= x_α). Hence π_α os is eventually in every nhd of x_α. In particular pa os is eventually in G. So there exists a positive integer m such that

$$(\pi_\alpha \text{ os})(n) = \pi_\alpha(s(n)) \in G \Rightarrow s(n) \in \pi_\alpha^{-1}[G]$$

for every $n \geq m$. Thus s is eventually in π_α^{-1} [G]. Since π_α^{-1} [G] was chosen to be an arbitrary member of B, containing x, it converges to x.

PRODUCT INVARIANT PROPERTIES

Theorem 1:

The product space $X = \times \{X_\lambda : \lambda \in \Lambda\}$ is T_1 iff each coordinate space X_λ is T_1.

Proof:

The 'iff' part : Let each Coordinate, space X_λ be T_1 and let $x = \{x_\lambda : \lambda \in \Lambda\}$ be any element of X. The $x_\lambda \in X_\lambda$ for every $\lambda \in \Lambda$. Since X_λ is T_1, (x_λ) is closed in X_λ for every $\lambda \in \Lambda$. Since projection map π_λ is continuous, it follows that π_λ^{-1} [{x_λ{] is closed in X for every $\lambda \in \Lambda$. Hence their intersection $\cap$ (π_λ^{-1} [{x_λ}] is also closed in X. But evidently $\cap\, \pi_\lambda^{-1}$ [{x_μ}] = {x}

{For $z \in \cap$ {x_λ^{-1} [{x_λ{] $\Rightarrow z \in \pi_\lambda^{-1}$ [{x_λ}] $\forall\, \lambda \in \Lambda$

$\Rightarrow$ $z_\lambda = x_\lambda \;\forall\, \lambda \in \Lambda \Rightarrow z = x$]

it follows that {x} is closed in X. Thus we have shown that every singleton subset of X is closed. Hence X is a T_1-space.

The 'only if' part : Let the product space X be T_1. Let $m \in \Lambda$ be arbitrary. We shall show that X_μ is T_1. Let a_μ and b_μ be any two distinct points of X_μ. Choose x and y such that x and y differ only in the μth coordinate and such that $x_\mu = a_\mu$ and $y_\mu = b_\mu$. Since X is T_1, there exists open sets G and H in X such that $x \in G$ but $y \notin G$ and $y \in H$ but $x \notin H$. Hence, by definition of base, there exists a basic open set $U = \times \{U_\lambda : \lambda \in \Lambda\}$ such that $x \in U \subset G$ but $y \notin U$. Similarly there exists a basic open set $V = \times (V_\lambda : \lambda \in \Lambda\}$ which contains y but not x. It follows that U_μ is open in X_μ and contains $x_\mu = a_\mu$ but not b_μ. Similarly V_μ is open in X_μ and contains b_μ but not a_μ. Hence X_μ is T_1. [Note that $U_\lambda = X_\lambda$ and $V_\lambda = X_\lambda$ for all but a finite number of λ' s but this is immaterial here].

Alternative proof of the 'only if' part : Let the product space X be T_1. We show that each coordinate space is homeomorphic to a certain subspace of X. Let $a = \{a_\lambda : \lambda \in \Lambda\}$ be a fixed element in X and for each μ consider the subset Am of X consisting of all points $\{x_\lambda : \lambda \in \Lambda\}$ such that $x_\lambda = a_\lambda$ if $\lambda \neq \mu$ while x_μ may be any point of X_λ. Let g_μ be the restriction of x_μ to A_μ. Then g_μ is a continuous mapping of A_μ onto X_μ. Also it is evident by the construction of A_μ that g_μ is one-one. We now show g_μ is open. Let B_μ be any basic open set B in X so that $B = \times \{G_\lambda : \lambda \in \Lambda\}$ where $G_\mu \in \mathfrak{I}_\lambda$ [Note that $G_\lambda = X$ for all but a finite number of λ's but this is immaterical here]. But $B_\mu = A_\mu \cap B$ is either empty or consists of those points of A_μ for which $x_\mu \in G_\mu$. Then image of B_μ under G_μ under Gm is either empty or G_μ and so open in either case. Now let U_μ be any open set in A_μ. Then U_μ is the union of a collection of base members. Hence (v), $g_\mu (U_\mu)$ is the union of images under g_μ of members of that collection and since such image is shown to be open, it follows that gm (U_μ) is an open set and consequently g_μ is an open mapping. Thus g_μ is homeomorphism of g_μ onto X_μ. But Am onto X_μ. But A_μ is T_1 .Hence its homeomorphic image X_μ is also T_1.

Theorem 2:

The product space $X = \times \{X_\lambda : \times \lambda \in L\}$ and y $\{y_\lambda : \lambda \in \mathfrak{I}\}$ be two distinct points of the product space X. Then $x_\mu \neq y_\mu$ for some $\mu \in \mathfrak{I}$ where $x_\mu \in X_\mu$ and y_μ. Since X_μ is Hausdorff,

Proof:

The 'if part' : Let each coordinate space X_λ,, be T_2 and let $x = \{x_\lambda : \lambda \in \Lambda\}$ and $y = \{y_\lambda : \lambda \in \Lambda\}$ be two distinct points of the product space X. Then x, $\neq y_\mu$ for some $\mu \in \Lambda$ where $x_\mu \in X_\mu$ and $y_\mu \in X_\mu$. Since X_μ is Hausdorff, there exist open sets G_μ and G_μ in X_μ such that

$$x_\mu \in G_\mu,\ y_\mu\ H_\mu \text{ and } G_\mu \cap H_\mu = \varnothing. \qquad ...(1)$$

Since $\pi_\mu (x) = x_\mu$ and $\pi_\mu (y) = y_\mu$, the relations (1) imply that $x \in \pi_\mu^{-1} [G_\mu]$, $y \in \pi_\mu^{-1} [H_\mu]$

and $\quad \pi_\mu^{-1} [G_\mu \cap H_\mu] = \pi_\mu^{-1} [\varnothing] = \varnothing$

or $\quad \pi_\mu^{-1} [G_\mu] \cap \pi_\mu^{-1} [H] = \varnothing.$

But π_μ^{-1} $[G_\mu]$ and π_μ^{-1} $[H_\mu]$ are open in X, being sub basic members of the product topology. Thus we have shown that to each pair x, y of distinct points of X, there exist two disjoint open sets one containing x and the other containing y. It follows that the product space X is Hausdorff.

The 'only if' part : Let the product space X be Hausdroff. We shall show that the coordinate space X_μ is Hausdorff for arbitrary $\mu \in L$. Let a_μ and b_μ be any two distinct points of X_μ. Choose x and y in X such that x and y differ only in the μth coordinate and such that $x_\mu = a_\mu$ and $y_\mu = b_\mu$. Since X is Hausdorff, there exist open sets G and H in X such that $x \in G$ and $y \in H$ and $G \cap H = \emptyset$. There exist basic open sets $U = \times \{U_\lambda : \lambda \in L\}$ and $V = \times \{V_\lambda : \lambda \in \mathfrak{I}\}$ such that $x \in U \subset H$ and $U \cap V = \emptyset$. It follows that U_μ and V_μ are open sets in X that $x_\mu = a_\mu \in U_\mu$ and $y_\mu = b_\mu \in V_\mu$ and $U_\mu \cap V_\mu = \emptyset$. Hence X_μ is Hausdorff.

Theorem 3:

Let $(X_\lambda : \mathfrak{I}l) : \lambda \in \Lambda\}$ be an arbitrary collection of topological sauces and let $(X, \mathfrak{I})$ be their product space where $X = \times \{X_\lambda : \lambda \in \Lambda\}$. Then $(X. \mathfrak{I})$ is first countable iff each of the coordinate spaces is first countable and all but a countable number of the coordinate spaces are indiscrete.

Proof:

The 'if' part. Let (X_λ, A_λ) be first countable for every $\lambda \in L$ and let all but a countable number of spaces (X_λ, A_λ) be indiscrete. Let $\Lambda' = \{\lambda : \lambda \in \Lambda$ and X_λ is not indiscrete$\}$. Then by hypothesis, Λ' is countable subset of Λ. Let $x \in X$. Then $x = \{x_\lambda : \lambda \in \mathfrak{I}\}$ where $x_\lambda \in X_\lambda$. Since each coordinate space X_λ is first countable, there exists a countable local base in x_λ for every $\in \Lambda$. Let X_λ be countable local base at x_λ. Since for every $\lambda \in \Lambda - \Lambda'$. X_λ is indiscrete. We have $B_\lambda = (X_\lambda)$ for each such λ. For every fixed $\lambda \in N$, let c_λ be the countable collection

$$\{\pi_\lambda^{-1} [G] : G \in B_\lambda\}. \text{ Let}$$

$$c = \{H : H \in c_\lambda \text{ for } \lambda \in \mathfrak{I}\} \cup \{c_\lambda : \lambda \in \Lambda\}$$

Then c is a countable collection, being a union of a countable collection of countable collections, Hence there are only a countable number of the finite subcollections of c. Then D is countable. It is easy to see that D is a local base at x. Hence X is first countable.

The 'only if part. Let $(X, \mathfrak{I})$ be first countable and let $x_\lambda \in X_\lambda$. Also let $x = \{x_\lambda : \lambda \in \Lambda\}$ where $x_\lambda \in X_\lambda$.

Then $x \in X$. Since is first countable, there exists a countable local base at x. say $E = \{E_n : n \in N\}$ where $E_n \in \mathfrak{I}$.

Theorem 4:

The product space $\times \{X_\lambda : l \in L\}$ is a second axiom space iff each coordinate space X_λ is second axiom and all but a countable number are indiscrete.

Proof:

Do yourself.

Theorem 5:

Let $\{X_\lambda : \lambda \in \Lambda\}$ be a non-empty collection of topological spaces. Then the non-empty product space $X = \times \{X_\lambda : \lambda \in L\}$ is regular if and only if each coordinate space X_λ is regular.

Proof:

Let each X_λ be regular. Let $x = \{x_\lambda : \lambda \in \sigma \Lambda\}$ be any point of the product space X and G any open set in X containing x. Then there exists a basic open set V in X such that $x \in V \subset G$. Now $V = \times \{V_\lambda : \lambda \in \Lambda\}$ where V_λ is open in X_λ. [Note that $V_\lambda = X_\lambda$ for all but a finite number of λ's but this is immaterial here]. Since each X_λ is regular and V_λ is an open set in X_λ containing x_λ, there exists an open set U_λ in X_λ. Such that $x_\lambda \in U_\lambda$ and $\overline{U_\lambda} \subset V_\lambda$. Let $W = \times \{U_\lambda : {}_\lambda \in L\}$. Then W is open in X and contains x. Also $\overline{W} = \times \overline{\{U_\lambda : \lambda \in \Lambda\}} = \times \{\overline{U_\lambda} : \lambda \in \Lambda\}$. Further, since $U_\mu \subset V_\mu$ for every λ, we have $\overline{W} = \times \{\overline{U_\lambda} : \lambda \in \Lambda\} \subset \times \{V_\lambda : \lambda \in V\} = V$. Thus, we have shown that for every point $x \in X$ and every open set G containing x there exists an open set W in X such that $x \in W$ and $\overline{W} \subset V \subset G$. It follows that X is regular.

Conversely, let X be a non-empty regular product space $\times \{X_\lambda : \lambda \in \Lambda\}$. Let $\mu \in \Im$ be arbitrary. We shall show that Xm is regular. Let am be any point of X_μ and G_μ any open set in X_μ containing a_μ.

Since X is non-empty, there is an element $x = \{x_\lambda : \lambda \in \Lambda\}$ in X such that x_μ. Since X is non-empty, there is an element $x = \{x_\lambda : \lambda \in \Lambda\}$ in X such that $x_\mu = a_\mu$. Let $G = \pi_\mu^{-1}[G]$. Then $x \in G$ and G is open in X (since π_μ is continuous).

Since X is a regular space, there exists a basic open set $V = \times \{V_\lambda : \lambda \in \Lambda\}$ such that $x \in V$ and $\overline{V} \subset G$. It follows that $x_\lambda = a_\lambda \in V_\mu$. It $\times \{V_\mu : \lambda \in \Im\} = \times \overline{\{V_\lambda : \lambda \in \Lambda\}}$. Hence $\overline{V} = \times \{\overline{V_\lambda} : \lambda \in \Lambda\} \subset G$ which implies that $\overline{V_\mu} \subset G_\mu$. Thus we have shown that to each am $\in X_\mu$ and exists an open set V_μ in X_μ each open set G_μ in X_μ containing a_μ, there exists an open set V_μ in X_μ such that $a_\mu \in V_\mu$ and $\overline{V_\mu} \subset G_\mu$. Hence X_μ is regular.

Corollary : *A non-empty product space* $\times \{X_\lambda : \lambda \in \Im\}$ *is a* T_3 *space if each coordinate space* X_λ *is* T_3.

Proof:

Before proving the main theorem we prove the following lemma.

Lemma : *Let B, be a subspace of topological space (X, ℑ). If for each U in B. and each* $p \in U$ *there is a continuous map of X into [0, 1] such that* $f(p) = 0$ *and* $f[X - U] \subset \{1\}$*, then X is completely regular.*

Proof of the Lemma : Let $p \in X$ and let G be a ℑ-open set containing p. Then (unless G = X which is a trivial case), there exist members $U_1, U_2, \ldots, U_n$ of B, such that

$$p \in U_1 \cap U_2 \cap \ldots \cap U_n \subset G.$$

[∵ every nhd of p must contain a basic nhd of p].

Then by hypothesis for each i = 1, 2,..., n, there is a continuous map.

$f_i : X \to [0, 1]$ such that $f_i(p) = 0$ and $f_i[X - U_3] = \{1\}$.

Now we define a map.

$f : X \to [0. 1]$

such that hypothesis for each i = 1, 2, ..., n}. Evidently f (p) = 0

and $f[X - G] = \{1\}$.

We now show that f is continuous as follows, For n = 2, we have

$$f(x) = f(x)\frac{|f_1(x) + f_2(x)|}{2} + \frac{|f_1(x) - f_2(x)|}{2} \quad \ldots(1)$$

It can be easily prove that the continuity of f_1, f_2 implies the continuity of $f_1 + f_2$ defined by $(f_1 + f_2)(x) = f_1(x) + f_2(x)$ and that the continuity of any function g implies the continuity of | g | defined by | g | (x) = | g (x). It then follows from (1) that f is a continuous map in case n = 2. Hence by induction f is continuous. It follows that X is completely regular.

Proof of the Main Theorem : The 'if' part. Let each coordinate space X_λ be completely regular. To prove that the product space is also completely regular. To prove this we shall use the above lemma. Let U be any member of the usual subbase for the product topology and let $x = \{x_\lambda : \lambda \in \Im\} \in U$. Then $U = \pi_\lambda^{-1}[G]$ here G is an open subset of X_λ containing x_λ for some $\lambda \in \Im$. Since X_λ is completely regular, there exists a continuous mapping

$f : X_\lambda \to [0, 1]$

such that $f(x_\lambda) = 0$ and $f(z) = 1$ if $z \in X_\lambda - G$. Since the projections are continuous of X that $f \circ \pi_\lambda$ is continuous mapping of X into [0, 1] such that

$$(f \circ \pi_\lambda)(x) = f(\pi_\lambda(x)) = \begin{cases} 0 \text{ if } x \in U \\ 1 \text{ if } x \in X - U \end{cases}$$

$$[\because x \in U = \pi_\lambda^{-1}[G] \Rightarrow \pi_\lambda(x) \in \pi_\lambda(\pi_\lambda^{-1}[G]) = G$$
$$\Rightarrow f(\pi_\lambda(x)) = 0 \text{ by def. off}$$
$$\text{and } x \in X \;\; U \Rightarrow x \in \pi_\lambda^{-1}[X_\lambda] - \pi_\lambda^{-1}[G]$$
$$\Rightarrow x \Rightarrow \pi_\lambda^{-1}[X_\lambda - G]$$
$$\Rightarrow \pi_\lambda(x) \in X_\lambda - G$$
$$\Rightarrow f(\pi_\lambda(x)) = 1 \text{ [by def. off]}$$

It then follows from the above lemma that the product space X is completely regular.

The 'only if part' : Since complete regularity is both a topological and a hereditary property, we advise the reader to write his own proof just as in the alternative proof of the 'only if', functions (f, g) we mean the function h defined by

$$h(x) = \begin{cases} f(x) & \text{if } f(x) \geq g(x) \\ -f(x) & \text{if } f(x) < g(x). \end{cases}$$

and min (f, g) defined likewise. The following identities are then evident.

$$\max \quad (f, g) = \frac{|f+g|}{2} + \frac{|f-g|}{2}$$

$$\text{and} \quad \min(f, g) = \frac{|f+g|}{2} - \frac{|f-g|}{2}.$$

This result can evidently be extended to any finite set of functions.

Theorem 6:

The product space $X = \times \{X_\lambda : \lambda \in \mathfrak{I}\}$ is a Tychonoff space iff each coordinate space X_λ is a Tychonoff space.

Proof:

Since a Tychonoff space is a completely regular T_1 space the proof.

Theorem 7:

The product space $X = \times \{X_\lambda : \lambda \in L\}$ is connected iff each coordinate space X_λ is connected.

Proof :

The 'if part' : Let each coordinate space X_λ be connected and a = $\{a_\lambda : \lambda \in \Lambda\}$ be a fixed point of X. Let C be the component of X to which a belongs. We shall prove that every point of X belongs to C. Let $\times \{G_\lambda : \lambda \in L\}$ be an arbitrary basic open set containing a point $x = \{x_\lambda : \lambda \in \Lambda\}$ of X where G_λ is open in X_λ and $G_\lambda = X_\lambda$ if $\lambda \neq \mu_1, \mu_2, ..., \mu_n$. The $A_{\mu_1} \times A_{\mu_2} \times ... \times A_{\mu_n}$, (that is, the set of all points $\{z_\lambda : \lambda \in \Im\}$ such that $z_\lambda = a_\lambda$ if $\lambda \neq \mu_1, \mu_2, ..., \mu_n$) is homeomorphic to $X_{\mu_1} \times X_{\mu_2} \times ... \times X_{\mu_n}$. But $X_{\mu_1} \times X_{\mu_2} \times ... \times X_{\mu_n}$ is connected by theorem. Hence its homeomorphic image $A_{\mu_1} \times A_{\mu_2} \times ... \times A_{\mu_n} = A$, say, is connected. Since by definition C is maximal connected subset of X, it follows that $A \subset C$. But the set A contains the point $\{z_\lambda : \lambda \in \Im\}$ for wish $z_\lambda = a_\lambda$ if $\lambda \neq \mu_1, \mu_2, ... \mu_n$ and $z_\lambda = x_\lambda$ for $\lambda = \mu_1, \mu_2, ..., \mu_n$. This point $\{z_\lambda : \lambda \in \Lambda\}$, however, lies in $\times \{G_\lambda : \lambda \in \Lambda\}$, which was an arbitrary basic open set containing $\{x_\lambda : \lambda \in \Lambda\}$. It follows that the point $x = \{x_\lambda : \lambda \in \Lambda\}$ is in the closure of C. Since we have shown that each basic nhd of x contains a point of A and hence a points of C. But C is closed so that $C = \bar{C}$. Hence $\{X_l, \lambda \in \Lambda\} \in C$. Thus every point of X belongs to C so that $X \subset C$. But surely $C \subset X$. Hence $X = C$. But by definition C is a connected set. It follows that X is connected.

Theorem 8:

The product space $X = \times \{X_\lambda : \lambda \in \Im\}$ is locally connected iff each coordinate space X_λ is locally connected and all but a finite number are connected.

Proof:

The 'if' part : Let the coordinate space X_λ be locally connected for every $\lambda \in L$ and connected for $\lambda \neq \mu_1, \mu_2, ... \mu_n$. To prove that the product space is locally connected. Let $x = \{x_\lambda : \lambda \in \Lambda\}$ be an arbitrary point of X and let x be contained in an arbitrary basic open set $G = \times \{G_\lambda : \lambda \in \Lambda\}$ where G_λ is open in X_λ and $G_\lambda = X_\lambda$ for $\lambda \neq \mu_1^*, \mu_2^*, ..., \mu_m^*$. Since each X_λ is locally connected and $x_\lambda \in G_\lambda$, there must exist a connected open set X_λ in X_λ such that $x_\lambda \in C_\lambda \subset X_\lambda$. Now consider the subset $H = \times \{H_\lambda : \lambda \in \Lambda\}$ of G where $H_\lambda = C_\lambda$ if $\lambda \neq \mu_1, \mu_2, ... \mu_n \mu_1^*, \mu_2^* ..., \mu_m$ and $H_\lambda = X_\lambda$ otherwise so that each H_λ is connected. It follows by the preceding theorem that H is connected. Also H is evidently a basic open set containing x and contained in G. Thus we have shown that to each basic open set G containing an arbitrary point x of X, there exists a basic open connected set H such that $x \in H \subset G$. It follows that the product space X is connected.

The 'only if' Part : Let the product space X be locally connected and let x_μ be an arbitrary point of X_μ contained in some open set G_λ in X_λ.

Consider the point $y = \{y_\lambda : \lambda \in \Lambda\}$ with $y_\mu = x_\mu$. Then evidently $y \in \pi_\mu^{-1} [G]$ which is an open subset of X (since the projections are continuous). By local connectedness of X, there must exist a connected open set C such that $y \in C \subset \pi^{-1}[G]$. Then

$$\pi_\mu(y) \in \pi_\mu[C] \subset \pi[\pi_\mu^{-1}[G_\mu]].$$

That is, $x_\mu = y_\mu \in \pi_\mu[C] \subset G_\mu$.

Since the projections are open mapping, $\pi_\mu[G]$ is an open subset of X_μ. Again since the projections are continuous $\pi_\mu[G]$ is connected. Thus we have shown that every open set G_μ in X_μ containing an arbitrary point x_μ contains a connected open set $\pi_\mu[C]$ containing. It follows that Xm is locally connected.

Now let a be an arbitrary point of the product space X. By local connectedness of X. There exists a connected open set C such that $y \in C$. Then there must exist basic open set

$$G = \times \{G_\lambda : \lambda \in \Lambda\}$$

where $G_\lambda \in \Im_\lambda$ for every $\lambda \in \Lambda$ and $G_\lambda = X_\lambda$ for all but a finite number of values of λ such that $y \in G \subset C$. Then $\pi_\lambda[G] = X_\lambda$ for all but a finite number of values of and consequently $\pi_\lambda[C] = X_\lambda$ for all but a finite number of l's. But $\pi_\lambda[C]$ is connected, being a continuous image of a connected set. Thus all but a since number of coordinate spaces are connected.

TYCHONOFF THEOREM

We now prove the most celebrated theorem of general topology known as. Tychonoff theorem. This theorem on the product of compact spaces is undoubtedly the most useful theorem on compactness. While looking forward for the proof, we first prove some preliminary theorems.

Definition : *Let X be a topological space. We say that a collection c of subsets of X is short iff c does not cover X. and we say that c is finitely short iff no finite subfamily of c covers X.*

Lemma 1:

Let X be a topological space and let M be a maximal finitely short family of open sets. If some member of zm contains $G_1 \cap G_2 \cap \dots \cap G_n$, where each G_1 is open then $G_m \in M$ for some M where $1 \leq m \leq n$.

Proof:

We shall prove the contrapositive of the lemma for n = 2. Suppose $G_1 \notin M$ and $G_2 \notin M$ where G_1, G_2 are open subsets of X. Then by the maximality of M, there exists members $E_1, E_2 \dots E_r$ of M such that

$$E_1 \cup E_2 \cup E_3 \cup \dots \cup E_3 = X. \qquad \dots(1)$$

Hence no open set containing G_1 belongs to M. Similarly, there are members $F_1, F_2, \dots, F_s$ of M such that

$$F_1 \cup F_2 \cup F_3 \cup \ldots \cup F_s = X \qquad \ldots(2)$$

From (1) and (2), it is easy to see that

$$(G_1 \cap G_2) \subset E_1 \cup E_2 \cup \ \ldots \cup E_r \cup \ F_1 \cup F_2 \cup \ldots \cup F_s = X \ldots(3)$$

Since M is finitely short, it follows from (3) that no member of M can contain $G_1 \cup G_2$. Thus, we have proved the contrapositive of the lemma for n = 2. Hence by finite induction, lemma is true for any positive integer n.

Lemma 2:

Let (X, $\mathfrak{I}$) be a topological space and let F be a finitely short subfamily of $\mathfrak{I}$. Then there exists a maximal finitely short subfamily. M of $\mathfrak{I}$ such that $F \subset M$.

Proof:

Let C be the class of all finitely short subfamilies of $\mathfrak{I}$, and let c be ordered by the inclusion relation $\subset$. Then $F \in c$. Now the singleton {F} is a trivial simply ordered set. Hence by.

Hausdorff Maximal Principle there exists a maximal simply ordered subclass D of c such that $\{F\} \subset D$, that is $F \in F$. Let $M = \cup D$. Then M is a collection of open sets since each member of D is such a collection. It will now be shown that

(i) M is finitely short and

(ii) M is a maximal finitely short subfamily of $\mathfrak{I}$.

To prove (i), let $A_1, A_2, \ldots, A_n \in M$. Then, for every i in N_k, there exists some c_j in D such that $A_i \in c_i$. Since D is simply ordered, one of these c_i' s, say c_j, contains each of the other c_i' s, and so $A_i \in c_j$ for each $i \in N_k$. Since c_j is finitely short, it follows that $A_1 \cup A_2 \cup \ldots \cup A_k \neq X$. Hence M is finitely short.

We now prove (ii). Suppose that (ii) does not hold, that is, suppose that there is some open set $G \notin M$ such that $M \cup \{G\}$ is still finitely short. Since M contains each member of D, $D \cup \{M \cup \{G\}\}$ would be simply ordered. Also since $G \notin M = \cup D$, $D \cup \{M \cup \ \{\{G\}\}$ will properly contain D. But this contradicts the maximality of D.

Hence (ii) must hold. Also $F \subset M$ since $f \in D$ and $M = \cup D$. Thus M is the required maximal finitely short subfamily of $\mathfrak{I}$ such that $F \subset N$.

Theorem 1: (Alexander, Subbase Lemma):

A topological space (X, $\mathfrak{I}$) is compact iff every subbasic open cover for X has a finite subcover.

Proof:

The 'only if' part follows immediately from the definition of compactness. For, if the space is compact, every open cover for X has a finite subcover and so in particular every sub-basic open cover must have a finite subcover.

We now prove the contrapositive of the 'if part', that is, we shall show that each finitely short subfamily of $\Im$ is short. Let F be any finitely short subfamily M of $\Im$-such that $F \subset M$. It will now be shown that M is short. Let B. Be a subbase for $\Im$. Since M is finitely short, the subfamily $M \cap B$. of M is also finitely short. Hence, by hypothesis, $M \cap B$, is short. This implies that $\cup (M \cap B,) \neq X$.

Now suppose $x \in \cup M$. Then $x \in$ for some $A \in M$. Since A is open and B, is a subbase

$$x \in B_1 \cap B_2 \cap \ldots \cap B_k \subset A.$$

for some B_i's in E, By lemma (5.13), $B_i \in M$ for some $j \in \{1, 2, \ldots k)$. Hence $x \in B_j$. But $B_j \in B$, and $B_j \in M$ implies that $B_j \in M \cap B$. It follows that $x \in \cap M \Rightarrow x \in \cup (M \cap B.)$.

Therefore $\cup M \subset \cap (M \cap B.)$. It then follows that M is short which certainly implies that F is short. Hence X is compact.

Another Statement : The above lemma can be stated contrapositively as follows : X is compact if easy finitely short subfamily of subbasic members is short.

Theorem 2: (Tychonoff):

Let $\{(X_\lambda, \Im_\lambda) : \lambda \in \Im\}$ be a collection of topological spaces and let $X = \times \{X_\lambda : \lambda \in \Im\}$. Then X is compact relative to the product topology iff each space X_λ is compact.

Proof:

Let X be compact. Since each projection π_λ, is a continuous mapping of X into X_λ. It follows that X_λ is a continuous image of the compact space X and hence each X_λ is compact.

To prove that converse, we shall make use of Alexander Subbase Lemma. Let B, the defining subbase for the product topology on X. that is, let

$$B = \{\pi_\lambda^{-1} [G_y] : \lambda \in \Im \text{ and } G_\lambda \in \Im_\lambda\}.$$

Then by Alexander subbase lemma, the space X will be compact if each finitely short subfamily A of B, is short, that is, if each subfamily A of B, such that no finite sub-family of A covers X fails to cover X. For each index $l \in \Im$, let B_λ be the family of open sets G_λ in X_λ such that $\{\pi^{-1} [G_\lambda], \in A\}$.

Let $B_\lambda = \{G_\lambda : G_\lambda \in \Im_\lambda, \pi_\lambda^{-1}[G] \in A\}$. Then B_λ is finitely short in X_λ otherwise A would not be finitely short in X. A Since X_λ is compact B_λ does not cover X_λ. This implies that there exist a point $x_\lambda \in X_\lambda$ such that $x_\lambda \notin G_\lambda$ for any $G_\lambda \in B_\lambda$. It follows that the point x whose λ^{th} coordinate is x_λ, belong to no member of A and hence A is not a cover of X, that is, A is short. Consequently, X is compact.

Theorem 3: (Generalized Heine Borel Theorem):

A subset of Euclidean n space is compact iff it is closed and bounded.

Proof:

Let A be a compact subset of Euclidean space E_n. Since E_n is a Hausdorff space (being a metric space), A is a closed subset of E_n. Since A is compact, it can be covered by a finite collection of follows that A is bounded.

Conversely, let A be a closed and bounded subset of E_n. Then π_i [A] is bounded in R for each i = 1, 2, ... n, there exists a closed interval li in R such that π_i [A] $\subset \lambda_i$. Now each I_i is compact. Hence by the preceding theorem, $\times \{I_i : i = 1, 2, \ldots n)$ is a compact subset of En (with due consideration of relative topologies and the relation between compact subset and compact subspaces). We now show that A is a closed subset of $\times \{I_i : i = 1, 2, \ldots n\}$. Let $A = A_1 \times A_2 \times \ldots \times \ldots A_n$ where each A_i is a bounded subset of R.

Then $\quad \pi_i[A] = A_i \subset I_i \qquad [\because \pi_i[A] \subset I_i]$.

Therefore $\quad A = A_1 \times A_2 \times \ldots \times A_n \subset I_1 \times I_2 \times \ldots I_n$.

Thus we have show that A is a closed subset of a compact space $\times \{I_i : i \in N_n\}$ and consequently A is compact.

TYCHONOFF CUBES

Theorem 1:

For every λ in an index set L, let X_λ, denote the closed unit interval

$$Q = \{x : x \in R, 0 \le x \le 1\} = [0, 1].$$

*with the relativized usual topology. Then the product space $\times \{X_\lambda : \lambda \in \Im\}$ is called a **Tychonoff cube,** and may be denoted by Q^N,. If L = N, then Q^N is called the **Hilbert cube.***

Theorem 2:

Every Tychonoff cube is normal.

Proof:

Let $X = \times \{X_\lambda : \lambda \in \Lambda\}$ where each X_λ is the closed unit interval Q = [0, 1] with the relativized usual topology so that X is a Tychonoff cube.

Follows that each X_λ is Hausdroff and consequently the product space X is also Hausdorff each X_λ is compact and hence by Tychonoff theorem the product space X is also compact. Thus, the Tychonoff cube is a compact. Hausdorff space and consequently, X is normal.

Theorem 3:

A non empty product space $X = \times \{X_\lambda : \lambda \in \mathfrak{I}\}$ where each Xl is a Hausdorff space is locally compact if and only if :

(i) each Xl is locally compact, and

(ii) all but a finitely many X_λ are compact.

Proof:

Let X be locally compact. Since each X_λ is a Hausdorff space and each projection π_λ is continuous and open such that $\pi_\lambda [X] = X_\lambda$. It we know that each X_λ is locally compact. This proves (i). We now prove (ii). Since X is locally compact, for $x \in X$ there exists a nhd N of x in X such that $G \subset N \subset \bar{N}$ and hence $\pi_\lambda [G] \subset \pi_\lambda [\bar{N}]$. But $\pi_\lambda [G] = X_\lambda$ for $\lambda \neq \lambda_1, \lambda_2 ... \lambda_n$. Thus $X_\lambda \subset \pi_\lambda [\bar{N}]$. But $\pi_\lambda [\bar{N}] \subset X_\lambda$ always. Hence $\pi_\lambda [\bar{N}] = X_\lambda$. Since N is compact and π_λ is continuous, it we know that $\pi_\lambda [\bar{N}] = X_\lambda$ is compact for $\lambda \neq \lambda_1$, λ_2,...λ_n,. Thus (ii) is also proved.

Conversely, let (i) and (ii) hold. We shall prove that X is locally compact. Let x be any point of X. By (ii). X_λ is compact for $\lambda \neq \underline{\lambda_1}, \lambda_1, \lambda_2 ... \lambda_n$ say. By (i), $\pi_\lambda (x) \in X_\lambda$ has an open nhd U_{λ_i} in X_{λ_i} such that $\bar{U}_\lambda$ is compact for i = 1, 2, ... n. Also X_λ is given to be compact for $\lambda \neq \lambda_1, \lambda_2 ... ,\lambda_n$. Now $U = \times \{U_\lambda : \lambda \in \Lambda\}$ where $U_\lambda = X_\lambda$ for $\lambda \neq \lambda_1, \lambda_2 ... \lambda_n$ is an open nhd of x in X. Also, $\overline{U} = \times \{\bar{U}_\lambda : \lambda \in \Lambda\}$ which is compact by Tychonoff theorem. Thus we have shown that each point $x \in X$ has a nhd whose closure is compact. Hence X is locally compact.

EMBEDDING THEOREMS AND METRIAZABILITY

Lemma 1:

For each λ in a set $\mathfrak{I}$ let f_λ be a mapping of a topological space X into a topological space Y_λ. Then the evaluation map e of X into the product space $\times \{Y_\lambda : \lambda \in \Lambda\}$ is continuous iff f_λ is continuous for each l in $\mathfrak{I}$.

Proof:

Let f_λ be continuous for each $\lambda \in L$ and let $\pi_\lambda^{-1} [G]$ be any subbasic open subset of the product space $\times \{Y_\lambda : \lambda \in \Lambda\}$ where G is open in Y.

$$\pi_\lambda \circ e = f_\lambda \text{ for all } \lambda \in \Lambda.$$

Hence $e^{-1}[\pi_\lambda^{-1}(G)] = (e^{-1} 0\, \pi_\lambda^{-1})[G] = \{\pi_\lambda \text{ o } e\}^{-1}[G]\, f_\lambda^{-1}[G]$ ...(1)

Since f_λ is continuous $f_\lambda^{-1}[G]$ is open in X. It follows from (1) that $e^{-1}[\pi_\lambda^{-1}(G)]$ is open in X. Thus we have shown that the inverse image under e^{-1} of every subasic open set of the product space $\times\{Y_\lambda : \lambda \in \Lambda\}$ is open in X. It follows that e is continuous. Thus arrives the 'if' part of theorem. Conversely, let the evaluation map e be continuous. By definition of the product space each projection π_λ of the product space $\times\{Y_\lambda : \lambda \in L\}$ onto the coordinate space Y_λ is continuous. It follows that the composition map π_λ o e is continuous. But π_λ o e = f_λ. Accordingly f_λ is continuous for every $\lambda \in \Lambda$.

Lemma 2:

For each $\lambda \in \Lambda$, let f_λ be a mapping of a topological space X into a topological space Y_λ. Then the evaluation map e of X into the product space $\times\{Y_\lambda : \lambda \in L\}$ is one-one iff the family $F = \{f_\lambda : \lambda \in \Im\}$ distinguishes points of X.

Proof:

Let e be one-one and let x and y be distinct points of X. Then

$x \neq y \Rightarrow e(y)$ [$\because$ e is one-one]

$\Rightarrow \quad \{f_\lambda(x)\, L : \in L\} \neq \{f_\lambda(y) : \lambda \in \Lambda\}$

$\Rightarrow \quad f_{\lambda_0}(x) \neq f_{\lambda_0}(y)$ for some $\lambda_0 \in \Lambda\}$.

It follows that the family $\{f_\lambda : \lambda \in L\}$ distinguishes points of X.

Conversely let $\{f_\lambda : \lambda \in \Im\}$ distinguish points of X, Then for each pair of distinct points x and y of X, there exists a $\lambda_0 \in \Im$ such that f_{λ_0}

$(x) \neq f_\lambda(y)$.

Thus $\quad x \neq y \Rightarrow f_{\lambda_0}$

$(x) \neq f_{\lambda_0}(y)$ for some $\lambda_0 \in \Lambda$

$\Rightarrow \quad \{f_\lambda(x) : \lambda \in \Lambda\} \neq \{f_\lambda(y) : \lambda \in \Lambda\}$

$\Rightarrow \quad e(x) \neq e(y)$.

This shows that e is one-one.

Lemma 3:

For each $\lambda \in \Lambda$, let f_λ be a mapping of a topological space X into a topological space Y_λ. Then the evaluation map e of X into the product space $Y = \times\{Y_\lambda \times \in \Lambda\}$ is an open map, of X onto the subspace e [X] of Y if the family $F = \{f_\lambda : \lambda \in \Lambda\}$ distinguishes points and closed sets.

Proof:

Let G be an open subset of X. Let $y \in e[G]$. Then there is some $x \in G$ such that $e(x) = y$. Let $G = X - G$. Then H is a closed subset of X such that $x \in X - H = G$. Since F distinguishes points and closed sets, there exists $f_\lambda \in F$ such that $f_\lambda(x) \notin \overline{(f_\lambda[H])}$ in Y_λ.

Let $V = Y_\lambda - (f_\lambda[H])$. Since $\overline{(f_\lambda[H])}$ is a closed subset of Y_λ, it follows that V is an open subset of Y_λ. Hence $\pi_1^{-1}[V]$ is open in Y being a subbasic member of Y. Let $W = e[X] \cap \pi_\lambda^{-1}[V]$. We clam that W is open set in the subspace e [X] such that $e(x) = y \in W$ and $W \subset e[G]$. For we have, $x \in X \Rightarrow e(x) \in e(X)$

and $\quad f_\lambda(x) \notin \overline{(f_\lambda[H])} \Rightarrow f_\lambda(x) \in Y_\lambda - \overline{f_\lambda[H]} = V$

$\Rightarrow \quad x \in f_\lambda^{-1}[V] = (\pi_\lambda oe)^{-1}[V] = e^{-1}\pi_\lambda^{-2}[V]]$

$\Rightarrow \quad e(x) \in \pi_\lambda^{-1}[V]$.

Thus $\quad e(x) \in e[X] \cap \pi_\lambda^{-1}[V]$

Further $V = Y = f_\lambda[H] \Rightarrow V \subset Y_\lambda - f_\lambda[H]$

$\Rightarrow \quad \pi_\lambda^{-1}[V] \subset \pi_\lambda^{-1}[Y_\lambda] - \pi_\lambda^{-1}[f_\lambda[H]] = Y - \pi_\lambda^{-1}[\pi_\lambda oe)[H]]$

$\Rightarrow \quad e[X] \cap \pi_\lambda^{-1}[V] \subset e[X] \cap [Y - e[H]] = e[X] - e[H]$

$[\because e[H] \subset e[X] \subset Y]$

$\Rightarrow \quad W \subset e[X\ H] = e[G] \qquad (\because e[X] - e[H] \subset e[X - H])$.

It follows the e [G] is an open subset of e [X]. Hence e is an open map of X onto e [X].

Theorem 1:

For each $\lambda \in L$, let f_λ be a continuous mapping of a topological space X into another space Y_λ If the family $F = \{F_\lambda : \lambda \in \Lambda\}$ distinguishes points as well as points and closed sets, then the evaluation map e of X into the product space $Y = \times\{Y_\lambda : \lambda \in \Lambda\}$ is a homeomorphism of X into the subspace [X] of Y. That is , the evaluation map e is embedding.

Proof:

Corollary : *If in addition, X is T_1-space then the result of the above theorem holds if F distinguishes points and closed sets.*

This follows form the fact that in a T_1-space singleton sets are closed.

Theorem 2: (Tychonoff Embedding Theorem):

Every Tychonoff space X can be embedded as a subspace of a cube.

Proof:

Recall that a Tychonoff space is a completely regular T_1-space. Let F denote the family of all continuous real functions $f_\lambda : X \to Q$ on the Tychonoff space X into the closed unit interval Q = [0, 1]. By using complete regularity of X, it can be easily proved that F distinguishes points and closed subsets of X. Again since X is also T_1, every singleton subset of X is closed and consequently F also distinguishes points. It follows by the preceding theorem that the evaluation map $e : X \to Q$ is an embedding. Hence every Tychonoff space can be embedding as a subspace of a cube.

Theorem 3:

A space X is Tychonoff iff X is homeomorphic to a subspace of a cube.

Proof:

For the 'only if' part of the theorem. To prove the 'if part' we observe that since each unit interval Q is compact. Hausdorff space, the cube Q^F is also a compact Hausdorff space by Tychonoff theorem. It follows that Q^F is a Tychonoff space (where $\mathfrak{I}$ is an arbitrary index set) and so is every subspace of Q^L,. Hence if X is homeomorphic to subspace of a cube, that X is 3 Tychonoff space.

Theorem 4:

For each $n \in N$, let (X_n, d_n) be a metric space such that $\delta_n(X) \leq 1$ and let $X = \times \{X_n : n \in N\}$. For x, $y \in X$

$$\text{define} \quad = d(x, y) \sum_{n=1}^{\infty} \{2^{-n} d_n(x_n, y_n)\}.$$

Then (i) d is metric for X, and (ii) the d-metric topology $\mathfrak{I}$ is the same as the product topology F for $X = \times \{X_n : n \in N\}$ where X_n has the d-metric topology.

Proof:

We first note the series for d (x, y) is convergent by comparison test for we have $2^{-n} d_n(x_n, y_n) \leq 2_n$ and the geometric series

$$\sum_{n=1}^{\infty} 2^{-n} \text{ is convergent.}$$

(i) We now prove that d is a metric for X. The axioms $[m_1]$, $[m_2]$ and $[m_3]$ immediately follow from the corresponding axioms for d_n. To prove $[m_4]$ (the triangle inequality), let x, y, z be arbitrary members of X. Then since each d_n is a metric, we have.

$$\sum_{n=1}^{n=k} 2^{-n} d_n(x_n, y_n) \le \sum_{n=1}^{n=k} 2^{-n} [d_n(x_n, a_n) + d_n(z_n, y_n)]$$

$$= \sum_{n=1}^{n=k} 2^{-n} d_n(x_n, z_n) + \sum_{n=1}^{n=k} 2^{-n} [d_n(z_n, y_n). \qquad ...(1)$$

proceeding to the limits when $k \to \infty$ we get from (1)

$$d(x, y) \le d(z, x) + d(z, y).$$

Hence d is a metric for X.

(ii) We first prove that $F \in \Im$. Let U be any member of the defining subbase for F. Then U consists of all points $x = \{x_n : n \in B]$ such that $x_k \in G_k$ for some open set G_k in X_k and $x_n \in X_n$ whenever $n = k$. Let $y \in U$.

Then $y_k \in G_k$. Since G_k is open in X_k, there exists $r > 0$ such that

$$D(y_k, r) \subset G_k$$

Let $V = S(y, \eta)$ where $n = 2^{-k} r$ and let $z \in V$. Then

$$2^{-k} d_k(z_k, y_k) \le d(z, y) \text{ by def. of } d < 2^{-k} r.$$

Therefore $d_k(z_k, y_k) < r$.

Then from (2), it follows that $z_k \in G_k$ and consequently $z \in U$ by definition of U. This we have shown that $z \in V \Rightarrow z \in U$. Hence $V \subset U$. Thus to each $y \in U$, there exists an open sphere centred at y such that $V \subset U$.

It follows that $U \in \Im$, that is, U is open with respect to the d-metric topology for X. Now since each subbase member of F belongs to $\Im$, It is easy to see that

$$F \subset \Im.$$

To prove $\Im \subset F$. let $z \in G$ where $G \in \Im$. Then there exists some m $\in N$ such that

$$S(z, 2^{-m}) \subset G.$$

Let $W = \{x : d_n(x_n, z_n) > 2^{-m-n-2}$ for each $n \in N_{m+2}\}$.

[Note that for convenience we write N_{m+2} for the set

$$\{1, 2, 3, .m + 2\}.$$

Then $W \subset G$ as shown below. Let $x \in W$. Then.

$$d(x, z) = \sum_{n=1}^{\infty} [2^{-n} d_n(x_n, z_n)].$$

$$= \sum_{n=1}^{n=m+2} 2^{-n} d_n(x_n, z_n) + \sum_{n=m+3}^{n=\infty} [2^{-n} d_n(x_n, z_n)]$$

$$= < \sum_{n=1}^{n=m+2} d_n (x_n, z_n) + \sum_{n=m+3}^{n=\infty} 2^{-n}$$

$$[\because 2^{-n} < 1 \text{ and } d_n (x_n, z_n) \leq 1]$$

$$< \sum_{n=1}^{m+2} 2^{-m-n-2} + \sum_{n=m+3}^{\infty} 2^{-n}$$

$$\frac{2^{-m-2} [1-(\frac{1}{2})^{m+2}]}{1-\frac{1}{2}} + \frac{2^{-m-3}}{1-\frac{.1}{2}}$$

$$= 2^{-m-2} [1-(\frac{1}{2})^{m+2}] + 2^{-m-2}$$

$$< 2^{-m-1} + 2^{-m-1} = 2^{-m}.$$

Hence $x \in S (z, 2^{-m}) \subset G$.

Thus it is shown that

$$x \in W \Rightarrow x \in G.$$

Hence $W \subset G$.

But evidently $z \in W$. Also it is easy to see that W is a base member of the product topology F. Thus we have shown that to each $z \in G$, there exists a base member W for the product topology F such that

$$z \in W \subset G.$$

It follows that G is F-open.

Hence $\mathfrak{I} \subset F$.

From (3) and (4) we have $\mathfrak{I} = F$.

Definition : *A topology space $(X, \mathfrak{I})$ is said to be Metrizable iff there exists a metric d for X which induces the same topology as the topology $\mathfrak{I}$.*

It immediately follows from this definition that a topological space is merrizable it is homeomorphic with a metric space.

Theorem 1:

Every subspace Y of a metrizable space X is metrizable, that is, metrizability is a herediatary property.

Proof:

Let $d : X \rightarrow R$ be a metric for X which induces the topology of X. Consider the restriction e of d to the subspace Y, that is, let, e=d/Y. Then e is evidently a metric for X which induces the relative topology of Y.

Theorem 2:

The space (R. U) is metrizable.

Proof:

Do yourself.

Corollary : *The unit interval Q = [0,1] is metrizable.*

Proof :

Do yourself.

Theorem 3:

The topological product of a countable family of metrizable spaces is metrizable.

Proof:

We first prove that the topological product $X_1 \times X_2 \times ... \times X_n$ of a finite number of metrizable spaces $X_1, X_2, ... X_n$ is metrizable. Let $d_i : X_i \times X_i \to R$ $(1 \leq i \leq n)$ be a metric for X_i which induces the topology of X_i. Consider the mapping.

$$d : X \times Y \to R$$

defined as follows. Let $x = (x_1, ..., x_n)$ and $y = (y_1, ... y_n)$ be any two points in X. Then d (x, y) is defined by

$$[d(x, y)] = \sum_{i=1}^{n} [d_i(x_i, y_i)]^2 .$$

It is easy to see that d is a metric for X which induces the product topology.

Now consider a denumerable collection $\{X_n : n \in N\}$ of metrizable spaces. Let d_i be a metric for x_i, i = 1, 2, 3,... Then there exists a metric for the cartesian product $X = \times \{X_n : n \in N\}$ which induces the product topology and consequently X is metrizable.

Corollary : *The Euclidean n-spaces R_n, the unit n-cube Q_n and the Hilbert cube Q_n are metrizable since R and Q = [0, 1] are metrizable.*

Theorem 4: (Urysohn Metrization Theorem):

Every regular T_1-space X satisfying the second axiom of countability is metrizable.

Proof:

If suffices to establish the existence of a countable family of closed function from X into Q = [0, 1], which distinguishes points and closed sets.

For then it will follow that X can be embedded as subspace of the Hilbert, cube Q^N which is metrizable and so X will be metrizable.

Let B be a countable base for the topology on X. This is possible since X is second countable. Consider the subcollection c of the cartesian product $B \times B$ consisting of pairs (U, V) such that $U \subset V$. Surely, c is countable. Let (U, V) be any member of c. Since X is regular $U \subset V$, we have $U \cap V = \varnothing$. Also $\bar{U}$ and B are closed in X, Since X is regular and second countable, it is normal. Hence by Urysohn's lemma there exists a continuous function f from X into Q such that $f(x) = 0$ if $x \in \bar{U}$ and $f(x) = 1$ if $x \in V$. Let F be the family of function so obtained. Then F is countable and it remains to show that F distinguishes points and closed sets. So let F be any closed subset of X and x any point in F′. Since F′ is an open neighbourhood of x, there exists a member V of B such that $x \in V \subset F'$. Again since X is regular, there is an open subset G of X such that $x \in G$ and $G \subset V$, Then by definition of base, we can choose. U in B such that $x \in U \subset G$ so that $x \in U \subset G$. Thus $x \in \bar{U} \subset V \subset F'$ and so (U. V) $\in$ c. If f be the corresponding member of F1 then

$$f(x) = 0 \varnothing \{1\} = f[V] \supset f[F] \qquad ...(1)$$

$$[\because V \subset F' \Rightarrow V' \supset F \Rightarrow f[V] \supset f[F]]$$

Since {1} is closed in Q, we have

$$\{1\} = \{1\} = f(V) \subset \overline{f[F]}.$$

It follows from (1) and (2) that $f(x) \varnothing f[F]$. Hence F distinguishes points and closed sets and so by our remark in the beginning of the proof, X is metrizable.

Theorem 5:

Let X be a T_1-space. Then the following statements are equivalent:

(i) *X is regular and second countable.*

(ii) *X is homeomorphic to a subspace of the Hilbert cube Q^N*

(iii) *X is metrizable and separable.*

Proof:

We shall show: (i) ⇒ (iii) ⇒ (i).

Now (i) ⇒ (ii) follows from the previous theorem. To prove (ii) ⇒ (iii), we first observe that Q^N is metrizable and second countable.

Since metrizability and second coutability, are hereditary properties, it follows that each subspace of Q^N is metrizable and second countable and is therefore separable.

Since by hypothesis X is homeomorphic to a subspace of Q^N, it follows that X is metrizable. Thus (ii) ⇒ (iii).

Finally assume that X is metrizable and separable. Then it X is regular and second countable and so (iii) ⇒ (i).

Theorem 6:

Let f be a mapping of a topological space (X, ℑ) onto a set Y. Then the collection v of all subsets G of Y such f^{1} [G] is open in X is the largest topology for Y such that f is ℑ-v- continuous. Further a subset F of Y is v-closed if and only if f^{1} [F] is ℑ-closed.

Proof:

We first show that v is a topology for Y.

[T1] : Since f^{-1} [∅] = ∅ is open in X, we have ∅ v. Also since f^{-1} [Y] = X is open in X, we have T ∈ 'v.

[T2] : Let G_1, G_2 ∈ v. Then f^{-1} [G_1] and f^{-1} are open in X by hypothesis and so f^{-1} [G_1] ∩ f^{-1} [G_1 ∩ G_2] is also open in X. Therefore $G_1 \cap G_2$ ∈ v.

[T3] : Let G_λ ∈ v for every λ ∈ L where L is an arbitrary index set. Then each f^{-1} [G_λ] is open in X by hypothesis and so their union ∪ {f^{-1} [G_λ] : λ ∈ L} = f^{-1} [∪ {G_λ : λ ∈ L} is also open in X and therefore ∪ {G_λ : λ ∈ Λ} ∈ v. Hence v is a topology for Y. It then follows that f is ℑ-v continuous.

Now let v be another topology for Y such that f is ℑ v' continuous. Then f^{-1} [H] ∈ v by hypothesis. Hence v' ⊂ v. In other words, v is the largest topology for Y such that f is ℑ-v continuous.

Finally, let F be subset of Y. Then F is V-closed ⇔ f^{-1} [Y – F] = X – F^{-1} [F] is ℑ-open {Theorem (1.8), ch. 4] ⇔ f^{-1} [F] is ℑ-closed.

QUOTIENT TOPOLOGY

Definition : *Let (X, ℑ) be a topology space, Y a set, and f a mapping of X onto. Then the largest topology v for Y such that f is ℑ-v continuous is called the quotient topology for Y (relative to f and ℑ) and shall be denoted by ℑf and the map f is called the quotient map.*

Theorem 1:

A subset G of Y is open in the quotient topology (relative to f : X → Y) if and only if f^{-1} [G] is an open subset of X.

Proof:

First let f be continuous and open. We want to show that 3t = bv where $ℑ_f$ denotes the quotient topology for Y. $ℑ_f$ is the largest topology for Y making f continuous. It follows that $v \subset \mathfrak{I}$. Conversely, let $G \in \mathfrak{I}\,\gamma$. Then f^{-1} [G] is-open *i.e.*, $G \in V$ and therefore $\mathfrak{I}f$ *i.e.*, V is the quotient topology for Y.

Now let f be continuous and closed. Then $V \subset \mathfrak{I}$ as before. Conversely, if $G \in \mathfrak{I}f$ then f^{-1} [G] is $\mathfrak{I}$-open and so $X = f^{-1}$ [G] is $\mathfrak{I}$-closed. But

$$X - f^{-1}[G] = f^{-1}[Y] - f^{-1}[G] = f^{-1}[Y - G].$$

Since f is a closed and onto map. $f[f^{-1}[Y - G]] = Y - G$ is V-closed and so $G \in V$. Therefore $\mathfrak{I}_f \subset V$. Thus, $\mathfrak{I}_f = V$ in this case also.

Theorem 2:

Let f a continuous map of a space X onto a space R and let Y have the quotient topology (relative to f). Then a mapping g on Y onto a topology space Z is continuous if and only if the composition map gof f is continuous.

Proof:

Suppose g o f is continuous and let G be open set in Z. Then $(gof)^{-1}$ $[G] = f^{-1}[g^{-1}[G]]$ is open. X and so g^{-1} [G] is open in Y. Hence g is a continuous map.

Conversely, let $g : Y \rightarrow Z$ be continuous. Since $f : X \rightarrow Y$ is given to be continuous, it follows that gof f is continuous map.

Remark : It can be shown that the quotient topology and the properties of open and closed maps have little to do with the range space. In fact if f is continuous map of a space X onto a space Y with quotient topology and the family of all sets of the from f^{-1} (y) with y in Y. The continuous can be described as follows. Let D be the collection of all subsets X of the from f^{-1}. | f (x) | for all $x \in X$. Define $g : Y \rightarrow D$ by $g(y) = f^{-1}(y)$. Then it is easy to see that g is a one-one map of Y onto D. Also go f = F g = 1 o F. If now we assign the quotient topology to D (relative to F), then shows that g and g^{-1} are continuous (since g o f = F and g^{-1} of = f). Hence g is a homeomorphism.

The range space Y is essentially extraneous to our discussion, and so the remaining theorems of this section will be formulated so as to display this fact.

The Decomposition (or partition) of a non-empty set X is the collection of non-empty disjoint subset of X whose union is X.

Theorem 3:

If f is a continuous map of a compact space (X, $\mathfrak{I}$) onto a Hausdorff space (Y, V), then V must be quotient topology for Y relative to a Hausdorff space (Y, V), then V must be quotient topology for Y relative to f.

Proof:

Let F be any $\Im$-closed subset of X. Then F is compact and so its continuous image f [F] is also a compact subset of Y. Since Y is Hausdorff, f [F] is a closed subset of Y and therefore f is a closed mapping. Hence V must be the quotient topology for Y.

QUOTIENT MAP

Definition : *Let X be non-empty set, and D be a decomposition of X. Then the mapping p from X onto D such that p (x) is the unique members of D to which x belongs is called the quotient map (or the projection map of canonical map).* ***X/R (called the quotient set of X modulo R) so that X/ R is the family of all equivalence.***

QUOTIENT SPACE

Definition : *Let (X, $\Im$) be a topological space and R an equivalence relation on X. Let f be the quotient map of X onto the quotient set X/R of X over R so that p (x) = [x] is the equivalence class to which x belongs. Then the quotient space is the family X/R with quotient topology (relative to p).*

Remark : (i) The quotient topology for X/R. is the strongest topology for X/R rendering the quotient map p is continuous on X. The quotient topology for X/R consists of all subsets G of X/R such that p^{-1} [G] is open in X.

(ii) If A is subset of X, then the set of all points which are R relative of points of A shall be denoted by [A] of R [A]. Thus R [A] [y : x, y) $\in$ R for some x $\in$ A}

In other words.

$$R [A] = \cup [D : D \subset X/R \text{ and } D \cap A \neq \emptyset].$$

If x is a point of X, then we abbreviate R [{x}] as R [x] or [{x}] as [x]. The set R [x] or [x] is the equivalence class to which s belongs and if p is the projection of X onto the decomposition, then p (x) = R [x].

(iii) If A $\subset$ X/R, then p^{-1} [A] = $\cup$ {A : A $\in$ $\Im$} and therefore Á is open (closed) relative to the quotient topology iff $\cup$ {A : A $\in$ $\Im$} is open (respectively closed) in X.

Theorem 1:

Let p be the quotient map of the topological space X onto the quotient space X/R. Then the following statements are equivalent.

(a) p is open (closed) in mapping

(b) If A is open (closed) in X, then R [A] is open (respectively closed).

(c) If B is a closed (open) subset of X, then the union of all members of X/R which are subsets of B is closed (respectively open).

Proof :

(a) ⇔ (b) : First observe that if A is a subset of X, then the set R [A] = π^{-1} [[A]].

Now let π be open map and let A be any open subset of X, Then by definition of an open map π [[A]] is open subset of X/R. Since π is continuous, π^{-1} [π [A]] is open in X and so by (1), R [A] is an open subset $2\pi^{-1}$ [p [A]] is open in X. Then π [A] is open in X/R, and consequently p is open mapping.

(b) ⇔ (c) : We first notice that the union of all these members of X/R which are subsets of B is X – R [X – B]. Now assume then R [A] be open for each open subset A of X and let B be any closed subset of X so that X – B is open in X. Hence by hypothesis R [X – B] is open X and consequently X – R [X – B] is closed in X, that is , the union of all those members of X/R which are subsets of B is closed. This proves (b) ⇒ (c). To prove (c) ⇒ (b), assume that for each closed subset B of X, the union of all those members of Y/R which are subsets of B is closed, that is assume that X – R [X – B] is closed. Let A be any open subset of X so that X – A is closed in X. Then and consequently R [A] is open.

If we interchange the words "open" and "closed" through, we get the proof of the dual proposition.

Theorem 2:

Let (X, ℑ) be a topological space. If X is compact, connected locally connected separable or Lindelof, the so is X/R..

Proof:

Since compactness, connectedness, separability and Lindelofness are preserved under continuous mapping and since X/R is a continuous image of X under the quotient map π, it follows that X/R also has these properties. To prove local connectedness of X/R, we shall use the fact that a topological space X is locally connected iff every component of an open set is open in X. So let A be an open set of an open set of X/R and let B be a component of A. Then, the set G = π^{-1} [A] is an open subset of X containing the set C = π^{-1} [B]. Let x be any point of C and let C, be the component of C containing x.

Now $x \in C \Rightarrow \pi(x) \in \pi[C] \Rightarrow \pi(x) \in \pi[\pi^{-1}(B)]$

$$\Rightarrow \pi(x) \in B \text{ and } x \in C_x \Rightarrow \pi(x) \in \pi[C_x].$$

By definition of component, C_x is connected and so its continuous image $\pi[C]$ is also connected. Also B, being a component, is connected. Since $\pi(x) \in B$ and $\pi(x) \in \pi[C_x]$, we have $B \cup \pi[C_x] \neq \varnothing$. It follows from theorem (2.24), ch. 5 that their union $B \cup \pi[C_x]$ is connected. Since $\pi\{C_x]$ is connected subset of A, it follows by the maximality of components that $\pi[C_x] \subset B$, and hence

$$\pi^{-1}[\pi[C_x]] \subset n^{-1}[B]$$

that is, $C_x \subset C$. Since X is locally connected, each component C_x must be open and hence C is a neighbourhood of x. Since x was an arbitrary point of C, we conclude that C is an open set in X. Finally, since $C = \pi^{-1}[B]$.B is an open subset of X/R and consequently X/R is locally connected.

Example:

Give an example of a Hausdorff space X which has a non-Hausdorff quotient space.

Solution:

Consider the space (R, U) where R denotes the set of all real numbers and U is usual topology for R. We know that this space is Hausdorff. Now consider the relation R on R consisting of all pairs (x, y) relation on R. Also it is evident that the quotient space R/R in this case is indiscrete and consequently non-Hausdorff

Theorem 3:

Let X be a topological space and let the quotient space X/R be Hausdorff. Then R is a closed subset of the product space X × X.

Proof:

We shall show that no point of X × X – R can be limit point of R. Let $(x, y) \in X \times X - R$. Then $(x, y) \notin R$ and hence $p(x) \neq \pi(y)$. Since X/R is Hausdorff, there exists open sets G and H inX/R such that

$$\pi(x) \in G, \pi(y) \in H$$

and $\quad G \cap H = \varnothing.$

$$[\pi^{-1}[G]] = G \text{ and } \pi[\pi^{-1}(H)] = H$$

are disjoint, no member of $\pi^{-1}[G]$ can be R-related to a member of $\pi^{-1}[H]$. Hence π^{-1} Hence $\pi^{-1}[G] \times \pi^{-1}[H]$ is a neighbourhood of (x, y) which does not contain any point of R and consequently (x, y) is not a limit point of R. It follows that R is closed..

Theorem 4:

Let X be a topological space and let X/R be the quotient space. If the quotient map p is an open map and R is closed in X × X, then X/R is a Hausdorff space.

Proof:

Let π (x) and π y) be any two members of X/R such that π (x) ≠ π (y). Then x is no R-related to y so that (x, y) ∉ R. Since R is closed, (x, y) can not be a limit point of R. Hence there exists a basic open subset G × H of the product space X × X containing (x, y) and disjoint from R. This implies that G and H are open neighbourhoods of x and y respectively such that no point of G is R-related to a point of H. Hence p [G] and p [H] are disjoint. Since p is given to be open map, it follows that π [G] and π [H] are open neighbourhoods of π (x) and π (y) respectively and consequently X/R is Hausdorff space.

UPPER SEMI-CONTINUOUS DECOMPOSITION

Definition: *Let X be a topological space. Then a decomposition D of X is said to be upper semi-continuous iff for every D ∈ D and every open set G containing D there exists an open set H such that D ⊂ H ⊂ G and H is the union of members of D.*

Theorem 1:

Let (X, ℑ) be a topological space and let D be an upper semi-continuous decomposition of X and let D have the quotient topology v. If G is a ℑ-open set containing a member A of D, then p [G] is a v-neighbourhood of A where p is the projection (or quotient) map.

Proof:

Since G is a ℑ-open set containing A ∈ D, by the definition of upper semi-continuity, there exists a ℑ-open set H which is a union of members of D such that

A ⊂ H ⊂ G.

Hence π [A] ⊂ π [H] ⊂ π [G]. ...(1)

Since A ∈ D, by the definition of π, we have π [A] = {A}, and consequently

A ∈ π [A]. ...(2)

It will now be shown that π [H] is V-open. By definition of H, we have

$$H = \cup \{A_i : A_i \in D' \subset D\}$$

Then

$$\pi [H] = \pi [\cup \{A_i : A_i \in D' \subset D\}]$$
$$= \cup \{\pi [A_i] : A_i \in D' \subset D\}$$
$$= \cup \{\{A_i\} : A_i \in D' \subset D\}$$

Therefore

$$\pi^{-1} [\pi [H]] = \pi^{-1} [\cup \{A_i : A_i \in D' \subset D\}]$$
$$= \cup \{\pi^{-1} [\{A_i\}] : A_i \in D' \subset D\}$$
$$= \cup \{A_i : A_i \in D' \subset D\} = H.$$

Since H is $\mathfrak{I}$-open, it follows from the definition of quotient topology that p [H] is a v-open set. Hence it follows from (1) and (2) that π [G] is a v-neighbourhood of A.

Theorem 2:

Let (X, $\mathfrak{I}$) be a topological space and let D be an upper semi continuous decomposition of X such that every member of D is a compact subset of X and let D have the quotient topology v. If X is :

(i) Hausdorff,

(ii) Regular,

(iii) Locally compact or

(iv) Satisfies the second axiom of countability, then D also has the corresponding properties.

Proof:

(i) Let X be a Hausdorff space and let A and B be distinct members of D. Since A and B are compact subsets of the Hausdorff space X, there exists disjoint $\mathfrak{I}$-nhds N and M of A and B respectively. Then π [N] and p [M] are disjoint V-nhds of A and B respectively and consequently D is a Hausdorff space.

(ii) Let X be regular. Let A be any point of D and let N be a V-nhd of A. Then the union G of members of N is a $\mathfrak{I}$-nhd of A. Since Y is regular, for each $x \in A$, there exists an open nhd U_x of x such that $U_x \subset G$. Now the collection $\{U_x : x \in A\}$ is an open cover of A. Since A is compact there exists a limit subcover $U_1, U_2, \ldots U_n$ of A of A. *Such* that $\overline{U}_i \subset G$ for each i so that $U_i \subset \cup \{A_i : i = 1, 2, \ldots n\}$. Let $V = \cup \{\overline{U} : i = 1, 2, \ldots n\}$. Then V is a $\mathfrak{I}$-closed neighbourhood of A such that $V \subset G$. It follows that π [V] is a V-closed neighbourhood of A such that $\pi [V] \subset \pi [G] = N$. Hence D is regular.

(iii) Let X be locally compact and A any member of D. Then by definition of local compactness, to each x ∈ A, there exists a ℑ-open compact nhd U_x of x. Now the collection $\{U_x : x \in A\}$ is an open cover of A. Since A is compact, there exists a finite subcover $U_1, U_2, \ldots, U_n$ so that

$A \subset \cup \{U_i : i = 1, 2. 3\ldots n)$. Let $V = \cup \{U_i : i = 1, 2, 3,\ldots n\}$. Also since V is the union of a finite member of compact neighbourhood of A. Since π is continuous, π [V] is compact. Also π [V] is a V-neighbourhood, and hence by definition of local compactness D is locally compact.

(iv) Let X be second countable and let B be a countable base for ℑ. The family H, let H of unions of finite subfamilies of B is countable. For every member H of H, let H* denote the union of all those members of D which are subsets of H and M be the family of all sets H* corresponding to every H ∈ H. Then it is easy to see that π [H*] is open in D for every H* = M. We will show that the countable collection {π [H*] : H* ∈ M} is a base for the quotient topology. For this purpose, it is sufficient to prove that for each A ∈ D and each neighbourhood N of A, there exists some H* ∈ M such that A ⊂ H* ⊂ N.

Since A is compact it may be covered by a finite number of members of B such that the union H of these members, which is members of H, is contained in N. Then H* ∈ M is such that

$$A \subset H^* \subset N.$$

This proves the theorem.

Theorem 3:

Let X be a topological space and let D be a decomposition of X. Then D is upper semi-continuous iff the projection p of X onto D is closed.

Proof:

We know that π is a closed map iff for each open subset G of X, the union H of all members of D which are subsets of G is an open set. Now let π be a closed map. Let D ∈ D and C be an open set containing D. If H is the union of all those members of D which are subsets of G, then H is an open set such that D ⊂ H ⊂ G. Hence D is upper semi-continuous. To prove the converse, let D be upper semi-continuous. Let G be an open subset of X and let H be the union of all members of D which are subsets of G. We shall show that H is open. Let x ∈ H. Then by definition of union, x ∈ D ⊂ G for some D ∈ D. Hence by upper semi-continuity, there exists an open set K which is the union of members of D such that D ⊂ K ⊂ G. It is follows that K ⊂ H. Thus we have shown that to each x ∈ H, there exists an open set K such that $x \in K \subset H$. Hence H is an open subset of X since it is a neighbourhood of each of its points. It follows that π is a closed map.

EXERCISES

1. Let $(X_1 ; \Im_1)$, $(X_2, \Im_2)$ and $(X_3, \Im_3)$ be topological spaces. Show that a mapping $f : X_3 \to X_1 \times X_2$ is continuous iff π_{x_1} of $: X_3 \to X_1$ and π_{x_2} of $: X_3 \to X_2$ are continuous.

2. (a) Let $\{X_i : i \in I\}$ be an indexed family of topological spaces and let X be its topological product. Prove that the projection function $\pi_i : X \to X_i$ is open for all $i \in I$.

   (b) Prove that a sequence $\{x_n\}$ in X above converges to a point x in X if an only if for every $i \in I$, $\{\pi_i (x_n)\}$ converges to π_i (x) in X_i.

3. Prove that the topological product is, upto a homeomorphism

   (i) a well defined

   (ii) commutative, and

   (iii) associative operation.

4. Let $(X, \Im)$ and (Y, v) be topological spaces and L and M be subbases for $\Im$ and v respectively. Show that the collection A of all subset of the from $\Lambda \times Y$ and $X \times M$ where $\Lambda \in \Lambda$ and $M \in M$ is a subbase for the product topology on $X \subseteq Y$.

5. Let X_1, X_2 be two topological spaces and let X be the product space. Let π be the projective defined by $\pi (x_1, x_2) = x$. Which of the following is a wrong statement:

   (a) p is onto, (b) p is continuous,

   (b) p is open, (d) p is closed.

6. Give an example of *a* partially ordered set which is not a lattice.

7. Let *L* be a lattice. If *x*, *y*, and *z* are elements of *Λ*, prove the following:

   $x \wedge x = x, x \vee x = x,$

   $x \wedge y = y \wedge x, x \wedge y = y \vee x, \ x \vee (y \wedge z) = (x \wedge y) \wedge z,$

   $x \vee (y \vee z) = (x \vee y) \vee z, (x \wedge y) \vee x = x,$

   $(x \vee y) \wedge x = x.$

   If *X* has the weak topology induced by a collection

   $\{f_\lambda \ \lambda \in \Lambda\}$

of functions $f_\lambda : X \to X_\lambda$, then $f : Y \to X$ is continuous iff f_λ of is continuous for each $\lambda \in \Lambda$.

8. Let f : X Y be onto, where Y is a topological space. If X is given the weak topology induce by f, show that f becomes a quotient map.
9. Show that each coordinate space X of a product space X × $\{X_\lambda : \lambda \in L\}$ has a quotient topology induced by the projection π_λ.

   [**Hint** : Since projections are continuous and open maps.]
10. Define quotient topology. If f is a continuous map of a topological space (X, μ) onto the space (Y, μ) such that f is either open or closed, then prove that m is a quotient topology.
11. We proved that a continuous, open or closed map of a space X onto another space is a quotient map. Give an example to show that an open quotient map is not closed an a closed quotient map which is not open.

    [**Hint** : The projection of R^2 onto R is an open quotient map]
12. Let X and Y be topological spaces and if X/R is the quotient space of X, show that any continuous function f : X $\xrightarrow{\text{onto}} Y$ induces a continuous functions g from X/R onto Y, provided (x, y) ∈ → f (x) = f (y), *i.e.*, R ⊂ {(x, y) : f (x) = f (y)}.

    [**Hint** : Define g : X/R → Y by g ([x]) = f (x) where [x] denotes the equivalence class in X/R containing x. Let [x] = [y]. Then (x, y) is in R and f (x) = f (y). Hence g ([x)] = g ([y]) and so g in one-one, Further f = g o π. Hence g is continuous].
13. A function g of a quotient space X/R into a space Y is continuous if and only if the compositive function g o π, where π denotes the quotient map is continuous on the space X.
14. Let f : g be maps from a set X onto two topological spaces Y_f and Y_g, and let X have the weak topology induced by {f, g}. Show that g need not be the quotient map.

    [**Hint** : Take Y_i descrete].
15. Show that a space X is. Hausdorff if and only if its diagonal L = {x, x) : x ∈ X} is a closed set in the product space X × X.
16. If F_λ is a proper closed set in X_λ, is it true that $F = \pi F_\lambda$ is closed in $X = \pi X_\lambda$ where λ is an infinite set ?

    $\lambda \in \mathfrak{I}$
17. Let X = × $\{X_\lambda : \lambda \in \Lambda\}$ be a product space. Prove that × $\{A_\lambda : \lambda \in \Lambda\}$ is dense in X iff A_λ is dense in X_λ for every $\lambda \in \Lambda$.

18. Let $X = \times \{X_\lambda : \in \Lambda\}$ be a product space. If $A_\lambda \subset X_\lambda$ for each $\lambda \in \Lambda$, show that $\times \{A_\lambda : \lambda \in \Lambda\}$ (where each A_λ has the subspace topology has the same topology as it is considered as a subspace of X.
19. Assuming that the product of two connected spaces is connected, prove that the product of any non-empty family of connected spaces is connected.
20. Show that the product space $X = \times \{X_\lambda : \lambda \in \Lambda\}$ is T_0-space iff each coordinate space X_λ is T_0.
21. Prove that a non-empty, product space of a countable number of separable spaces is separable.
22. Let X_n (n = 1, 2, 3, ...) be discrete topological spaces consisting of the elements. Investigate whether their product is :

    (i) Compact

    (ii) Locally compact

    (iii) Discrete.
23. State necessary and sufficient conditions under which a product space $X = \Pi X_\lambda$ is locally compact. Prove sufficiency of the condition.
24. Prove that the product of each family of locally compact spaces is locally compact.
25. Explain why metrizability is only a countably product invariant property.
26. Prove that every metrizable space X has a bounded distance function $d : X^2 \to [0.1]$ which defines its topology.
27. Prove that the continuous image of a compact metrizable space in any Hausdorff space is metrizable.
28. Prove that every discrete space is metrizable and hence a metrizable space any fail to be a second countable space.
29. Let f be a closed subset of $X \times Y$ and Y be compact. Show that f is continuous mapping of X into Y.
30. Show that the product of two regular spaces is a regular space.
31. Show that the product space of two first axiom space is a first axiom space.
32. Let $f : X \to Y$ continuous and let Y be. Hausdorff. Show that f is a closed subset of $X \times Y$.
33. Let X be any topological space and let Y be a compact space. Show that the projection mapping π_x is a closed mapping of $X \times Y$ into Y.

34. Show that each projection map π_{x_i} of the product space is a continuous map onto X_i.

35. Show that the product space of two T0-spaces is a T_0-space.

36. Show that the product space of two completely regular spaces is a completely regular space.

37. Let $(x_i\ \Im_i)$, i = 1, 2, ..., n be n topological spaces. Show that the collection $B = \{G_1 \times G_2 \times \ldots \times G_n : G_i \in \Im_1, i = 1, 2 \ldots, n\}$ is a base for a topology for $X_1 \times X_2 \times \ldots \times X_n$,

    [The topology W generated by B in exercise 10 above is called the product topology for the set $X_1 \times X_2 \times \ldots X_n = X$, say and (X, W) is product space of the given spaces. The mapping $\pi_{x_i} : X \to X_i$: $\pi_{x_i}(x) = x_i \ \forall\ x \in X$ where $X = X_1{\times}X_2{\ldots}{\times}X_n$ is called the i-th projection mapping or the i-th projection].

38. Show that the product space of two separable spaces is a separable space.

39. Suppose X and Y are compact spaces and v is an open cover of X × Y. Prove that for every y ∈ Y there exists an open set V_y in Y containing Y such that, X × V_y is covered by finitely many members of U. Hence show that X × Y is compact.

40. Prove that X is Hausdroff if and only if the diagonal in X is a closed subset of X × X.

41. Prove that the product of two locally connected spaces is locally connected.

42. Show that the product topology W is the smallest topology for the set $X_1 \times X_2 \times \ldots \times X_n$ for which each projection π_1 is continuous.

43. Show that the product space of two T_1-spaces is a T_1-space.

2

HILBERT SPACE

INTRODUCTION

One of the principal applications of the theory of Banach algebras developed the study of operators on Hilbert spaces. Our purpose in this chapter is to present enough of the elementary theory of Hilbert spaces and their operators to provide an adequate foundation for the deeper theory discussed in these later chapters.

We shall see from the formal definition given in the next section that a Hilbert space is a special type of Banach space, one which possesses additional structure enabling us to tell when two vectors are orthogonal (or perpendicular). The first part of the chapter is concerned solely with the geometric implications of this additional structure.

As we have said before, the objects of greatest interest in connection with any linear space are the linear transformations on that space. In our treatment of Banach spaces, we took advantage of the metric structure of such a space by focusing our attention on its operators. A Banach space, however, is still a bit too general to yield a really rich theory of operators. One fact that did emerge, which is of great significance for our present work, is that with each operator T on a Banach space B there is associated an operator T* (its conjugate) on the conjugate space B*. We shall see below that one of the central properties of a Hilbert space H is that there is a natural correspondence between H and its conjugate space H*. If T is an operator on H, this correspondence makes it possible to regard the conjugate T* as acting on H itself (instead of H*), where it can be compared with T. These ideas lead to the concept of the adjoint of an operator on a Hilbert space, and they make it easy to understand the importance of operators (such as self-adjoint and normal operators) which are related in simple ways to their ad joints.

THE DEFINITION AND SOME SIMPLE PROPERTIES

The Banach spaces studied in the previous chapter are little more than linear spaces provided with a reasonable notion of the length of a vector. The main geometric concept missing in an abstract space of this type is that of the angle between two vectors. The theory of Hilbert spaces does not hinge on angles in general, but rather on some means of telling when two vectors are orthogonal.

In order to see how to introduce this concept, we begin by considering the three-dimensional Euclidean space R^3. A vector in R^3 is of course an ordered triple $x = (x_1, x_2, x_3)$ of real numbers, and its norm is denned by

$$\| x \| = (| x_1 |^2 + | x_2 |^2 + | x_3 |^2)^{1/2}.$$

In elementary vector algebra, the inner product of x and another vector $y = (y_1, y_2, y_3)$ is denned by

$$(x, y) = x_1y_1 + x_2y_2 + x_3y_3,$$

and this inner product is related to the norm by

$$(x, x) = \| x \|^2.$$

We assume that the reader is familiar with the equation

$$(x, y) = \| x \| \| y \| \cos \theta,$$

where θ is the angle between z and y, and also with the fact that z and y are orthogonal precisely when $(x, y) = 0$.

Most of these ideas can readily be adapted to the three-dimensional unitary space C^3. For any two vectors $x = (x_1, x_2, x_3)$ and $y = (y_1, y_2, y_3)$ in this space, we define their inner product by

$$(x, y) = x_1\bar{y}_1 + x_2\bar{y}_2 + x_3\bar{y}_3. \quad ...(1)$$

Complex conjugates are introduced here to guarantee that the relation

$$(x, x) = \| x \|^2$$

remains true. It is clear that the inner product defined by (1) is linear as a function of z for each fixed y, and is also conjugate-symmetric, in the sense that $\overline{(x, y)} = (y, x)$. In this case, it is no longer possible to think of (x, y) as representing the product of the norms of x and y and the cosine of the angle between them, for (x, y) is in general a complex number. Nevertheless, if the condition $(x, y) = 0$ is taken as the definition of orthogonality, then this concept is just as useful here as it is in the real case.

With these ideas as a background, we are now in a position to give our basic definition. A *Hilbert space* is a complex Banach space whose norm arises from an *inner product,* that is, in which there is defined a complex function (x, y) of vectors x and y with the following properties:

(1) $(\alpha x + \beta y, z) = \alpha (x, z) + \alpha(y, z)$;

(2) $\overline{(x, y)} = (y, x)$;

(3) $(x, x) = \| x \|^2$.

It is evident that the further relation

$$(x, \alpha y + \beta z) = \bar{\alpha}(x, y) + \bar{\beta}(x, z)$$

is a direct consequence of properties (1) and (2).

The reader may wonder why we restrict our attention to complex spaces. Why not consider real spaces as well ? As a matter of fact, we could easily do so, and many writers adopt this approach. There are a few places in this chapter where complex scalars are necessary, but the theorems involved are not crucial, and we could get along with real scalars without too much difficulty. It is only in the complex case, however, that the theory of operators on a Hilbert space assumes a really satisfactory form. This will appear with particular clarity in the next chapter, where we make essential use of the fact that every polynomial equation of the nth degree with complex coefficients has exactly n complex roots (some of which, of course, may be repeated). For this and other reasons, we limit ourselves to the complex case throughout the rest of this book.

The following are the main examples of Hilbert spaces. In accordance with the above remarks, the scalars in each example are understood to be the complex numbers.

Example 1:

The space l_2^n, with the inner product of two vectors

$$x = (x_1, x_2, \ldots, x_n) \text{ and}$$
$$y = (y_1, y_2, \ldots, y_n) \text{ defined by}$$

$$(x, y) = \sum_{i=1}^{n} x_i \bar{y}.$$

It is obvious that conditions (1) to (3) are satisfied.

Example 2:

The space l_2, with the inner product of the vectors $x = \{x_1, x_2, \ldots, x_n \ldots\}$ and $y = y_1 i\, y_2, \ldots, y_n, \ldots\}$ defined by

$$(x. y) = \sum_{n=1}^{\infty} x_n, \bar{y}_n$$

The fact that this series converges—and thus defines a complex number —for each x and y in l_2 is an easy consequence of Cauchy's inequality.

Example 3:

The space L_2 associated with a measure space X with measure m, with the inner product of two functions f and g denned by

$$(f, g) = \int f(x)\overline{g(x)}\, dm(x).$$

This Hilbert space is of course not part of the official content of this book, but we mention it anyway in case the reader has some knowledge of these matters.

As our first theorem, we prove a fundamental relation known as the Schwarz inequality.

Theorem 1:

If x and y are any two vectors in a Hilbert space, then $\|(x, y)| \leq \|x\|\,\|y\|$.

Proof:

When y = 0, the result is clear, for both sides vanish. When $y \neq 0$, the inequality is equivalent to $(x, y/\|y\|) \leq \|x\,x\|$. We may therefore confine our attention to proving that if $\|y\| = 1$, then we have $|(x, y)| \leq \|x\|$ for all x. This is a direct consequence of the fact that

$$0 \leq \|x - (x, y)\,y\|^2 = (x - (x, y)\,y, x - (x, y)\,y)$$

$$= (x, x) - (x, y)\,\overline{(x, y)} - (x, y)\overline{(x, y)} + (x, y)\,\overline{(x, y)}\,(y, y)$$

$$= (x, x) - (x, y)\,\overline{(x, y)} = \|x\|^2 - |(x, y)|^2.$$

An inspection of Fig. 2.1 will reveal the geometric motivation for this computation.

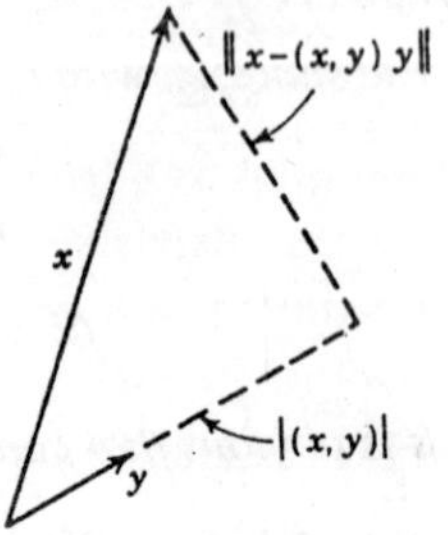

It follows easily from Schwarz's inequality that the inner product in a Hilbert space is jointly continuous.

$$x_n \to x \text{ and } y_n \to y \Rightarrow (x_n, y_n) \to (x, y).$$

To prove this, it suffices to observe that

$$|x_n, y_n - (x, y)| = |x_n, y_n) - (x_n, y)$$
$$- (x, y) \leq |(x_n, y_n) - (x_n, y)|$$
$$+ |x_n, y) - (x, y|$$
$$= |(x_n, y_n - y)| + |x_n - x, y)\ | \leq \|x_n\|\,\|y_n - y\|$$
$$+ \|x_n - x\|\ y\|.$$

Fig. 2.1 : Schwarz's inequality.

A well-known theorem of elementary geometry states that the sum of the squares of the sides of a parallelogram equals the sum of the squares of its diagonals. This fact has an analogue in the present context, for in any Hilbert space the so-called *parallelogram law* holds:

$$\| x + y \|^2 + \| x - y \|^2 = 2 \| x \|^2 + 2 \| y \|^2.$$

This is readily proved by writing out the expression on the left in terms of inner products:

$$\begin{aligned} \| x + y \|^2 + \| x - y \|^2 &= (x + y, x + y) + (x - y, x - y \\ &= (x, x) + (x, y) + (y, x) + (y, y) \\ &\quad + (x, x) - (x, y) - (y, x) + (y, y) \\ &= 2 (x, x) + 2 (y, y) = 2 \| x \|^2 + 2 \| y \|^2. \end{aligned}$$

The parallelogram law has the following important consequence for our work in the next section.

Theorem 2:

A closed convex subset C of a Hilbert space H contains a unique vector of smallest norm.

Proof:

We recall from the definition in since C is convex, it is non-empty and contains $(x + y)/2$ whenever it contains x and y. Let $d = \inf \{\| x \| : x \varepsilon C\}$. There clearly exists a sequence $\{x_n\}$ of vectors in C such that $\| x_n\| \to d$. By the convexity of C, $(x_m + x_n)/2$ is in C and $\| (x_m + x_n)/2\| \geq d$, so $\| x_m + x_n \| \geq 2d$. Using the parallelogram law, we obtain

$$\begin{aligned} \| x_m - x_n \|^2 &= 2 \| x_m \|^2 + 2 \| x_n \|^2 - \| x_m \mid x_n \|^2 \\ &\leq 2 \| x_m \|^2 + 2 \| x_n \|^2 - 4d^2. \end{aligned}$$

and since $2\| x_m \|^2 + 2 \| x_n \|^2 - 4d^2 \to 2d^2 + 2d^2 - 4d^2 = 0$, it follows that $\{x_n\}$ is a Cauchy sequence in C. Since H is complete and C is closed, C is complete, and there exists a vector a; in C such that $x_n \to x$. It is clear by the fact that $\| x \| = \| \lim x_n \| = \lim \| x_n \| = d$ that x is a vector in C with smallest norm. To see that x is unique, suppose that x′ is a vector in C other than x which also has norm d. Then $(x + x')/2$ is also in C, and another application of the parallelogram law yields

$$\begin{aligned} \left\| \frac{x + x'}{2} \right\|^2 &= \frac{\| x' \|^2}{2} + \frac{\| x' \|^2}{2} - \left\| \frac{x - x'}{2} \right\|^2 \\ &< \frac{\| x \|^2}{2} + \frac{\| x' \|^2}{2} \; d^2, \end{aligned}$$

which contradicts the definition of d.

The parallelogram law has another interesting application, which depends on the fact that in any Hilbert space the inner product is related the norm by the following identity:

$$4\,(x, y) = \|x + y\|^2 - \|x - y\|^2 + i\,\|x + iy\|^2 - i\,\|x - iy\|^2 \quad ...(2)$$

This is easily verified by converting the expression on the right into inner products.

Theorem 3:

If B is a complex Banach space whose norm obeys the parallelogram law, and if an inner product is defined on B by (2), *then B is a Hilbert space.*

Proof:

All that is necessary is to make sure that the inner product defined by (2) has the three properties required by the definition of a Hilbert space. This is easy in the case is best treated by splitting it into two parts:

$$(x + y, e) = (z, z) + (y, z),$$

and $(\alpha x, y) = \alpha\,(\xi, \psi)$. The first requires the parallelogram law, and the second follows from the first. We ask the reader to work out the details.

This result has no implications at all for our future work. However, it does provide a satisfying geometric insight into the place Hilbert spaces occupy among all complex Banach spaces: they are precisely those in which the parallelogram law is true.

EXERCISES

1. Show that the series which defines the inner product in Example 2 is convergent.
2. The *Hilbert cube* is the subset of l_2 consisting of all sequences $x = \{x_1, x_2, \ldots, x_n, \ldots\}$ such that $|x_n| \le 1/n$ for all *n*. Show that this set is compact as a subspace of l_2.
3. For the special Hilbert space l_2^n, use Cauchy's inequality to prove Schwarz's inequality.
4. Show that the parallelogram law is not true in l_1^n $(n > 1)$.
5. In a Hilbert space, show that if $\|x\| = \|y\| - 1$, and if $e > 0$ is given, then there exists $\delta > 0$ such that $\|(x + y/)/2\| > 1 - \delta \Rightarrow \|x - y\| < \in$. A Banach space with this property is said to be *uniformly convex.*
6. Give a detailed proof of Theorem 3.

ORTHOGONAL COMPLEMENTS

Two vectors *x* and *y* in a Hilbert space *H* are said to be *orthogonal* (written $x \perp y$) if $(x, y) = 0$. The symbol $\pm$ is often pronounced "perp." Since $\overline{(x, y)} = (y, x)$, we have $x \perp y \Leftrightarrow y \perp x$. It is also clear that $x \perp 0$ for every *x*, and (x, x)

$= \| x \|^2$ shows that 0 is the only vector orthogonal to itself. One of the simplest geometric facts about orthogonal vectors is the *Pythagorean theorem:*

$$x \perp y \Rightarrow \| x + y \|^2 = \| x - y \|^2 = \| x \|^2 + \| y \|^2.$$

A vector x is said to be orthogonal to a non-empty set S (written $x \perp S$) if $x \perp y$ for every y in S, and the *orthogonal complement* of S – denoted by $S^\perp$ — is the set of all vectors orthogonal to S. The following statements are easy consequences of the definition:

$$\{0\}^\perp = H; H^\perp = \{0\};$$

$$S \cap S^\perp \subseteq \{0\};$$

$$S_1 \subseteq S_2 \Rightarrow S_1^\perp \supseteq S_2^\perp;$$

($S^\perp$ is a closed linear subspace of H.

It is customary to write $(S^\perp)^\perp$ in the form $S^{\perp\perp}$. Clearly, $S \subseteq S^{\perp\perp}$. Let M be a closed linear subspace of H. We know that M-1 is also a closed linear subspace, and that M and Mx are disjoint in the sense that they have only the zero vector in common. Our aim in this section is to prove that $H = M \rightarrow M^\perp$, and each of our theorems is a step in this direction.

Theorem 1:

Let M be a closed linear subspace of a Hilbert space H, let x be a vector not in M, and let d be the distance from x to M. Then there exists a unique vector y_0 in M such that $\| x — y_o \| = d$.

Proof:

The set $C = x + M$ is a closed convex set, and d is the distance from the origin to C. By Theorem 2, there exists a unique vector z_0 in C such that $\| z_0 \| = d$. The vector $y_0 = x - z_0$ is easily seen to be in M, and $\| x — y_o \| = \| z_o \| =$ d. The uniqueness of j/n follows from the fact that if y_i is a vector in M such that $y_1 \neq y_0$ and $\| x - y_1 \| = d$, then $z_1 = x - y_1$ is a vector in C such that $z_1 \neq z_0$ and $\| z_1 \| = d$, which contradicts the uniqueness of z_o.

We use this result to prove

Theorem 2:

If M is a proper closed linear subspace of a Hilbert space H, then there exists a non-zero vector z_0 in H such that $z_0 \perp M$.

Proof:

Let x be a vector not in M, and let d be the distance from x to M. By Theorem 1, there exists a vector y_0 in M such that $\| x - y_0 \| = d$. We define z_0 by $z_0 = x - y_0$ (Fig. 2.2), and we observe that since $d > 0$, z_0 is a non-zero vector. We conclude the proof by showing that if y is an arbitrary vector in M, then $z_0 \perp y$. For any scalar a, we have

$$\| z_0 - \alpha y \| = \| x - (y_0 + \alpha y)\| \geq d = \| z_0 \|,$$

so

$$\| z_0 - \alpha y \|^2 - \| z_0 \|^2 \geq 0.$$

and

$$-\bar{\alpha}(z_0, y) - \alpha\overline{(z_0, y)} | | \alpha |^2 \| y \|^2 \geq 0 \quad ...(3)$$

If we put $\alpha = \beta\ (z_0, y)$ for an arbitrary real number β, then (1) becomes

$$-2\beta\, |(z_0, y)|^2 + \beta^2\, |(z_0, y)|^2 \| y \|^2 \geq 0. \quad ...(4)$$

If we now put $\alpha = |(z_0, y)|^2$ and $\beta = \| y \|^2$, we obtain

$$-2\beta\alpha + \beta^2\alpha\beta \geq 0$$

so

$$\beta a\,(\beta b - 2) \geq 0.$$

For all real β. However, if $a > 0$, then (2) is obviously false for all sufficiently small positive β. We see from this that $a = 0$, which means that $z_0 \perp y$.

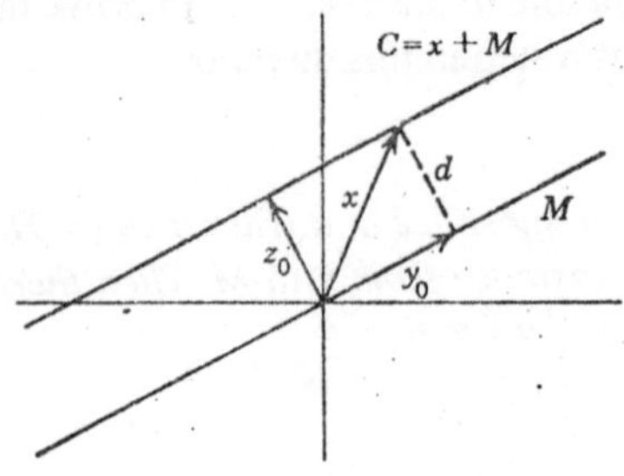

Fig. 2.2 : Schwarz's inequality.

This proof of Theorem 2 may strike the reader as being excessively dependent on ingenious computations. If so, he will be pleased to learn that the ideas developed in the next section can be used to provide another proof which is free of computation.

In order to state our next theorem, we need the following additional concept. Two non-empty subsets *S1 and* S_2 of a Hilbert space are said to be orthogonal (written $S_1 \perp S_2$) if $x \perp y$ for all x in S_1 and y in S_2.

Theorem 3:

If M and N are closed linear subspaces *of a Hilbert space H such that M* $\perp$ *N, then the linear subspace M + N is also closed.*

Proof:

Let z be a limit point of M + N. It suffices to show that z is in M + *N*. There certainly exists a sequence $\{z_n\}$ in M + *N* such that $z_n \to z$. By the assumption that $M \perp N$, we see that M and *N* are disjoint, so each z_n can be written uniquely in the form $z_n = x_n + y_n$ where x_n is in *M* and *yn* is in *N*. The Pythagorean theorem shows that $\| z_m - z_n \|^2 = \| x_m = \| x_m - x_n \|^2 + \| y_m - y_n \|^2$, so $\{x_n\}$ and $\{y_n\}$ are Cauchy sequences in M and N. M and N are closed, and therefore complete, so there exist vectors x and y in M and N such that $x_n \to x$ and $y_n \to y$. Since $x + y$ is in *M* + *N*, our conclusion follows from the fact that $z = \lim z_n = \lim (x_n + y_n) = \lim x_n + \lim y_n = x + y$.

The way is now clear for the proof of our principal theorem.

Theorem 4:

If *M is a closed linear subspace of a Hilbert space H, then* $H = M \oplus M^{\perp}$.

Proof:

Since M and $M^{\perp}$ are orthogonal closed linear subspaces of H, Theorem C shows that $M + M^{\perp}$ is also a closed linear subspace of *H*. We prove that $M + M^{\perp}$ equals *H*. If this is not so, then by Theorem 2 there exists a vector $z_0 \perp 0$ such that $z_0 \perp (M + M^{\perp})$. This non-zero vector must evidently lie in $M^{\perp} \cap M^{\perp\perp}$; and since this is impossible, we infer that $H = M + M^{\perp}$. To conclude the proof, it suffices to observe that since M and $M^{\perp}$ are disjoint, the statement that $H = M + M^{\perp}$ can be strengthened to $H = M \oplus M^{\perp}$.

The main effect of this theorem is to guarantee that a Hilbert space is always rich in projections. In fact, if M is an arbitrary closed linear subspace of a Hilbert space H, then Theorem 4 shows that there exists a projection defined on H whose range is M and whose null space is $M^{\perp}$. This satisfactory state of affairs is to be contrasted with the situation in a general Banach space, as explained in the remarks following Theorem 4.

EXERCISES

1. If *S* is a non-empty subset of a Hilbert space, show that $S^{\perp} = S^{\perp\perp\perp}$.
2. If *M* is a linear subspace of a Hilbert space, show that *M* is closed $\Leftrightarrow M = M^{\perp\perp}$.
3. If S is a non-empty subset of a Hilbert space *H*, show that the set of all linear combinations of vectors in *S* is dense in $H \Leftrightarrow S^{\perp} = \{0\}$.
4. If *S* is a non-empty subset of a Hilbert space *H*, show that $S^{\perp\perp}$ is the closure of the set of all linear combinations of vectors in *S*. This

is usually expressed by saying that $S^{\perp\perp}$ is the smallest closed linear subspace of *H* which contains *S*.

ORTHONORMAL SETS

An *orthonormal set* in a Hilbert space *H* is a non-empty subset of *H* which consists of mutually orthogonal unit vectors; that is, it is a non-empty subset $\{e_i\}$ of *H* with the following properties:

(1) $i \neq j \Rightarrow e_i \perp e_{ji}$,

(2) $\| e_i \| = 1$ for every i.

If H contains only the zero vector, then it has no orthonormal sets. If H contains a non-zero vector x, and if we normalize x by considering $e = x/\| z \|$, then the single-element set $\{e\}$ *is* clearly an orthonormal set. More generally, if $\{x_i\}$ is a non-empty set of mutually orthogonal non-zero vectors in H, and if the x_i's are normalized by replacing each of them by $e_i = x_i/\| x_i \|$, then the resulting set $\{e_i\}$ is an orthonormal set.

Example 1:

The subset $\{e_1, e_2, \ldots, e_n\}$ of l_2^n, where e, is the n-tuple with 1 in the ith place and 0's elsewhere, is evidently an orthonormal set in this space.

Example 2:

Similarly, if en is the sequence with 1 in the nth place and 0's elsewhere, then $\{e_1, e_2, \ldots, e_n, \ldots\}$ is an orthonormal set in l_2.

At the end of this section, we give some additional examples taken from the field of analysis.

Every aspect of the theory of orthonormal sets depends in one way or another on our first theorem.

Theorem 1:

Let $[e_1, e_2, \ldots, e_n]$ be a finite orthonormal set in a Hilbert space H. If x is any vector in H, then

$$\sum_{i=1}^{n} |(x, e_i)|^2 \leq \| x \|^2 \qquad \text{...(5)}$$

$$\textit{further, } x - \sum_{i=1}^{n} (x, e_i)e_i \perp e_j \textit{ for each } j. \qquad \text{...(6)}$$

Proof:

The follows from a computation similar to that used in proving Schwarz's inequality:

$$0 \leq \| x - \sum_{i=1}^{n} (x, e_i) e_i \|^2$$

$$= (x - \sum_{i=1}^{n} (x, e_i) e_i, x - \sum_{j=1}^{n} (x, e_j) e_j)$$

$$= (x, x) - (x - \sum_{i=1}^{n} (x, e_i) \overline{(x, e_i)}, - \sum_{j=1}^{n} (x, e_j) \overline{(x, e_i)}$$

$$+ \sum_{i=1}^{n} \sum_{j=1}^{n} (x, e_i) \overline{(x, e_j)} (e_i, e_j)$$

$$= \| x \|^2 - \sum_{i=1}^{n} | (x, e_i) |^2$$

To conclude the proof, we observe that

$$(x - \sum_{i=1}^{n} (x, e_i) e_i, e_j) = (x, e_j) - \sum_{i=1}^{n} (x, e_i)(e_i, e_j) = (x, e_i) - (x, e_j) = 0$$

From which statement follows at once.

The reader should note that the inequality can be given the following loose but illuminating geometric interpretation: the sum of the squares of the components of a vector in various perpendicular directions does not exceed the square of the length of the vector itself. This is usually called *Bessel's inequality,* though, as we shall see below, it is only a special case of a more general inequality with the same name. In a similar vein, relation (2) says that if we subtract from a vector its components in several perpendicular directions, then the result has no component left in any of these directions.

Our next task is to prove that both parts of Theorem 1 generalize to the case of an arbitrary orthonormal set. The main problem here is to show that the sums in (1) and (2) can be defined in a reasonable way when no restriction is placed on the number of e_i's under consideration. The key to this problem lies in the following theorem.

Theorem 2:

If $\{e_i\}$ is an orthonormal set in a Hilbert space H, and if x is any vector in H, then the set $S = \{e_i : (x, e_i) \neq 0\}$ is either empty or countable.

Proof:

For each positive integer *n,* consider the set

$$S_n = \{e_i : |(x, e_i)|^2 > \| x \|^2/n\}.$$

By Bessel's inequality, *Sn* contains at most *n — I* vectors. The conclusion now follows from the fact that $S = \bigcup_{n=1}^{\infty} S_n$.

As our first application of this result, we prove the general form of Bessel's inequality.

Theorem 3:

(Bessel's Inequality) : If $\{e_i\}$ *is an orthonormal set in a Hilbert space H, then*

$$\Sigma \mid (x_i, e_i)^2 \mid \leq \| x \|^2 \quad ...(7)$$

for every vector x in H.

Proof:

Our basic obligation here is to explain what is meant by the sum on the left of (3). Once this is clearly understood, the proof is easy. As in the preceding theorem, we write $S = \{e_i:(x,e_i) \neq 0\}$. If S is empty, we define $S\,|(x, e_i)|^2$ to be the number 0; and in this case, (3) is obviously true. We now assume that *S* is non-empty, and we see from Theorem B that it must be finite or countably infinite. If S is finite, it can be written in the form $S = [e_1, e_2, \ldots, e_n\}$ for some positive integer *n.* In this case, we define $\Sigma\,|(x,ei)\,|^2$ to be $\Sigma_{i=1}^{n} |(x, e_i)|^2$, which is clearly independent of the order in which the elements of *S* are arranged. The inequality (3) now reduces to (1), which has already been proved. All that remains is to consider the case in which *S* is countably infinite. Let the vectors in *S* be arranged in a definite order:

$$\Sigma = \{e_1, e_2, \ldots, e_n, \ldots\}.$$

By the theory of absolutely convergent series, if $\Sigma_{i=1}^{\infty} |(x, e_n)|^2$ converges, then every series obtained from this by rearranging its terms also converges, and all such series have the same sum. We therefore define $\Sigma|(x, e_i)|^2$ to be $\Sigma_{n=1}^{\infty} |(x, e_n)|^2$, and it follows from the above remark that $\Sigma|(x, e_i)|^2$ is a non-negative extended real number which depends only m *S,* and not on the arrangement of its vectors. We conclude the proof ay observing that in this case, (3) reduces to the assertion that

$$\sum_{n=1}^{\infty} |(x, e_n)|^2 \leq \|x\|^2$$

and since it follows from (1) that no partial sum of the series on the left of (4) can exceed $\| x \|^2$, it is clear that (4) itself is true.

The second part of Theorem A is generalized in essentially the same way.

Theorem 4:

If $\{e_i\}$ *is an orthonormal set in a Hilbert space H, and if x is an arbitrary in H, then*

$$x - \Sigma (x,e_i) e_i \perp e_j \qquad ...(8)$$

for each j.

Proof:

As in the above proof, we define S $(x,e_i)e_i$, for each of the various cases, and we prove (5) as we go along. We again write

$$\Sigma = \{e_i : (x,e_i) \neq 0\}.$$

When *S* is empty, we define $\Sigma (x,e_i)e_i$ to be the vector 0, and we observe that (5) reduces to the statement that $x - 0 = x$ is orthogonal to each e_j, which is precisely what is meant by saying that *S* is empty. When S is non-empty and finite, and can be written in the form

$$\Sigma = \{e_1, e_2, \ldots, e_n\},$$

We define $\Sigma (x,e_i) e_i$ to be $\Sigma_{n=1}^{n} (x,e_i) e_i$; and in this case, (5) reduces to (2), which has already been proved.

By Theorem B, we may assume for the remainder of the proof that *S* is countably infinite. Let the vectors in S be listed in a definite order: $S = \{e_1, e_2, \ldots, e_n, \ldots\}$. We put $s_n = \Sigma_{n=1}^{n} (x,e_i) e_i$, and we note that for m > n we have

$$\| x_m - s_n \|^2 = \| \sum_{i=n+1}^{m} (x,e_i) e_i \|^2 = \sum_{i=n+1}^{m} |(x,e_i)|^2$$

Bessel's inequality shows that the series $\Sigma_{n=1}^{\infty} |(x, e_n)|^2$ converges, so $\{s_n\}$ is a Cauchy sequence in H; and since H is complete, this sequence converges to a vector s, which we write in the form $s = \Sigma_{n=1}^{\infty} (x, e_n) e_n$. We now define $\Sigma (x,e_i) e_i$ to be $\Sigma_{n=1}^{\infty} (x, e_n) e_n$, and—deferring for a moment the question of what happens when the vectors in S are rearranged—we observe that (5) follows from (2) and the continuity of the inner product :

$$\begin{aligned}(x - \Sigma(x,e_i)e_i, e_j) &= (x - s_1, e_j) = (x, e_j) - (x,e_j,) - (\lim s_n\, e_j) \\ &= (x, e_j) - \lim (s_n, e_j) = (x,e_j) - (x,e_j) = 0.\end{aligned}$$

All that remains is to show that this definition of $\Sigma(x,e_i) e_i$ is valid, in the sense that it does not depend on the arrangement of the vectors in *S.* Let the vectors in *S* be rearranged in any manner:

$$S = \{f_1, f_2, \ldots, f_n, \ldots\}.$$

We put $s'_n = \Sigma_{i=1}^{\infty} (x, f_i) f_i$, and we see — as above — that the sequence *[$',,}* converges to a limit s', which we write in the form $s' = \Sigma_{n=1}^{\infty} (x, f_n) f_n$. We conclude the proof by showing that *s'* equals s. Let $\varepsilon > 0$ be given, and let n_0 be a positive integer so large that if $n \geq n_0$, then $\| s_n - s \| < \varepsilon$, $\| s'_n - s' \| <$

ε, and $\Sigma_{i=n_0+1}^{\infty} |(x, e_i)|^2 < \varepsilon^2$. For some positive integer $m_0 > n_0$, all terms of s_{n_0}, occur among those of s'_{m_0}, so s'_{m_0} sum of terms of the form $(x, e_i)\, e_i$ for $i = n_0 + 1, n_o + 2$, yields $\| s'_{m_0} - s_{n_0} \|^2 \leq \Sigma_{i=n_0+1}^{\infty} |x, e_i)|^2 < \varepsilon^2$, so $\| s'_{m_0} - s_{n_0} \|$ < and

$$\| s' - s \| \geq \| s' - s'_{m_0} \| + \| s'_{m_0} - s \| < \varepsilon + \varepsilon = 3e.$$

Since is arbitrary, this shows that s ' = s.

Let *H* be a non-zero Hilbert space, so that the class of all its orthonormal sets is non-empty. This class is clearly a partially ordered set with respect to set inclusion. An orthonormal set $\{e_i\}$ in *H* is said to be *complete* if it is maximal in this partially ordered set, that is, if it is impossible to adjoin a vector *e* to $\{e_i\}$ in such a way that $\{e_i, e)$ is an orthonormal set which properly contains $\{e_i\}$.

Theorem 5:

Every non-zero Hilbert space contains a complete orthonormal mal set.

Proof:

The statement follows at once from Zorn's lemma, since the union of any chain of orthonormal sets is clearly an upper bound for the chain in the partially ordered set of all orthonormal sets.

Orthonormal sets are truly interesting only when they are complete. The reasons for this are presented in our next theorem.

Theorem 6:

Let H be a Hilbert space, and let $\{e_i\}$ *be an orthonormal set in H. Then the following conditions are all equivalent to one another:*

(1) $\{e_i\}$ *is complete;*

(2) $x \perp \{e_i\}$ =>* = 0;

(3) *if x is an arbitrary vector in H, then x* — Σ (x, e_i) e_i;

(4) *if x is an arbitrary vector in H, then* $\| x \|^2 = \Sigma |(x, e_i)|^2$.

Proof.

We prove that each of the conditions (1), (2), and (3) implies the one following it and that (4) implies (1).

(1) ⇔ (2). If (2) is not true, there exists a vector $x \neq 0$ such that $x \perp \{e_i\}$. We now define *e* by $e = x / \| x \|$, and we observe that $\{e_i\, e\}$ is an orthonormal set which properly contains $\{e_i\}$. This contradicts the completeness of $\{e_i\}$.

(2) ⇒ (3). By Theorem D, $x - \Sigma\, (x, e_i)\, e_i = 0$ is orthogonal to $\{e_i\}$, so (2) implies that $x - S\, (x, e_i)\, e_i = 0$, or equivalently, that $y = 2\, (x, e_i)\, e_i$.

(3) $\Rightarrow$ (4). By the joint continuity of the inner product, the expression in (3) yields

$$\|x\|^2 = (x, x) = (\Sigma(x, e_i)\, e_i\, \Sigma\,(x, e_j)\, e_j) = \Sigma\,(x, e_i)\,\overline{(x, e_i)} \not= \Sigma\,|\,(x, e_i)|^2.$$

(4) $\Rightarrow$ (1). If $\{e_i\}$ is not complete, it is a proper subset of an orthonormal set $\{e_i, e\}$. Since e is orthogonal to all the e_i's, (4) yields $\|e\|^2 = \Sigma|(e, e_i)|^2 = 0$, and this contradicts the fact that e is a unit vector.

There is some standard terminology which is often used in connection with this theorem. Let $\{e_i\}$ be a complete orthonormal set in a Hilbert space *H*, and let x be an arbitrary vector in *H*. The numbers (x, e_i) are called the *Fourier coefficients* of x, the expression $x = \Sigma\,(x, e_i)\, e_i$ is called the *Fourier expansion* of x, and the equation $\|x, e_i|\} = \Sigma\,|(x,e_i)|^2$ is called *Parseval's equation*—all with respect to the particular complete orthonormal set $(e_i\}$ under consideration. These terms come from the classical theory of Fourier series, as indicated in our next example.

Example 3:

Consider the Hilbert space L_2 associated with the measure space $[0, 2\pi]$, where measure is Lebesgue measure and integrals are Lebesgue integrals. This space essentially consists of all complex functions f defined on $[0,2\pi]$ which are Lebesgue measurable and square integrable, in the sense that

$$\int_0^{2\pi} |f(x)|^2\, dx < \infty$$

Its norm and inner product are defined by

$$\|f\| = \left(\int_0^{2\pi} |f(x)|^2\, dx\right)^{1/2}$$

and
$$(f, g) = \int_0^{2\pi} f(x)\overline{g(x)}\, dx\,.$$

A simple computation shows that the functions e^{inx}, for

$$n = 0, \pm 1, \pm 2, \ldots,$$

are mutually orthogonal in L_2:

$$\int_0^{2\pi} e^{imx}\, e^{-inx}\, dx = \begin{cases} 0 & m \neq n \\ 2\pi & m = n \end{cases}$$

It follows from this that the functions e_n $(n = 0, \pm 1, \pm 2, \ldots)$ defined by $e_n(x) = e^{inx} / \sqrt{2\pi}$ form an orthonormal set in L_2. For any function f in L_2, the numbers

$$c_n = (f, e_n) = \frac{1}{\sqrt{2\pi}} \int_0^{2\pi} f(x)\, e^{-inx}\, dx \qquad \ldots(9)$$

are its classical *Fourier coefficients,* and *Bessel's inequality* takes the form

$$\sum_{n=-\infty}^{\infty} |c_n|^2 \leq \int_0^{2\pi} |f(x)|^2 \, dx .$$

It is a fact of very great importance in the theory of Fourier series that the orthonormal set $\{e_n\}$ is complete in L_2. As we have seen in Theorem F, the completeness of $\{e_n\}$ is equivalent to the assertion that for every f in L_2, Bessel's inequality can be strengthened to *Parseval's equation:*

$$\sum_{n=-\infty}^{\infty} |c_n|^2 = \int_0^{2\pi} |f(x)|^2 \, dx$$

Theorem F also tells us that the completeness of $\{e_n\}$ is equivalent to the statement that each / in L_2 has a
Fourier expansion:

$$f(x) = \frac{1}{\sqrt{2\pi}} \sum_{n=-\infty}^{\infty} c_n e^{inx} \qquad ...(10)$$

It must be emphasized that this expansion is not to be interpreted as saying that the series converges pointwise to the function. The meaning of (9) is that the partial sums of the series, that is, the vectors f_n in L_2 denned by

$$f_n(x) = \frac{1}{\sqrt{2\pi}} \sum_{k=-n}^{\infty} c_k e^{ikx} \qquad ...(11)$$

Converge to the vector f in the sense of L_2

$$\| f_n - f \| \to 0.$$

This situation is often expressed by saying that f is the *limit in the mean* of the fn's. We add one final remark to our description of this portion of the theory of Fourier series. If f is an arbitrary function in L_2 with Fourier coefficients c_n defined by (6), then Bessel's inequality tells us that the series $\Sigma_{n=-\infty}^{\infty} |c_n|^2$ converges. The celebrated *Riesz-Fischer theorem* asserts the converse: if c, (n = 0, ± 1, ± 2, . . .) are given complex numbers for which $\Sigma_{n=-\infty}^{\infty} |c_n|^2$ converges, then there exists a function f in L_2whose Fourier coefficients are the c_n's. If we grant the completeness of L_2 as a metric space, this is very easy to prove. All that is necessary is to use the c_n's to define a sequence of f_n's in accordance with (10) . The functions $e^{inx}/\sqrt{2\pi}$ form an orthonormal set, so for m > n we have

$$\|f_m - f_n\|^2 = \sum_{|k|=n+1}^{m} |c_k|^2$$

By the convergence of $\Sigma_{n=-\infty}^{\infty} |c_n|^2$, the sum on the right of (9) can be made as small as we please for all sufficiently large n and all m > n. This tells us that the f_n's form a Cauchy sequence in L_2; and since L_2 is complete, there exists a function f in L_2 such that $f_n \to f$. This function f is given by (7), and the c_n's are clearly its Fourier coefficients. It is apparent from these remarks that the essence of the Riesz-Fischer theorem lies in the completeness of L_2 as a metric space.

We shall have use for one further item in the general theory of orthonormal sets, namely, the Gram-Schmidt orthogonalization process. Suppose that $\{x_1, x_2, \ldots, x_n, \ldots\}$ is a linearly independent set in a Hilbert space H. The problem is to exhibit a constructive procedure for converting this set into a corresponding orthonormal set $\{x_1, x_2, \ldots, x_n, \ldots\}$ with the property that for each n the linear subspace of H spanned by {ei, e2, . . . , en) is the same as that spanned by $\{x_1, x_2, \ldots, x_n, \ldots\}$. Our first step is to normalize x_1 — which is necessarily nonzero — by putting

$$e_1 = \frac{x_1}{\|x_1\|}$$

The next step is to subtract from x_2 its component in the direction of e_1 to obtain the vector $x_2 — (x_2,e_1)\, e_1$ orthogonal to e_1, and then to normalize this by putting

$$e_2 = \frac{x_2 - (x_2, e_1)e_1}{\|x_2 - (x_2, e_1)e_1\|}$$

We observe that since x_2 is not a scalar multiple of x_1, the vector $x_2 - (x_2,e_1)\, e_1$ is not zero, so the definition of e_2 is valid. Also, it is clear that e_2 is a linear combination of x_1 and x_2, and that x_2 is a linear combination of e_1 and e_2. The next step is to subtract from x_3 its components in the directions of e_1 and e_2 to obtain a vector orthogonal to e_1 and e_2, and then to normalize this by putting

$$e_3 = \frac{x_3 - (x_3, e_1)e_1 - (x_3, e_2)e_2}{\|x_3 - (x_3, e_1)e_1 - (x_3, e_2)e_2\|}$$

If this process is continued in the same way, it clearly produces an orthonormal set $\{e_1, e_2, \ldots, e_n, \ldots)$ with the required property.

Example 4:

Many orthonormal sets of great interest and importance in analysis can be obtained conveniently by applying the Gram-Schmidt process to sequences of simple functions.

(a) In the space L_2 associated with the interval [—1,1], the functions x^n (n = 0, 1, 2, . . .) are linearly independent. If we take these functions to be the

x_n's in the Gram-Schmidt process, then the e_n's are the normalized Legendre polynomials.

(b) Consider the space L_2 over the entire real line. If the x_n's here are taken to be the functions $x_n e^{-x^2/2}$ (n = 0, 1, 2, . . .), then the corresponding e_n's are the normalized Hermite functions.

(c) Consider the space L_2 associated with the interval $[0, + \infty)$. If the x_n's are the functions $x^n e^{-x}$ (n = 0, 1, 2, . . .), then the e_n's are the normalized Laguerre functions.

Each of the orthonormal sets described in the above example can be shown to be complete in its corresponding Hilbert space. The analysis involved in a detailed study of these matters is quite complicated and has no proper place in the present book. The reader should recognize, however—and this is our only reason for mentioning the material in Examples 3 and 4—that the theory of Hilbert spaces does have significant contacts with many solid topics in analysis.

EXERCISES

1. Let $\{e_2, e_2, \ldots, e_n\}$ be a finite orthonormal set in a Hilbert space *H*, and let z be a vector in *H*. If $a_1, a_2, \ldots, a_n$ are arbitrary scalars, show that $\|x - \Sigma_{i=1}^{n} \alpha_i e_i\|^2$ attains its minimum value $\Leftrightarrow a_i = (x, e_i)$ for each *i*. (*Hint:* expand $\|x - \Sigma_{i=1}^{n} \alpha_i e_i\|^2$, add and subtract $\Sigma_{i=1}^{n} |x, e_i)|^2$, and obtain an expression of the form $\Sigma_{i=1}^{n} |x, e_i) - \alpha_i|^2$ in the result.)
2. Show that the orthonormal sets described in Examples 1 and 2 are complete.
3. Show that every orthonormal set in a Hilbert space is contained in some complete orthonormal set, and use this fact to give an alternative proof.
4. Prove that a Hilbert space *H* is separable *of* every orthonormal set in *H* is countable.
5. Show that an orthonormal set in a Hilbert space is linearly independent, and use this to prove that a Hilbert space is finite-dimensional $\Leftrightarrow$ every complete orthonormal set is a basis.
6. Prove that any two complete orthonormal sets in a Hilbert space *H* have the same cardinal number. This cardinal number is called the *orthogonal dimension* of *H* (if *H* has no complete orthonormal sets, into orthogonal dimension is said to be 0).
7. If *H* and *H'* are Hilbert spaces, prove that *H* is isometrically isomorphic to *H'* «=> they have the same orthogonal dimension. (*Hint:*

by Eq. (2), an isometric isomorphism T preserves inner products,] in the sense that $(T(x),T(y)) = (x, y)$.)

8. Let S be a non-empty set, and let l_2 *(S)* be the set of all complex functions f defined on S with the following two properties:
   (1) $\{s : f(s) \neq 0\}$ is empty or countable;
   (2) $\Sigma \mid f(s) \mid\mid^2 < \infty$.

   These functions clearly form a complex linear space with respect to pointwise addition and scalar multiplication. Show that l_2 (S) becomes a Hilbert space if the norm and inner product are defined by $\| f \| = (S \mid f(s) \mid^2)^{1/2}$ and $(f, g) = \Sigma f(s)\overline{g(s)}$. Show also that the set of all functions defined on S which have the value 1 at a single point and are 0 elsewhere is a complete orthonormal set in $l_2(S)$. We shall see in the next problem that Hilbert spaces of the type described here are universal models for all non-zero Hilbert spaces.

9. Let $S = .\{e_i)$ be a complete orthonormal set in a Hilbert space H. Each vector x in H determines a function f defined on S by

   $$f(e_i) = (x, e_i),$$

   and Theorems 2 and 3 tell us that f is in l_2 (S). Show that the mapping $x \to f$ is an isometric isomorphism of H onto l_2 (S).

THE CONJUGATE SPACE H*

We pointed out in the introduction to this one of the fundamental properties of a Hilbert space H is the fact that there is a natural correspondence between the vectors in H and the functionals in H*. Our purpose in this section is to develop the features of this correspondence which are relevant to our work with operators in the rest of t chapter.

Let y be a fixed vector in H, and consider the function f_y defined on H by $f_y(x) = (x, y)$. It is easy to see that f_y is linear, for.

$$\begin{aligned} f_y(x_1 + x_2) &= (x_1 + x_2, y) \\ &= (x_1, y) + (x_2, y) \\ &= f_y(x_1) + f_y(x_2) \end{aligned}$$

and

$$\begin{aligned} f_y(\alpha x) &= (\alpha x, y) \\ &= \alpha (x, y) \\ &= \alpha f_y(x). \end{aligned}$$

Further, f_y is continuous and is therefore a functional, for Schwarz's inequality gives

$$| f_y(x) | = | x, y) | \leq \| x \| \| y \|,$$

which shows that $\| fy \| \leq \| y \|$. Even more, equality is attained here, that is, $\| f_y \| = \| y \|$. This is clear if $y = 0$; and if $y \neq 0$, it follows from

$$\| fy \| = \sup | \{f_y(x) | : \| x \| = 1\}$$

$$\geq \left| f_y\left(\frac{y}{\|y\|}\right) \right|$$

$$= \left| \left(\frac{y}{\|y\|}, y\right) \right| = \| y \|.$$

To summarize, we have seen that $y \to f_y$ is a norm-preserving mapping of H into H*. This observation would be of no more than passing interest if it were not for the fact that every functional in H* arises in just this way.

Theorem:

Let H be a Hilbert space, and let f be an arbitrary functional in H. Then there exists a unique vector y in H such that*

$$f(x) = (x, y) \qquad ...(12)$$

for every x in H.

Proof:

It is easy to see that if such a y exists, then it is necessarily unique. For if we also have f(x) = (x, y') for all x, then (x, y') = (x, y) and (x, y' — y) = 0 for all x; and since 0 is the only vector orthogonal to every vector, this implies that y' – y – 0 or y' = y.

We now turn to the problem of showing that y does exist. If f = 0, then it clearly suffices to choose y = 0. We may therefore assume that $f \neq 0$. The null space M of f is thus a proper closed linear subspace of H, and by theorem there exists a non-zero vector y_0 which is orthogonal to M . We show that if a is a suitably chosen scalar, then the vector $y = \alpha y_0$ meets our requirements. We first observe that no matter what a may be, (1) is true for every x in M; for f(x) = 0 for such an x, and since x is orthogonal to y_0, we also have (x, y) = 0. This allows us to focus our attention on choosing a in such a way that (1) is true for $x = y_0$. The condition this imposes on a is that

$$f(y_0) = (y_0, \alpha y_0) = \bar{\alpha} \ \| y0 \|^2.$$

We therefore choose α to be $\overline{f(y_0)} / \|y_0\|^2$, and it follows that (1) is true for every x in M and for $x = y_0$. It is easily seen that each x in H can be written in the form $x = m + \beta y_0$ with m in M: all that is necessary is to choose β in such a way that $f(x - \beta y_0) = f(x) - f(x) - \beta f(y_0) = 0$, and this is accomplished by putting $\beta = f(x)/f(y_0)$. Our conclusion that (1) is true for every x in H now follows at once from

$f(x) = f(m + \beta y_0) = f(m) + \beta f(y_0) = (m, y) + \beta (y_0, y) = (m + \beta y_0, y) = (x, y)$.

This result tells us that the norm-preserving mapping of H into H* defined by

$$y \to f_y \text{ where } f_y(x) = (x, y), \qquad ...(13)$$

is actually a mapping of H onto H*. It would be pleasant if (2) were also a linear mapping. This is not quite true, however, for

$$f_{y_1} + y_2 = f_{y_1} + f_{y_2} \text{ and } f_{\alpha y} = \bar{\alpha} f_y. \qquad ...(14)$$

It is an easy consequence of (3) that the mapping (2) is an isometry, for $\| f_x - fy \| = \| f_{x-y} \| = \| x - y \|$. We state several interesting additional facts about this mapping (and what it enables us to do) in the problems, and we leave their verification to the reader. It should be remembered, however, that the real significance of this entire circle of ideas lies in its influence on the theory of the operators on H. We begin the treatment of these matters in the next section.

EXERCISES

1. Verify relations (3).
2. Let H be a Hilbert space, and show that H* is also a Hilbert space with respect to the inner product defined by $(f_x, f_y) = (y, x)$. In just the same way, the fact that H* is a Hilbert space implies that H** is a Hilbert space whose inner product is-given by $(F_f, F_0) = (g, f)$.
3. Let H be a Hilbert space. We have two natural mappings of H into H**, the second of which is onto: the Banach space natural embedding $x \to F_x$, where $F_x(f) = f(x)$, and the product mapping $x \to fx \to F_{f_x}$, where $f_x(y) = (y, x)$ and $F_{y_x}(f) = (f, f_x)$. Show that these mappings are equal, and conclude that H is reflexive, Show also that $(F_x, F_y) = (x, y)$.

THE ADJOINT OF AN OPERATOR

Throughout the rest of this chapter, we focus our attention on a fixed but arbitrary Hilbert space H, and unless we specifically state otherwise, it is to be understood that H is the context for all our discussions and theorems.

Let T be an operator on H. That T gives rise to an operator T* (its conjugate) on H*, where T* is denned by

$$(T^*f)\,x = f(T_x)\,1.$$

We also saw that the mapping T → T* is an isometric isomorphism of $\mathcal{B}$(H) into $\mathcal{B}$(H*) which reverses products and preserves the identity transformation. In the same way, T* gives rise to an operator T** on H**; and since H is reflexive, it follows that T** = T when H** is identified with H by means of the natural imbedding. These statements depend only on the fact that H is a reflexive Banach space. We now bring its Hilbert space character into the picture, and we use the natural correspondence between H and H* discussed in the previous section to pull T* down to H. The details of this procedure are as follows (Fig. 2.3). Let y be a vector in H, and f_y, its corresponding functional in H*; operate with T* on f_y, to obtain a functional $f_y = T^*f_y$, and return to its corresponding vector z in H. There are three mappings under consideration here, and we are forming their product:

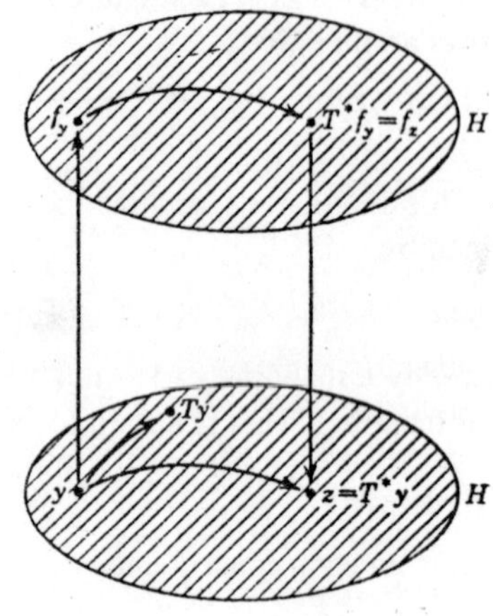

Fig. 2.3 : The conjugate and the adjoint of T.

$$y \to f_y \to T^*f_y = f_x \to z. \qquad ...(15)$$

We write z = T*y, and we call this new mapping T* of H into itself the adjoint of T. The same symbol is used for the adjoint of T as for its conjugate because these two mappings are actually the same if H and H * are identified by means of the natural correspondence. It is easy to keep track of whether T* signifies the conjugate or the adjoint of T by noticing whether it operates on functionals or on vectors. The action of the adjoint can be linked more closely to the structure of H by observing that for every vector x we have $(T^*f_y)\,x = f_y\,(Tx) = (Tx, y)$ and $(T^*f_y)\,x = f_x\,(x) = (x, z) = (x, T^*y)$, so that

$$(Tx, y) = (x, T^*y) \qquad ...(16)$$

for all x and y. Eqn. (16) is much more than merely a property of the adjoint of T, for it uniquely determines this adjoint. The proof is simple, if T′ is any mapping of H into itself such that (Tx, y) = (x,T′ y) or all x and y, then (x,T′y) = (x,T*y) for all z, so T′y = T*y;[1] and since the latter is true for all y, T′ = T*.

Our remarks in the above paragraph have shown that to each operator T on H there corresponds a unique mapping T* of H into itself (called the adjoint of T) which satisfies relation for all x and y. There is a more direct but less natural approach to these ideas, one which avoids any reference to the conjugate of T. If y is fixed, it is clear that the expression (Tx, y) is a scalar-valued continuous linear function of x. By Theorem 1, there exists a unique vector z such that (Tx, y) = (x, z) for all x. We now write z = T*y, and since y is arbitrary, we again have relation (2) for all x and y. The fact that T* is uniquely determined by (2) follows just as before.

The principal value of our approach to the definition of the adjoint (as opposed to that just mentioned) lies in the motivation it provides for considering adjoints at all. We can express this by emphasizing that an operator on a Banach space always has a conjugate which operates on the conjugate space; and when the Banach space happens to be a Hilbert space, then, as we have seen, the natural correspondence discussed in the previous section makes it almost inevitable that we regard the conjugate as an operator on the space itself. Once the definition of the adjoint is j fully understood, however, there is no further need to mention conjugates. All our future work with adjoints will be based on Eq. (16), I and from this point on, the symbol T* will always signify the adjoint of T (and never its conjugate).

As our first step in exploring the properties of adjoints, we verify that T* actually is an operator on H (all we know so far is that it maps H into itself). For any y and 2, and for all x, we have

$$(x, T^*(y + z)) = (Tx, y + z) = (Tx, y) + (Tz, z)$$
$$= (x, T^*y) + (x, T^*z) = (x, T^*y + T^*z),$$

so $$T^*(y + z) = T^*y + T^*z.$$

The relation $T^*(\alpha y) = \alpha T^*y$

is proved similarly, so T* is linear. It remains to be seen that T* is continuous; and to prove this, we note that

$$\| T^*y \|^2 = (T^*y, T^*y)$$
$$= (TT^*y, y) \leq \| TT^*y \| \| y \| \leq \| T \| \| T^*y \|$$

implies that $\| T^*y \| \leq \| T \| \| y \|$ for all y, so

$$\| T^* \| \leq \| \ \|.$$

These facts tell us that $T \to T^*$ is a mapping of $\mathcal{B}$(H) into itself. This mapping is called the *adjoint operation* on *$\mathcal{B}$(H)*.

Theorem:

The adjoint operation $T \to T^*$ on (B(H) has the following properties:

(1) $(T_1 + T_2)^* = T_{1'}^* + T_{2'}^*,$

(2) $(\alpha T)^* = \bar{\alpha}\, T^*,$

(3) $(T_1 T_2)^* = T_2^* T_1 V^*,$

(4) $T^{**} = T,$

(5) $\| T^* \| = \| T \|$

(6) $\| T^* \| = \| T \|^2.$

Proof:

The arguments used in proving (1) to (4) are all essentially the same. As an illustration of the method, we observe that (3) follows from the fact that for all x and y we have

$$(x, (T_1T_2)^*y) = (T_1T_2x, y) = (T_2x, T_1^*y) = (x, T_2^*T_1^*y)$$

To prove (5), we note that we already have $\| T^* \| < \| T \|$, and if we apply this to T* instead of T and use (4), we obtain $\| T \| = \| T^{**} \| \leq \| T^* \|$. Half of (6) follows from (5) and the inequality (5), for

$$\| T^*T \| \leq \| T^* \| \, \| T \| \, \| T \| = \| T \|^2,$$

and the fact that $\| T \|^2 \leq \| T^*T \|$ is an immediate consequence of

$$\| Tx \|^2 = (Tx, Tx) = (T^*Tx, x) \leq \| T^*Tx \| \, \| x \| \leq \| T^*T \| \, \| x \|^2$$

The presence of the adjoint operation is what distinguishes the theory of the operators on H from the more general theory of the operators on a reflexive Banach space. In the next three sections, we use this operation as a tool by means of which we single out for special study certain types of operators on *H* whose theory is particularly complete and satisfying.

EXERCISES

1. Prove parts (1), (2), and (4) of Theorem 1.
2. Show that the adjoint operation is one-to-one onto as a mapping of $\mathcal{B}$(H) into itself.
3. Show that 0* = 0 and 7* = I. Use the latter to show that if T is non-singular, then T* is also non-singular, and that in this case $(T^*)^{-1} = (T^{-1})^*$.
4. Show that $\| TT^* \| = \| T |^2$.

SELF-ADJOINT OPERATORS

There is an interesting analogy between the set $\mathcal{B}$(H) of all operators on our Hilbert space H and the set C of all complex numbers. This can be summarized by observing that each is a complex algebra together with a mapping of the algebra onto itself ($T \rightarrow T^*$ and $z \rightarrow \bar{z}$) and that these mappings have similar properties. We shall see that this analogy is quite useful as an intuitive guide to the study of the operators on H. The most significant difference between these systems is that multiplication in the algebra $\mathcal{B}$(H) is in general non-commutative. and it will become clear as we proceed that this is the primary source of the much greater structural complexity of $\mathcal{B}(H)$.

The most important subsystem of the complex plane is the real line, which is characterized by the relation $z = \bar{z}$. By analogy, we consider those operators A on H which equal their adjoints, that is, which satisfy the condition $A = A^*$. Such an operator is said to be self-adjoint. The self-adjoint operators on H are evidently those which are related in the simplest possible way to their adjoints.

We know that $0^* = 0$ and $I^* = I$, so 0 and I are self-adjoint. If A_1 and A_2 are self-adjoint, and if α and β are real numbers, then

$$(\alpha A_1 + \beta A_2)^* = \bar{\alpha} A_1^* + \bar{\beta} A_2^* = \alpha A_1 + \beta A_2$$

shows that $\alpha A_1 + \beta A_2$ is also self-adjoint. Further, if $\{A_n\}$ is a sequence of self -adjoint operators which converges to an operator A, then it is easy to see that A is also self -ad joint; for

$$\| A - A^* \| \leq \| A - A_n \| + \| A_n - A_n^* \| + \| (A_n{}^* - A)^* \| = \| A - A_n \|$$
$$+ \| (A_n - A)^* \| = \| A - A_n \| + \| A_n - A \| = 2\| A_n - A \| \to 0$$

shows that $A - A^* = 0$, so $A = A^*$. These remarks yield our first theorem.

Theorem 1:

The self-adjoint operators in $\mathcal{B}(H)$ form a closed real linear subspace of $\mathcal{B}(H)$–and therefore a real Banach space–which contains the identity transformation.

The reader will notice that we have said nothing here about the product of two self-adjoint operators. Very little is known about such products, and the following simple result represents almost the extent of our information.

Theorem 2:

If A_1 AND A_2 are self-adjoint operators on H, then their product A_1A_2 is self-adjoint $\Leftrightarrow A_1A_2 = A_2A_1$.

Proof:

This is an obvious consequence of

$(A_1A_2)^* = A_2^*A_1^* = A_2A_1$.

The order properties of self-adjoint operators are more interesting, and we devote the remainder of the section to establishing some of the simpler facts in this direction.

If T is an arbitrary operator on H, it is easy to see that

$$T = 0 \Leftrightarrow (Tx, y) = 0$$

for all x and y. It is also clear that $T = 0 \Rightarrow (Tx, x) = 0$ for all x. We shall need the converse of this implication.

Theorem 3:

If T is an operator on H for which (Tx, x) = 0 for all x, then T = 0.

Proof:

It suffices to show that (Tx, y) = 0 for any x and y, and the proof of this depends on the following easily verified identity:

$$(T(\alpha x + \beta y), \alpha x + \beta y) - |\alpha|^2 (Tx, x) - |\beta|^2 (Ty, y)$$

$$= \alpha\bar{\beta}\ (Tx, y) + \bar{\alpha}\beta\ (Ty, x). \qquad ...(17)$$

We first observe that by our hypothesis, the left side of (1)—and therefore the right side as well–equals 0 for all α and β. If we put $\alpha = 1$ and $\beta = 1$, then (1) becomes

$$(Tx, y) + (Ty, x) = 0, \qquad ...(18)$$

and if we put $\alpha = i$ and $\beta = 1$, we get

$$i\ (Tx, y) - i\ (Ty, x) = 0. \qquad ...(19)$$

Dividing (3) by i and adding the result to (2) yields 2(Tx, y) = 0, so (Tx, y) = 0 and the proof is complete.

It is worth emphasizing that this proof makes essential use of the fact that the scalars are the complex numbers (and not merely the real numbers).

We now apply this result to proving our next theorem, which indicates that self-adjoint operators are linked to real numbers by stronger ties than might be suspected from the loose analogy that led to their definition.

Theorem 4:

An operator T on H is self-adjoint $\Leftrightarrow$ (Tx, x) is real far all x.

Proof:

If T is self-adjoint, then

$$\overline{(Tx, x)} = (x, Tx) = (x, T^*x) = (Tx, x)$$

shows that (Tx, x) is real for all x. On the other hand, if (Tx, x) is real for all z, then (Tx, x) = (Tx, x) = (x, T*x) = (T*x, x) or

$$([T - T^*]x, x) = 0$$

for all x. By Theorem C, this implies that T – T* = 0, so T = T*.

This theorem enables us to define a respectable and useful order relation on the set of all self-adjoint operators. If A_1 and A_2 are self-adjoint, we write $A_1 \leq A_2$ if $(A_1x, x) \leq (A_2x, x)$ for all x. The main elementary facts about this relation are summarized in

Theorem 5:

The real Banach space of all self-adjoint operators on H is a partially ordered set whose linear structure and order structure are related by the following properties:

(1) *if $A_1 \leq A_2$, then $A_1 + A \leq A_2 + A$ for every A;*

(2) *if $A_1 \leq A_2$ and $\alpha \geq 0$, then $\alpha A_1 \leq \alpha A_2$.*

Proof:

The relation in question is obviously reflexive and transitive. To show that it is also antisymmetric, we assume that $A_1 \leq A_2$ and $_2 \leq A_1$. This implies at once that $([A_1 - A_2,]x, x) = 0$ for all x, so by Theorem C, $A_1 - A_2 = 0$ and $A_1 = A_2$. The proofs of properties (1) and (2) are easy. For instance, if $A_1 \leq A_2$, so that $(A_1x, x) \leq (A_2x, x)$ for all *x,* then $(A_1x, x) + (Ax, x) \leq (A_2x, x) + (Ax, x)$ or $([A_1 + A]x, x) \leq ([A_2 + A]x, x)$ for all x, so $A_1 + A \leq A_2 + A$. The proof of (2) is similar.

A self-adjoint operator A is said to be *positive* if $A \geq 0$, that is, if $(Ax, x) \geq 0$ for all x. It is clear that 0 and I are positive, as are T*T and TT* for an arbitrary operator T.

Theorem 6:

*If A is a positive operator on H, then I + A is non-singular. In particular, I + T*T and I + TT* are non-singular for an arbitrary operator T on H.*

Proof:

We must show that I + A is one-to-one onto as a mapping of H into itself. First, it is one-to-one, for

$$(I + A)x = 0 \Rightarrow Ax = -x$$

$$\Rightarrow \quad (Ax, x) = (-x, x) = -\|x\|^2 \geq 0 \Rightarrow x = 0.$$

We next show that the range M of I + A is closed. It follows from $\|(I+A)x\|^2 = \|x\|^2 + \|Ax\|^2 + 2(Ax, x)$–and the assumption that A is positive–that $\|z\| \leq \|(I + A)x\|$. By this inequality and the completeness of H, M is complete and therefore closed. We conclude the proof by observing that M = H, for otherwise there would exist a nonzero vector x_0 orthogonal to M, and this would contradict the fact that $(x_0, [I + A]x_0) = 0 \Rightarrow \|x_0\|^2 = -(Ax_0, x_0) \leq 0 \Rightarrow x_0 = 0$.

If the reader wonders why we fail to show that the partially ordered set of all self-adjoint operators is a lattice, the reason is simple: it isn't true. As a matter of fact, this system is about as far from being a lattice as a partially ordered set can be, for it can be shown that two operators in the set have a greatest lower bound ⇔ they are comparable. This whole situation is intimately related to questions of commutativity for algebras of operators and is too complicated for us to explore here. For further details.

EXERCISES

1. Define a new operation of "multiplication" for self-adjoint operators by $A_1 \circ A_2 = (A_1A_2 + A_2A_1)/2$, and note that $A_1 \circ A_2$ is always self-adjoint and that it equals A_1A_2 whenever A_1 and A_2 commute. Show that this operation has the following properties:

$$A_1 \circ A_2 = A_2 \circ A_1,$$

$$A_1 o (A_2 + A_3) = A_1 o A_2 + A_1 o A_3,$$

$$\alpha (A_1 \circ A_2) = (\alpha A_1) o A_2 = A_1 o (\alpha A_2),$$

and $A I = I o A = A$. Show also that $A_1 \circ (A_2 \circ A_3) = (A_1 o A_2) 0 A_3$ whenever A_1 and A_3 commute.

2. If T is any operator on H, it is clear that $|(Tx, x)| \leq \| Tx \| \| x \| \leq \| T \| \| x \|^2$ so if $H \neq \{0\}$, we have $\sup \{|(Tx, z)|/\| z \|^2 x \neq 0\} \leq \| T \|$. Prove that if T is self-adjoint, then equality holds here. (Hint: write $o = \sup \|(Tx, x)|/\| x \|^2 x \neq 0\} = \sup [|(Tx, x)| : \| x \| = 1\}$,and show that $\| Tx \| \leq a$ whenever $\| x \| = 1$ by putting $b = \| Tx \|^{1/2}$— f $Tx \neq 0$ – and considering

$$4 \| Tx \| = (T (bx + b^{-1} Tx), bx + b^{-1}Tx)$$

$$- (T (bx - b^{-1}Tx), bx - b^{-1}Tx) \leq a$$

$$[\| bx + b^{-1}Tx \|^2 + \| bx - b^{-1}Tx \|^2]$$

$$= 4a \| Tx \|.)$$

NORMAL AND UNITARY OPERATORS

An operator N on H is said to be normal if it commutes with its adjoint, that is, if $NN^* = N^*N$. The reason for the importance of normal operators will not become clear until the next chapter. We shall see that they are the most general operators on H for which a simple and revealing structure theory is possible. Our purpose in this section is to resent a few of their more elementary properties which are necessary or our later work.

It is obvious that every self-adjoint operator is normal, and that if y is normal and a is any scalar, then aN is also normal. Further, the limit N of any convergent sequence {Nk} of normal operators is normal; 'or we know that N** → N*, so

$$\| NN^* - N^*N \| \leq \| NN^* - N_kN_k^* \| + \| N^*N_k^* - N_k^*N_k \| + \| N_k^*N_k - N^*N \| = \| NN^* - N_kN_k^* \| + \| N_k^*N_k - N^*N \| \to 0,$$

which implies that NN* – N*N = 0. These remarks prove

Theorem 1:

The set of all normal operators on H is a closed subset of (&(H) which contains the set of all self-adjoint operators and is closed under scalar multiplication.

It is natural to wonder whether the sum and product of two normal operators are necessarily normal. They are not, but nevertheless, we can say a little in this direction.

Theorem 2:

If N_1 and N_2 are normal operators on H with the property that either commutes with the adjoint of the other, then $N_1 + N_2$ and $N_1 + N_2$ are normal.

Proof:

It is clear by taking adjoints that

$$N_1N_2^* = N_2^* N_1 \Leftrightarrow N_2 N_1^* = N_1^*N_2$$

so the assumption implies that each commutes with the adjoint of the other. To show that $N_1 + N_2$ is normal under the stated conditions, we have only to compare the results of the following computations:

$$(N_1 + N_2)(N_1 + N_1)^* = (N_1 + N_2)(N_1^* + N_1^*)$$
$$= N_1N_1^* + N_1N_1^* + N_2N_2^*$$

and $$(N_1 + N_2)^*(N_1 + N. = (N_1^* + N_2^*)(N_1 + N_2)$$
$$= N_1^*N_1 + N_1^*N_2^* + N_2^* N_1 + N_2^* N_2$$

The fact that N_1N_2 is normal follows similarly from

By definition, a self-ad joint operator A is one which satisfies the identity A*x = Ax. Many properties of self -adjoint operators do not depend on this, but only on the weaker identity || A*x || = || Ax ||. Our next theorem shows that all such properties are shared by normal operators.

Theorem 3:

An operator T on H is normal $\Leftrightarrow \| T^*x \| = \| Tx \|$ *for every x.*

Proof:

In view of Theorem 3, this is implied by the fact that

$\| T^*x \| = \| Tx \| \Leftrightarrow \| T^*x \|^2 = \| Tx \|^2 \Leftrightarrow (T^*x, T^*x)$

$= (Tx, Tx) \Leftrightarrow (TT^*x, x) = (T^*Tx, x) \Leftrightarrow ([TT^* - T^*T]\, x, x) = 0.$

The following consequence of this result will be useful in our later work.

Theorem 4:

If N is a normal operator on H, then $\| N^2 \| = \| N \|^2$.

Proof:

The preceding theorem shows that

$$\| N^2x \| = \| NNx \| = \| N^*Nx \|$$

for every x, and this implies that $\| N^2 \| = \|N^*N \|$. By theorem A, we have $\| N^*N \| = \| N \|^2$, so the proof is complete.

We know that any complex number z can be expressed uniquely in the form $z = a + ib$ where a and b are real numbers, and that these real numbers are called the real and imaginary parts of z and are given by $a = (z + \bar{z})/2$ and $b = (z - \bar{z})/2i$. The analogy between general operators and complex numbers, and between self-adjoint operators and real numbers, suggests that for an arbitrary operator T on H we form $A_1 = (T + T^*)/2$ and $A_2 = (T - T^*)/2i$. A_1 and A_2 are clearly self-adjoint, and they have the property that $T - A_1 + i\, A_2$. The uniqueness of this expression for T follows at once from the fact that

$$T^* = A_1 - i\, A_2.$$

The self-adjoint operators A_1 and A_2 are called the real part and the imaginary part of T.

We emphasized earlier that the complicated structure of $\mathcal{B}(H)$ is due in large part to the fact that operator multiplication is in general non-commutative. Since our future work will be focused mainly on normal operators, it is of interest to see–as the following theorem shows–that the existence of non-normal operators can be traced directly to the non-commutativity of self-adjoint operators.

Theorem 5:

If T is an operator on H, then T is normal $\Leftrightarrow$ *its real and imaginary parts commute.*

Proof:

If A_1 and A_2 are the real and imaginary parts of *T*, so that $T = A_1 + iA_2$ and $T^* = A_1 - iA_2$, then

$TT^* = (A_1 + iA_2)(A1 - iA_2) = A_1^2 + A_2^2 + i(A_2A_1 - A_1A_2)$

and $T^*T = (A_1 - iA_2)(A_1 + iA_2) = A_1^2 + A_2^2 + i(A_1A_2 - A_2A_1).) =$

It is clear that if $A_1A_2 = A_2A_1$, then $TT^* = T^*T$, Conversely, if $TT^* = T^*T$, then $A_1A_2 - A_2A_1 - A_1A_2$, so $2A_1A_2 = 2A_2A_1$ and $A_1A_2 = A_2A_1$.

Perhaps the most important subsystem of the complex plane after the real line is the unit circle, which is characterized by either of the equivalent identities $|z| - 1$ or $z\bar{z} = \bar{z}z = 1$. An operator U on H which satisfies the equation $UU^* = U^*U = I$ is said to be unitary. Unitary operators–which are obviously normal–are thus the natural analogues of complex numbers of absolute value 1. It is clear from the definition that the unitary operators on H are precisely the non-singular operators whose inverses equal their adjoints. The geometric significance of these operators is best understood in the light of our next theorem.

Theorem 6:

If T is an operator on H, then the following conditions are all equivalent to one another:

(1) $T^*T = I$,

(2) *(Tx, Ty) = (x, y) for all x and y,*

(3) $\| Tx \| = \| x \|$| *for all x.*

Proof:

If (1) is true, then $(T^*Tx, y) = (x, y)$ or $(Tx, Ty) = (x, y)$ for all x and y, so (2) is true; and if (2) is true, then by taking $y = x$ we obtain $(Tx, Tx) = (x, x)$ or $\| Tx \|^2 = \| x \|^2$ for all x, so (3) is true. The fact that (3) implies (1) is a consequence of Theorem 3 and the following chain of implications:

$$\| Tx \| = \| x \| \Rightarrow \| Tx \|^2 = \| x \|^2 \Rightarrow (Tx, Tx) = (x, x) \Rightarrow \Rightarrow (T^*Tx, x)$$
$$= (x, x) \Rightarrow ([T^*T - I]\, x, x) = 0.$$

An operator on H with this theorem is simply an isometric isomorphism of H into itself. That an operator of this kind need not be unitary is easily seen by considering the operator on h defined by

$T\{x_1, x_2, \ldots\} = \{0, x_1, x_2, \ldots\}$,

which preserves norms but has no inverse. These ideas lead at once to

Theorem 7:

An operator T on H is unitary ⇔ it is an isometric isomorphism of H onto itself.

Proof:

If T is unitary, then we know from the definition that it is onto; and since by Theorem F it preserves norms, it is an isometric isomorphism of H onto itself. Conversely, if T is an isometric isomorphism of H onto itself, then T^{-1} exists, and by Theorem F we have T*T = I. It now follows that (T*T) T^{-1}, so $T^* = T^{-1}$ and TT* = T*T = I, which shows that T is unitary. This theorem makes quite clear the nature of unitary operators: they are precisely those one-to-one mappings of H onto itself which preserve all structure—the linear operations, the norm, and the inner product.

EXERCISES

1. If T is an arbitrary operator on.H, and if α and β are scalars such that $|a| = |\beta|$, show that $\alpha T + \beta T^*$ is normal.
2. If H is finite-dimensional, show that every isometric isomorphism of H into itself is unitary.
3. Show that an operator T on H is unitary ⇔ T({e_i}) is a complete orthonormal set whenever {e_i} is.
4. Show that the unitary operators on H form a group.

THEOREMS OF COMPLETENESS METRIC SPACE

Theorem 1:

Let A be a subset of a metric space (X, d).

Then A is non-dense in X ⇔ $(\overline{A}) = X - \overline{A}$ is dense in X.

Or

A closed set is nowhere dense ⇔ its complement is everywhere dense.

Proof :

Recall that a is dense in X if $\overline{A} = x$ and A is nowhere dense in X if int $(\overline{A}) = \phi$.

A is non-dense (or no where dense) in X

⇔ int $(\overline{A}) = \phi$ ⇔ ∃ no open nbd N of any point of $\overline{A}$ s.t. $N \subset \overline{A}$

⇔ $\overline{A}$ contains no nbd. ⇔ $S_r(x) \not\subset \overline{A}$ ∀ r > 0 & ∀ x ∈ X

⇔ $S_r(x) \cap (\overline{A})' \neq \phi$

⇔ x is adherent point of $(\overline{A})'$ ∀ x ∈ X

$\Leftrightarrow [(\overline{A})'] = X = \Leftrightarrow (\overline{A})'$ is dense in X.

Theorem 2:

Let A and B be disjoint closed sets in a metric space (X, d). Then $\exists$ disjoint open sets G, H s.t. $A \subset G$, $B \subset H$.

Proof:

If either A or B is empty, say $A = \phi$, then $\exists$ disjoint open sets ϕ, X s.t.

$A \subset \phi$, $B \subset X$.

Hence we may assume $A \neq \phi$, $B \neq \phi$.

Let $a \in A$, $b \in B$ arbitrary.

$$A \cap B = \phi,$$

$$\Rightarrow \quad a \notin B,\ b \notin A$$

$$\Rightarrow \quad d(a, B) > 0,\ d(b, A) > 0.$$

Take $d(a, B) = \delta_a > 0$ and $d(b, B)\ \delta_b > 0$.

Set $\quad S_a = S(a, 1/3\ \delta_a)$, $s_b = (b, 1/3\ \delta_b)$.

Evidently $\quad a \in S_a$, $b \in S_b$.

Write $\quad G = \cap \{S_a : a \in A\}$, $H = \cap \{S_b : a \in B\}$

Being arbitrary union of open spheres, G and H are open sets.

$a \in S_a \Rightarrow A \subset G$ and $b \in S_b \Rightarrow B \subset H$.

Now we claim $G \cap H = \phi$.

If possible let $p \in G \cap H$, then $p \in G$, $p \in H$.

Consequently $\exists\ a_0 \in A$, $b_0 \in B$ s.t.

$p \in S(a_0, 1/3\ \delta_{a0})$,

$p \in S(b_0, 1/3\ \delta_{b0})$.

Let $d(a_0, b_0) = \varepsilon > 0$.

Then $d(a_0, B) = \delta_{a0} \leq \varepsilon$, $d(b_0, A) = \delta_{b0} \leq \varepsilon$

$\varepsilon = d(a_0, b_0) \leq d(a_0, p) + d(p, b_0)$

$$< \frac{1}{3}\delta_{a0} + \frac{1}{3}\delta_{b0} < \frac{\varepsilon}{3} + \frac{3}{3} = \frac{2\varepsilon}{3} \text{ or } \varepsilon < \frac{2\varepsilon}{3}.$$

which is not possible. Hence $G \cap H = \phi$.

Theorem 2:

Every non empty open set $G \subset R$ is the union of a countable collection of pairwise disjoint open intervals.

Proof:

Let G be a non-empty open subset of R and $x \in G$ be arbitrary. Then $\exists$ an interval I_x with centre x s.t. $x \in G$.

Let C_x denote the union of all such open intervals then $C_x = \cup \{I_x : x \in I_x \subset G\}$. Evidently C_x is the largest open interval containing x s.t. $x \in C_x \subset G$.

Moreover this also suggests that $G = \cup C_x$.

$\forall$ x, $y \in C_y$ we have either

$C_x \cap C_y = \phi$ or $C_x \cap C_y \neq \phi$

Consider the case $C_x \cap C_y \neq \phi$.

Evidently $C_x \cap C_y$ is an open interval containing both x and y

s.t. $C_x \subset C_x \cup C_y \subset G$,

$$C_y \subset C_x \cup C_y \subset G.$$

By def. of C_x, and C_y, $C_x \cup C_y \subset C_x$,

$$C_x \cup C_y \subset C_y$$

Hence $\quad C_x = C_x \cup C_y = C_y$

or $\quad C_x = C_y$

$\therefore \forall$ x, $y \in G$ we have either $C_x \cap C_y = \phi$ or $C_x = C_y$

s.t. $\quad G = \cup C_x$.

Second Part: It remains to show that the intervals C_x form a countable collection. Let $G = \{C_x\}$ be a collection of open intervals s.t. $C_n \neq \Phi\ \forall\ n$ and $C_n \cap C_m = \phi$ for $n \neq m$.

To prove that G is countable.

We know that each C_n contains an infinite number of rationals. Let K be the set of all rationals belonging to $\cup C_n$. Define a map $f : K \to G$ s.t.

$f(x) = C_x$ where $C_x \in G$ s.t. $x \in C_x$.

For each $x \in K$, there is unique open interval C_x containing x, the rational number. Hence f is a well defined map. For each $C_a \in G$, has at least one rational $a \in K$ s.t. $f(a) = C_a$, showing thereby f is onto.

Also $\quad f(x) = f(y) \Rightarrow C_x = C_y$

where $\quad x \in C_x$, $y \in C_y \Rightarrow x = y$

Now f is one-one onto. Hence $K \sim G$. But K is countable. Hence G is countable.

For $K \subset Q$, Q is countable $\Rightarrow$ K is countable.

Theorem 3:

Let X and Y be metric spaces and f a mapping of X into Y. Then f is continuous $\Leftrightarrow f^{-1}(G)$ *is open in X whenever G is open in Y.*

Proof:

We first assume that f is continuous. If G is an open set in Y, we must show that $f^{-1}(G)$ is open in X. $f^{-1}(G)$ is open if it is empty, so we may assume that it is non-empty. Let x be a point in $f^{-1}(G)$. Then f(x) is in G, and since G is open, there exists an open sphere $S_\epsilon(f(x))$ centered on f(x) and contained in G. By the definition of continuity, there exists an open sphere $S_\delta(x)$ such that $f(S_\delta(x)) \subset S_\epsilon(f(x))$. Since $S_\epsilon(f(x)) \subseteq G$, we also have $f(S_\delta(x)) \subseteq G$, and from this we see that $S_\delta(x) \subseteq f^{-1}(G)$. $S_\delta(x)$ is therefore an open sphere centered on x and contained in $f^{-1}(G)$, so $f^{-1}(G)$ is open.

We now assume that $f^{-1}(G)$ is open whenever G is, and we show that f is continuous. We show that f is continuous at an arbitrary point x in X. Let $S_\epsilon(f(x))$ be an open sphere centered on f(x). This open sphere is an open set, so its inverse image is an open set which contains x. By this, there exists an open sphere $S_\delta(x)$ which is contained in this inverse image. It is clear that $f(S_\delta(x))$ is contained in $S_\epsilon(f(x))$, so f is continuous at x. Finally, since x was taken to be an arbitrary point in X, f is continuous.

The fact just established—that continuous mappings are precisely those which pull open sets back to open sets—will be of great importance for all our work.

We now come to the useful concept of *uniform continuity*. In order to explain what this is, we examine the definition of continuity expressed in condition (1) at the beginning of this section. Let X and Y be metric spaces with metrics d_1 and d_2, and let f be a mapping of X into Y. We assume that f is continuous, that is, that for each point x_0 in X the following is true: given $\epsilon > 0$, a number $\delta > 0$ can be found such that $d_1(x, x_0) < \delta \Rightarrow d_2(f(x), f(x_0)) < \epsilon$. The reader is no doubt familiar with the idea that if xa is held fixed and e is made smaller, then, in general, δ has to be made correspondingly smaller. Thus, in the case of the real function f defined by f(x) = 2x, δ can always be chosen as any positive number $\leq \epsilon/2$, and no larger δ will do. In general, therefore, δ depends on ϵ. Let us return to our examination of the definition. It says that for our given ϵ, a δ can be found which "works" in the above sense at the particular point x_0 under consideration. But if we hold ϵ fixed and move to another point x_0, then it may happen that this δ no longer works; that is, it may be necessary to take a smaller δ to satisfy the requirement of the definition. We see in this way that δ nay well depend, in general, not only on ϵ but also on x_0. Uniform continuity is essentially continuity plus the added condition that for each $\in$ we can find a δ which works *uniformly over the entire space* X, in the sense that it does not depend on x_0. The formal definition is as follows. If X and Y are metric spaces with metrics d_1 and d_2, then a mapping f of X into Y is said to be *uniformly continuous* if for each $\in > 0$ there exists $\delta > 0$ such that $d_1(x, x') < \delta \Rightarrow d_2(f(x), f(x')) < \in$. It is clear that any uniformly continuous mapping is automatically continuous. The reader will observe that the above real function f defined on the entire real line R by $f(x) = 2x$ is uniformly continuous. On the other hand, the function g defined on R by $g(x) = x^2$ is continuous but not uniformly continuous. Similarly, the continuous function h defined on (0, 1) by $h(x) = 1/x$ is not uniformly continuous.

Uniformly continuous mappings—as opposed to those which are merely continuous—are of particular significance in analysis. The following theorem expresses a property of these mappings which is often useful.

Theorem 4:

Let (X, d) be a metric space. Suppose A is a closed subset of X and x is a point of X s.t. $x \notin A$. Then $\exists$ disjoint open subsets U and V of X s.t. $x \in U$, $A \subset V$.

Proof:

$x \notin A \Rightarrow \exists\ \delta > 0$ s.t. $S_\delta(x) \cap A = \phi$.

Choose ε, η with $0 < \varepsilon < \eta < \delta$.

write $U = S_\varepsilon(x)$, $V = X - S_\eta[x]$.

Since each open sphere is an open set and each closed sphere is a closed set. Also complement of closed set is open set. These prove that U and V are open sets s.t. $U \cap V = \phi$.

For $U = S_\varepsilon(x) \subset S_\eta(x) \subset S_\delta(x)$.

Also $A \subset V$, by construction of V.

Theorem 5:

Let $<f_n(x)>$ be a sequence of continuous functions defined on a metric space (X,d). Let this sequence converge uniformly to f on X. Then f(x) is continuous on X.

Or

Uniform limit of a sequence of continuous function is continuous.

Proof:

Since $<f_n(x)>$ converges uniformly to f on (X, d). Hence given $\varepsilon > 0$, $\exists\ n_0 \in N$ independent of $x \in X$ s.t. $n \geq n_0$.

$$\Rightarrow d\ (f_n(x), f(x) < \varepsilon/3 \qquad ...(1)$$

Let a $\in$ X be arbitrary. To prove that f is continuous on X; we have to prove that f is continuous at x = a. For this we have to show that given $\varepsilon > 0$, $\exists\ \delta > 0$, 0 s.t. d (x, a) < δ.

$$\Rightarrow d\ (f(x), f(a) < \varepsilon \qquad ...(2)$$

Continuity of f_n at a $\in$ X

$$\Rightarrow d\ (f_n(x), f(x) < \varepsilon/3 \text{ for } d\ (x, a)\ ,\ \delta \qquad ...(3)$$

By (1), d $(f_n(a), f(a)) < \varepsilon/3\ \forall\ n \geq n_0$ ...(4)

If d (x, a) < δ, then

$$d\ (f(x), f(a)) \leq d\ [f(x), f_n(\dot{x})] + d\ [f_n(x), f_n(a) + d\ [f_n(a), f(a)]$$

$$< \frac{\varepsilon}{3} + \frac{3}{3} = \frac{\varepsilon}{3} o = \varepsilon \text{ by (1), (3), \& (4)}$$

or, d (f(x), f(a)) < ε for d (x, a) < δ. Hence the result (2).

Theorem 6:

Frechet Space: *Let F be the set of infinite sequences of real numbers.*

x, y, z $\in$ F, then

$$x = < x_n > = < x_1, x_2 \ldots >,$$

$$y = < y_n > ,\ z = < z_n >$$

where $\quad x_n, y_n, z_n \in R$

Proof:

We define a map

$$d : F \times F : \rightarrow R \text{ s.t.}$$

$$d\left(x,y\right) = \sum_{n=1}^{\infty} \frac{1}{2^n} \frac{\left(x_n - y_n\right)}{\left[1 + \left|x_{n} - y_n\right|\right]}$$

To show that d is metric on F.

(i) d (x, y) $\geq$ 0 For $|x_n - y_n| \geq 0\ \forall_n$

(ii) d (x, y) = 0 $\Leftrightarrow$ x = y

For $d\left(x,y\right) = 0 \Leftrightarrow \sum_{n=1}^{\infty} \frac{1}{2^n} \frac{\left(x_n - y_n\right)}{\left[1 + \left|x_{n} - y_n\right|\right]} = 0$

$$\Leftrightarrow \frac{1}{2^n} \frac{\left(x_n - y_n\right)}{\left[1 + \left|x_{n} - y_n\right|\right]} = 0\ \forall\ n$$

$$\Leftrightarrow |x_n + y_n| = 0\ \forall\ n$$

$\Leftrightarrow |x_n + y_n| = \forall\ n$

$\Leftrightarrow x = y$

(iii) d (x, y) = d (y, x)

For $|x_n + y_n| = |y_n + x_n|$

(iv) d (x, y) ≥ d (x, z) + d (z, y)

Here we use the fact that

$$\frac{|\alpha+\beta|}{1+|\alpha+\beta|} \leq \frac{|\alpha|}{1+|\alpha|} + \frac{|\beta|}{1+|\beta|}$$

In view of this we have

$$d(x,y) = \sum_{n=1}^{\infty} \frac{|x_n - y_n|}{1+|x_n - y_n|} \cdot \frac{1}{2^n}$$

$$= \sum_{n=1}^{\infty} \frac{1}{2^n} \cdot \frac{|(x_n - z_n) + (z_n - y_n)|}{1+|(x_n - z_n) + (z_n - y_n)|}$$

$$\geq \sum_{n=1}^{\infty} \frac{1}{2^n} \cdot \frac{|x_n - z_n|}{1+|x_n - z_n|} + \sum_{n=1}^{\infty} \frac{1}{2^n} \cdot \frac{|z_n - y_n|}{1+|z_n - y_n|}$$

= d (x, z) + d (z, y).

Thus d is metric on F. The pair (F, d) is a metric space and this metric space is called *Frechet space.*

Theorem 7:

Let X is a metric space. Then (1) any union of open sets in X is open; and (2) any finite intersection of open sets in X is open.

Proof:

To prove (1), let $(G_1\}$ be an arbitrary class of open sets in X. We must show that $G = \cup_i G_i$ is open. If $\{G_i\}$ is empty, then G is empty, and by Theorem A, G is open. Suppose that $\{G_i\}$ is non-empty. By Theorem C, each G, (being an open set) is a union of open spheres; G is the union of all the open spheres which arise in this way; and by another application of Theorem C, G is open.

To prove (2), let $\{G_i\}$ be a finite class of open sets in X. We must show that $G = \cap_i G_i$ is open. If (G",} is empty, then G = X; and by Theorem A, G is open. Suppose that $\{G_i\}$ is non-empty and that $\{G_i\} = \{G_1, G_2, ... G_n\}$ for some positive integer n. If G happens to be empty, then it is open by Theorem A, so we may assume that G is non-empty. Let a; be a point in G. Since x is in each G_i, and each G_i is open, for each i there is a positive real number r, such that $S_{ri}(x) \subseteq G_i$. Let r be the smallest number in the set

$\{r_1, r_2, \ldots, R_n\}$. This number r is a positive real number such that $S_r(x) \subseteq S_{r_i}(x)$ for each i, so $S_r(x) \subseteq G_i$ for each i, and therefore $S_r(x) \subseteq G$. Since $S_r(x)$ is an open sphere centered on x and contained in G, G is open.

The above theorem says that the class of all open sets in a metric space is closed under the formation of arbitrary unions and finite intersections. The reader should clearly understand that Theorem A is an immediate consequence of this statement, since the empty set is the union of the empty class of open sets and the full space is its intersection. The limitation to *finite* intersections in this theorem is essential. To see this, it suffices to consider the following sequence of open intervals on the real line:

$$(-1,1)\left(-\frac{1}{2},\frac{1}{2}\right),\left(-\frac{1}{3},\frac{1}{3}\right)$$

The intersection of these open sets is the set $\{0\}$ consisting of the single point 0, and this set is not open.

In an arbitrary metric space, the structure of the open sets can be very complicated indeed. Theorem C contains the best information available in the general case: each open set is a union of open spheres. In the case of the real line, however, a description can be given of the open sets which is fairly explicit and reasonably satisfying to the intuition.

Theorem 8:

In a metric space (X, d), prove that

F is closed $\Leftrightarrow D(F) \subset F$.

Or

Prove that a subset F of a metric space X contains all its limit points iff X – F is open.

Proof :

Let (X, d) be a metric space and $F \subset X$.

We know that F is closed $\Leftrightarrow$ X – F is open.

Step I : Let X – F be open so that F is closed.

Aim. $D(F) \subset F$.

Let $x \in X - F$ be arbitrary. The X – F is an open set containing x s.t. $(X - F) \cap F = \phi$.

This $\Rightarrow$ is not a limit point of F

$\Rightarrow x \notin D(F) \Rightarrow x \in X - D(F)$

Thus, " $x \in X - F \Rightarrow x \in X - D(F)$

$\therefore \quad X - F \subset X' - D(F)$

or, $\quad D(F) \subset F$

Step II : Given $\quad D(F) \subset F$ ...(1)

To prove F is closed.

Let $y \in X - F$, then $y \notin F$

$y \notin F,\ D(F) \subset F \Rightarrow y \notin D(F)$

$\Rightarrow \exists$ open set G with $y \in G$ s.t.

$(G - \{y\}) \cap F = \phi$

$\Rightarrow G \cap F = \phi$ as $y \notin F$

$\Rightarrow G \subset X - F$

Thus we have shown that

any $y \in X - F \Rightarrow \exists$ open set G with $y \in G$ s.t. $G \subset X - F$

This $\Rightarrow X - F$ is open $\Rightarrow$ is closed.

Theorem 9:

If a_i, b_i $(i = 1, 2 \ldots n)$ are real or complex numbers, then

$$\left(\int_a^b fg\,dx \le \left(\int_a^b f^p\,dx\right)^{1/p} \left(\int_a^b g^q\,dx\right)^{1/q}\right)$$

$$\sum_{i=1}^{n} |a_i b_i| \le \left[\sum_{i=1}^{n} |a_i|^2\right]^{1/2} \left[\sum_{i=1}^{n} |b_i|^2\right]^{1/2}$$

Proof :

If we replace a_i, b_i, p, q by $|a_i|$, $|b_i|$, 2, 2 respectively, then we get this inequality. An independent proof is given below :

Write $A = (a_1, a_2, a_3, \ldots a_n)$, $b = (b_1, b_2, \ldots, b_n)$.

The magnitudes of these vectors are given by

$$\|a\|^2 = \sum_{i=1}^{n} |a_i|^2 \quad \|b\|^2 \sum_{i=1}^{n} |b_i|^2$$

If $a = 0$ or $b = 0$, then the inequality reduces to equality.

So we suppose that $a \ne 0$. $b \ne 0$. We know that

Geometric $\le$ mean Arithmetic mean.

This$\Rightarrow \sqrt{(xy)} \le \dfrac{x+y}{2}$

Set $\sqrt{x} = \dfrac{|a_i|}{\|a\|}, \sqrt{y} = \dfrac{|b_i|}{\|b\|}$

Then $$\frac{2|a_i|.|b_i|}{\|a\|.\|b\|} \le \frac{|a_i|^2}{\|a\|^2} + \frac{|b_i|^2}{\|b\|^2}$$

or $$2\sum_{i=1}^{n}\frac{|a_i|.|b_i|}{\|a\|.\|b\|} \le \frac{\sum|a_i|^2}{\|a\|^2} + \frac{\sum|b_i|^2}{\|b\|^2} = 1+1=2$$

or $$\sum_i |a_i|.|b_i| \le \|a\|.\|b\|$$

or $$\sum_{i=1}^{n} |a_i b_i| \le \left(\sum_{i=1}^{n}|a_i|^2\right)^{1/2} \left(\sum_{i=1}^{n}|b_i|^2\right)^{1/2}$$

If a_i , b_i are real numbers, then

$$\sum_i |a_i\, b_i| \le \left(\sum_{i=1}^{n} a_i^{\,2}\right)^{1/2} \left(\sum_{i=1}^{n} b_i^{\,2}\right)^{1/2}$$

Theorem 10:

If $p > 1$ and a_i , b_i are non-negative real numbers, then

$$\left[\sum_{i=1}^{n}(a_i + b_i)^p\right]^{1/p} \le \left(\sum_{i=1}^{n} a_i^p\right)^{1/p} + \left(\sum_{i=1}^{n} b_i^p\right)^{1/p}$$

Proof :

By Holder's inequality,

$$\sum_{i=1}^{n} a_i + b_i \le \left(\sum_{i=1}^{n} a_i^p\right)^{1/p} \left(\sum_{i=1}^{n} b_i^q\right)^{1/q} \quad \text{... (1)}$$

with $$\frac{1}{p} + \frac{1}{q} = 1 \quad \text{... (2)}$$

By (2), $$q\,(p-1) = p \quad \text{... (3)}$$

$$\sum_{i=1}^{n}(a_i + b_i)^p = \sum_{i=1}^{n}(a_i + b_i)(a_i + b_i)^{p-1}$$

where $\sum_{i=1}^{n} = \sum$

$$= \sum a_i\,(a_i + b_i)^{p-1} + \sum b_i\,(a_i + b_i)^{p-1}$$

Using (1) for each member of R.H.S.,

$$\sum\,(a_i + b_i)^p \le \left(\sum a_i^p\right)^{1/p} \left[\sum(a_i , b_i)^{q(p-1)}\right]^{1/q}$$

$$+ \left(\sum b_i^p\right)^{1/p} \left[\sum(a_i + b_i)^{q(p-1)}\right]^{1/q}$$

Using (3), $\sum (a_i + b_i)^p \le \left(\sum a_i^p\right)^{1/p} \left[\sum (a_i + b_i)^p\right]^{1/q}$

$$+ \left(\sum b_i^p\right)^{1/p} \left[\sum (a_i + b_i)^p\right]^{1/q}$$

Dividing by $\left[\sum (a_i + b_i)^p\right]^{1/q}$ we get

$$\left[\sum (a_i + b_i)^p\right]^{1-(1/q)} \le \left(\sum a_i^p\right)^{1/p} + \left(\sum b_i^p\right)^{1/p}$$

or, $\left[\sum_{i=1}^{m} (a_i + b_i)^p\right]^{1/p} \le \left(\sum a_i^p\right)^{1/p} + \left(\sum b_i^p\right)^{1/p}$

Note : Minkowski's inequality for integrals

$$\left[\int_a^b (f+g)^p\right] \le \left(\int_a^b f^p\right)^{1/p} + \left(\int_a^b g^p\right)^{1/p}$$

where $p > 1$.

Theorem 11:

Let $a = (a_1, a_2, \ldots, a_n)$, $b = b_1, b_2, \ldots, b_n)$ be two n tuples or real or complex numbers, then

$$\|a+b\| \le \|a\| + \|b\|$$

$$\left[\sum_{i=1}^{n} |a_i + b_i|^2\right]^{1/2} \le \left(\sum_{i=1}^{n} |a_i|^2\right)^{1/2} + \left(\sum_{i=1}^{n} |b_i|^2\right)^{1/2}$$

Proof:

We have $\|a+b\|^2 = \sum_i |a_i + b_i|^2$,

or $\|a\|^2 = \sum_i |a_i|^2$

$$\|b\|^2 = \sum_i |b_i|^2$$

where $\sum_i$ *means* $\sum_{i=1}^{n}$

$$\|a+b\|^2 = \sum_i |a_i + b_i|^2 = \sum_i |a_i + b_i| . |a_i + b_i|$$

$$\le \sum_i |a_i + b_i| . (|a_i + b_i|)$$

$$\leq \sum_i |a_i + b_i| . |a_i| + \sum_i |a_i + b_i| . |b_i|$$

$$\leq \sum_i |(a_i + b_i) a_i| + \sum_i |(a_i + b_i) b_i|$$

Applying Cauchy Schwarz inequality to each sigma or R.H.S., we get

$$\|a + b\|^2 \leq \left(\sum_i |a_i + b_i|^2 \right)^{1/2} \left(\sum_i |a_i^2| \right)^{1/2}$$

$$+ \left(\sum_i |a_i + b_i|^2 \right)^{1/2} . \left(\sum_i |b_i^2| \right)^{1/2}$$

$$= \|a+b\| . \|a\| + \|a+b\| . \|b\|$$

Dividing by $\|a+b\|$, we get

$$\|a+b\| \leq \|a\| + \|b\|$$

Theorem 12:

If a_i, b_i (i = 1, 2 ... n) are non- negative real numbers and $p > 1$, then

$$\sum_{i=1}^{n} (a_i + b_i)^p \leq \sum_{i=1}^{n} a_i^p + \sum_{i=1}^{n} b_i^p$$

Proof :

Let $f(x) = 1 + x^p - (1 + x)^p$.

Then taking derivative ; it can be shown that

$f(x) \geq 0$ for $x \geq 0$.

or $\quad 1 + x^p - (1 + x)^p \geq 0$.

or $\quad (1 + x)^p \leq 1 + x^p$

put $x = a_i/b_i$, we get

$$\left(1 + \frac{a_i}{b_i}\right)^p \leq 1 + \left(\frac{a_i}{b_i}\right)^p$$

or, $(a_i + b_i)^p \leq (a_i)^p + (b_i)^p$

or $$\sum_{i=1}^{n} (a_i + b_i)^p \leq \sum_{i=1}^{n} a_i^p + \sum_{i=1}^{n} b_i^p$$

Theorem 13:

(or l_2-space) : Let H be the set of infinite real sequences $< x_n >$ s.t. $\sum_{i=1}^{n} x_i^{\,2} < \infty$.

Let x , y ∈ H s. t x = < x_n >, y = < y_n > . Define a map

d : H × H → R s.t. $d(x, y) = \left[\sum_{i=1}^{\infty}(x_i - y_i)^2\right]^{1/2}$

Aim : d is a metric in H.

(1) To prove that d is well defined map

$$d(x, y) = [\Sigma \{x_i + (-y_i)\}^2]^{1/2}$$
$$\leq (\Sigma x_i^2)^{1/2} + (\Sigma (-y_i)^2)^{1/2} \text{ by Minkowski inequality}$$
$$= \Sigma x_i^2 + \Sigma y_i^2 < \infty \text{, by given condition,}$$

or, $d(x, y) < \infty$,

meaning thereby d s a finite real number. Hence d is well defined map.

(2) $d(x, y) \geq 0$

For x_i , y_i, are real

⇒ $(x_i - y_i)^2 \geq 0$

⇒ $d(x, y) \geq 0$

(3) $d(x, y) = d(y, x)$

For $(x_1 - y_i)^2 = (y_i - x_i)^2$

⇒ $\{\Sigma (x_1 - y_i)^2\}^{1/2} = \{\Sigma (y_i - x_i)^2\}^{1/2}$

⇒ $d(x, y) = d(y, x)$

(4) $d(x, z) \leq d(x, y) + d(y, z)$.

where z = < z_n > ∈ H.

$$d(x,z) = \left[\sum_{i=1}^{\infty}(x_i - z_i)^2\right]^{1/2}$$
$$= [\Sigma \{(x_i - y_i) + (y_i - z_i)\}^2]^{1/2}$$
$$\leq [\Sigma (x_i - y_i)^2]^{1/2} + [\Sigma (y_i - z_i)^2]^{1/2},$$

by Minkowski inequality

$$= d(x, y) + d(y, z).$$

The above facts prove that d is a metric on H. The space (H, d) is called Hilbert space or l_2-space.

Theorem 14:

Let (x, ρ) be a metric space, Show that ∅ and X are open sets in X.

Proof :

Let (X, ρ) be a metric space.

To prove that ∅ and X are open sets .

To prove that ∅ is open in X, it suffices to show that

any $x \in \Phi \Rightarrow \exists\ \varepsilon > 0$, s.t. $S_{\varepsilon\,(x)} \subset \varnothing$.

Since ∅ does not contain any element and hence this condition is automatically fulfilled.

Corresponding to every point x ∈ X, ∃ an open sphere with its centre at x which is contained in X, showing there by X is an open .

Example : *Is the subset [0, 3] open in X = [0, 3] under the metric d given by d (x, y) = | x – y?*

Solution:

Since Φ and X both are open sets in any metric space (X, d).

∴ [0, 3] is open in x = [0, 3] .

Theorem 15:

In a metric space (X, ρ), ∅ and X are closed sets.

Proof :

Let (X, ρ) be a metric space.

To prove that ∅ and X are closed sets.

∵ $D(\varnothing) = \varnothing \subset \varnothing$

∴ $D(\varnothing) \subset \varnothing$

∵ This ⇒ ∅ is a closed set.

All the limiting points of X belong ... . For X is the universal set.

i.e., any $x \in D(X) \Rightarrow x \in X$

∵ $D(X) \subset X$.

This ⇒ X is a closed set.

Theorem 16:

Let (X, d) be a metric space. Show that F ⊂ X, F is closed ⇔ F' is open.

Proof:

Let (X, d) be a metric space .

Let C be a closed subset of X, so that D (F) ⊂ F.

To prove that F' is open in X.

Let X ∈ F' be arbitrary. Then X ∉ F.

$$D(F) \subset F.\ x \notin F \Rightarrow x \notin D(F)$$

$$\Rightarrow (S_{r(x)} - \{x\}) \cap F = \varnothing \text{ for some } r > 0.$$

$$\Rightarrow S_{r(x)} \cap F = \varnothing \qquad (\because x \notin F)$$

$$\Rightarrow S_{r(x)} \subset X - F$$

$$\Rightarrow S_{r(x)} \subset F'.$$

∴ Given $x \in F'$, ∃ any an open sphere $S_{r(x)}$ s.t.

$$S_{r(x)} \subset F'.$$

By definition, this proves that F' is open.

Conversely suppose that F' is open in X.

To prove that F is closed in X.

Let $x \in F'$ be arbitrary, then $x \notin F$.

∵ F' is open, ∃ $r \subset R^+$ s.t., $S_{r(x)} \subset F'$

$$\text{This} \Rightarrow S_{r(x)} \cap F = \varnothing$$

$$\Rightarrow (S_{r(x)} - \{x\}) \cap F = \varnothing$$

$$\Rightarrow x \notin D(F).$$

Thus, any $x \in F' \Rightarrow x \notin D(F)$.

i.e., any $x \in X - F$

$$\Rightarrow x \in X - D(F)$$

This ⇒ $X - F \subset X - D(F)$

or $D(F) \subset F$.

This ⇒ F is closed.

Theorem 17:

In any metric space (X, d) each open sphere is an open set.

Proof:

Let (X, d) be a metric space. Let $S_{r0\,(x0)}$ be an open sphere in X.

To prove that $S_{r0\,(x0)}$ is an open set.

Let $x \in S_{r0\,(x0)}$ be arbitrary, then $d(x, x_0) < r_0$

Write $r = r_0 - d(x, x_0)$, ...(1)

By definition, $S_{r0\,(x0)} = \{y \in X : d(y, x_0) < r_0\}$

$$S_{r(x)} = \{y \in X : d(y, x) < r\}.$$

We claim $S_{r(x)} \subset S_{r0(x0)}$

Let $y \in S_{r(x)}$ be arbitrary.

Then $d(x, y) < r$

$$d(y, x_0) \le d(y, x) + d(x, x_0)$$

$$< r + d(x, x_0) = r_0.$$

$\therefore$ $d(y, x_0) < r_0$ on using (1)

$\Rightarrow$ $y \in S_{r0(x0)}$

and $y \in S_{r(x)}$

$\Rightarrow$ $y \in S_{r0(x0)}$

$\Rightarrow$ $S_{r(x)} \subset S_{r0(x0)}$

Thus we have shown that

given any $x \in S_{r0(x0)}$, $\exists\, r > 0$ s.t. $S_{r(x)} \subset S_{r0(x0)}$.

By definition, this proves that $S_{r0(x0)}$ is an open set.

Theorem 18:

Let (X, d) be a metric space. Let $\langle x_n \rangle$ be a Cauchy sequence in X and $\langle y_n \rangle$ be a sequence in X s.t.

$$d(x_n, y_n) < \frac{1}{n} \ \forall\, n \in N.$$

Then show that

(i) $\langle y_n \rangle$ is a Cauchy sequence,

(ii) $\langle y_n \rangle$ converges to x iff $\langle x_n \rangle$ converges to x.

Proof :

Let $\langle x_n \rangle$ be a Cauchy sequence. Then given $\varepsilon > 0$.

$\exists\, n_0 \in N$ s.t. $m, n \ge n_0$

$\Rightarrow$ $d(x_n, x_m) < \varepsilon/3$ ...(1)

Choose $m_0 \in N$ s.t. $\frac{1}{m_0} < \frac{\varepsilon}{3}$.

Then $n \ge m_0 \Rightarrow \frac{1}{n} \le \frac{1}{m_0} < \frac{\varepsilon}{3}$.

or, $d(x_n; y_n) < \frac{1}{n} \le \frac{1}{m_0} < \frac{\varepsilon}{3}$.

or, $d(x_n, y_n) \le \frac{\varepsilon}{3} \ \forall\, n \ge m_0$ ...(2)

Choose $k_0 = \max \{n_0, m_0\}$. Then $\forall$ n, m $\geq k_0$, we have

$$d(y_n, y_m) \leq d(y_n, x_n) + d(x_n, x_m) + d(x_m, y_m)$$

$$< \frac{\varepsilon}{3} + \frac{\varepsilon}{3} + \frac{\varepsilon}{3} \text{ by (1) \& (2)}$$

or, $\quad d(y_n, y_m) < \varepsilon,$

Showing that $\langle y_n \rangle$ is a Cauchy sequence.

(ii) Let $\lim x_n = x$, then $\exists\, m_1 \in N$ s.t.

$$d(x_n, x) < \varepsilon/3 \;\forall\, n \geq m_1 \quad ...(3)$$

But $\quad d(y_n, x) \leq d(y_n, x_n) + d(x_n, x) \quad ...(4)$

Choose $\quad m_2 = \max \{m_1, m_0\}$

Using (2) and (3), in (4).

$$d(y_n, x) < \frac{\varepsilon}{3} + \frac{\varepsilon}{3} = \frac{2\varepsilon}{3} < \varepsilon \;\forall\, n \geq m_2$$

This $\Rightarrow \lim y_n = x$

Similarly, $y_n \to x \Rightarrow x_n \to x$.

Theorem 19:

Any sequence in a metric space converges to the unique point.

Or, A sequence in a metric space can converge to almost one point.

Or, If $\langle a_n \rangle$ is a sequence in a metric space (X, ρ) s.t. $a_n \to a_0$, $a_n \to a_0'$, then $a_0 = a_0'$.

Proof :

Let $\langle a_n : n \in N \rangle$ be a sequence in metric space (X, ρ) s.t.

$a_n \to a_0$, $a_n \to a_0'$,

To prove that $a_0 = a_0'$.

Suppose the contrary.

Then $a_0 \neq a_0'$.

Let $\rho(a_0, a_0') = 2r$.

Then $\quad S_r(a_0) \not\in S_r(a_0') = \phi$.

$\because \quad a_n \to a_0,\; a_n \to a_0'$

$\therefore$ Given any $r > 0$, $\exists\, n_0, n_0' \in N$ s.t.

$$\forall\, n \geq n_0 \Rightarrow a_n \in S_r(a_0)$$

$$\forall\, n \geq n_0' \Rightarrow a_n \in S_r(a_0')$$

Taking $k_0 \Rightarrow \max \{n_0, n_0'\}$, we obtain

$$\forall\, n \geq k_0 \Rightarrow a_n \in S_r(a_0)\ a_n \in S_r(a_0')$$
$$\Rightarrow a_n \in S_r(a_0) \cap S_r(a_0')$$
$$\Rightarrow S_r(a_0) \cap S_r(a_0') \neq \phi. \text{ A contradiction.}$$

Hence our initial assumption $a_0 \neq a_0'$ is wrong. Consequently

$$a_0 = a_0'.$$

Theorem 20:

Let $\langle x_n \rangle$ be a sequence in a metric space (X, d) and the sets A_n be defined by

$$A_1 = \{x_1, x_2, x_3, \ldots\}$$
$$A_2 = \{x_2, x_3, \ldots\},$$
$$A_n = \{x_n, x_{n+1}, \ldots\},\ \forall n \in N.$$

Then the sequence $< x_n >$ is a Cauchy sequence iff $d(A_n) = 0$ as $n \to \infty$.

Proof :

Let $\langle x_n \rangle$ be a sequence in a metric space (X, d) and let

$$A_n = \{x_n, x_{n+1}, \ldots\},\ \forall\, n \in N \text{ so that}$$

and
$$\left.\begin{array}{l} n < m \Rightarrow A_m \subset A_n \\ r \geq n \Rightarrow x_r \in A_n \end{array}\right\} \quad \ldots(1)$$

Step (i) : Let $\langle x_n \rangle$ be a Cauchy sequence so that given any

$$\varepsilon > 0,\ \exists\ n_0 \in N \text{ s.t.}$$
$$n, m \geq n_0 \Rightarrow d(x_n, x_m) < \varepsilon \quad \ldots(2)$$

To prove that $d(A_n) = 0$ as $n \to \infty$.

Let $n \geq n_0$, then

$$d(A_n) = \sup \{d(x_r, x_s) : x_r, x_s \in A_n\}$$
$$= \sup \{d(x_r, x_s) : r, s \geq n\} < \varepsilon, \text{ by (1), (2)}$$

$$\therefore \quad d(A_n) < \varepsilon\ \forall\ n \geq n_0$$

This$\Rightarrow \lim_{n \to \infty} d(A_n) = 0$

Step (ii) : Let us suppose $\lim_{n \to \infty} d(A_n) = 0$, so that given any

$$\varepsilon > 0,\ \exists\ n_0 \in N \text{ s.t.}$$
$$n \geq n_0 \Rightarrow d(A_n) < \varepsilon$$

$\Rightarrow \quad d(A_{n0}) < \varepsilon.$

To prove that $< x_n >$ is a Cauchy sequence, it is enough to prove that

$$d(x_r, x_s) < \varepsilon \ \forall \ r, s \geq n_0.$$

Now, r, s $\geq n_0$

$\Rightarrow \quad x_r, x_s \in A_{n0}$ by (1)

$\Rightarrow \quad d(x_r, x_s) < \varepsilon$

$\because \quad d(A_{n0}) < \varepsilon$

Theorem 21:

In any metric space, any closed sphere is a closed set.

Proof:

Let $Sr_{0\,[x0]}$ denote a closed sphere in a metric space (X, d).

To prove that $Sr_{0\,[x0]}$ is a closed set.

For this we must show that $S'\,r_0\,[x_0]$ is open in X.

Let $x \in S'\,r_{0\,[x0]}$ be arbitrary,

$x \in S'\,r_{0\,[x0]} \Rightarrow x \notin S'\,r_{0\,[x0]}$

$\Rightarrow d(x, x_0)\ , <\ , \neq r_0,$

$\because Sr_{0\,[x0]} = \{ y \in X : d(y, x_0) \leq r_0 \}$

$\Rightarrow d(x, x_0) > r_0$

$\Rightarrow d(x, x_0) - r_0 > 0$

$\Rightarrow r > 0$, on taking $r = d(x, x_0) - r_0$ ...(1)

We claim $S_{r\,(x)} \subset S'\,r_{0\,[x0]}$.

Let $y \in S_{r\,(x)}$ be arbitrary, so that, $d(y, x) < r$.

$\therefore \quad d(x, x_0) \leq d(x, y) + d(y, x_0).$

$\therefore \quad d(y, x_0) \geq d(x, x_0) - d(x, y).$

$> d(x, x_0) - r = r_0$ on using (1)

$\therefore \quad d(y, x_0) > r_0$

$\Rightarrow y \notin Sr_{0\,[x0]}$

Thus, any $y \in S_{r\,(x)} \Rightarrow y \in S'r_{0\,[x0]}$

This$\Rightarrow$ $S_{r\,(x)} \subset S'r_{0\,[x0]}$

$\therefore$ Given any $x \in S'r_{0\,[x0]}$, $\exists\, r > 0$ s.t. $S_{r\,(x)} \subset S'r_{0\,[x0]}$

This proves that $S'r_{0\,[x0]}$ is open in x.

Theorem 22:

In any metric space X, the empty set ϕ and the full space X are open sets.

Proof:

To show that ϕ is open, we must show that each point in ϕ is the center of an open sphere contained in ϕ; but since there are no points in ϕ, this requirement is automatically satisfied. X is clearly open, since every open sphere centered on each of its points is contained in X.

We have seen that [0, 1) is not open as a subset of the real line. However, if we consider [0, 1) as a metric space X in its own right, as a subspace of the real line, then [0, 1) is open as a subset of X, since from this point of view it is the full space.

This apparent paradox disappears when we realize that points outside of a given metric space have no relevance to any discussion taking place within the context of that space. A set is open or not open only with respect to a specific metric space containing it, never on its own.

Our next theorem justifies the adjective in the expression "open sphere."

Theorem 23:

Let X be a metric space. A subset G of X is open $\Leftrightarrow$ it is a union of open spheres.

Proof:

We assume first that G is open, and we show that it is a union of open spheres. If G is empty, it is the union of the empty class of open spheres. If G is non-empty, then since it is open, each of its points is the center of an open sphere contained in it, and it is the union of all the open spheres contained in it.

We now assume that G is the union of a class S of open spheres. We must show that G is open. If S is empty, then G is also empty, and by Theorem A, G is open. Suppose that S is non-empty. G is also non-empty.

Let a; be a point in G. Since G is the union of the open spheres in S, x belongs to an open sphere $S_r(x_0)$ in S. By Theorem B, x is the center of an open sphere $S_{r_1}(x) \subseteq S_r(x_0)$. Since $S_r(x_0) \subseteq G$, $S_{r_1}(x) \subseteq G$ and we have an open sphere centered on x and contained in G. G is therefore open.

The fundamental properties of the open sets in a metric space are those stated in

Theorem 24:

Let (X, d) be a metric space and let $G \subset X$ be an arbitrary set. Show that G is open $\Leftrightarrow$ it is a union of open spheres.

Proof:

Let (X, d) be a metric space. Let $G \subset X$ be an open set.

To prove that G is a union of open spheres.

If $G = \emptyset$, the proof is trivial,

If $G \neq \emptyset$, then by definition of an open set

any $\quad x \in G \Rightarrow \exists\, r \in R^+$ s.t. $S_{r\,(x)} \subset G$ ...(1)

In view of this statement we are justified to write.

$$G = \bigcup_i \{S_{ri(xi)} : x_i \in G\}$$

where each $r_i \in R^+$ is s.t. $S_{ri(xi)}$ satisfies the condition (1).

Conversely suppose that (X, d) is a metric space. Let $G \subset X$ be expressible as a union of open spheres, *i.e.*, $G = \bigcup_i S_{ri(xi)}$.

To prove that G is an open set.

Let $x \in G$ be arbitrary.

Then x is a centre of some open sphere, say.

$$S_{(x,\, r)} \in \{S_{ri(xi)}\}$$

Then $\quad S_{r\,(x)} \subset \bigcup_i S_{ri\,(xi)} = G,$

or $\quad S_{r\,(x)} \subset G,$

$\therefore$ any $\quad x \in G \Rightarrow \exists\, r \in R^+$ s.t. $S_{r\,(x)} \subset G$.

By definition this proves that G is an open set.

Theorem 25:

Let (X, d) be a metric space. Prove that

(i) an arbitrary union of open sets is open in X.

(ii) any finite intersection of open sets is open.

Proof:

Let (X, d) be a metric space.

Let $G_i \subset X$ be open X $\forall\, i \in N$

Let $G = \bigcap_{i=1}^{\infty} G_i,\ G = \bigcap_{i=1}^{n} G_i$

To prove that

(i) G is open in X.

(ii) G is open in X.

(i) By Theorem 7, each G_i is a union of open spheres and hence G is the union of all open spheres arising in this way.

Hence g is open in X.

(ii) If G $= \Phi$, then G is open in X. For Φ is open in X.

If $G \neq \Phi$, then $\exists\ x \in \bigcap_{i=1}^{n} G_i$ so that

$x \in G_i$ for $1 \leq i \leq n$.

$x \in G_i$, G_i is open

$\Rightarrow \quad \exists\ r_i \in R^+$ s.t. $S_{r\,(x)} \subset G_i$ ...(1)

This is true for each i (= 1, 2, ..., n).

Take $\quad r = \min \{r_1, r_2 \ldots, r_n\}$

Then $\quad r \leq r_i$ for $1 \leq i \leq n$.

This $\quad \Rightarrow S_{r\,(x)} \subset S_{ri\,(x)}$ for $1 \leq i \leq n$

$\Rightarrow S_{r\,(x)} \subset S_{ri\,(x)} \subset G_i$ by (1)

$\Rightarrow S_{r\,(x)} \subset G_i$

$\Rightarrow S_{r\,(x)} \subset G = \bigcap_{i=1}^{n} G_i = G \Rightarrow S_{r\,(x)} \subset G$

$\therefore$ any $x \in G \Rightarrow \exists\ r \in R^+$ s.t. $S_{r\,(x)} \subset G$

By definition, this $\Rightarrow$ G is open in X.

Remark: The intersection of an infinite number of open sets may or may not be open. If we take an open set, $A_n = (-1/n, 1/n)$, then

$$\bigcap_{n=1}^{\infty} A_n = \{0\} = \text{closed set}]$$

Theorem 26:

In any metric space (X, d), show that

(i) an arbitrary intersection of closed sets is closed.

(ii) any; finite union on closed sets is closed.

Proof:

Let (X, d) be a metric space

Let $F_i \subset X$ be a closed set $\forall\ i \in N$

Let $\quad F = \bigcap_{i=1}^{\infty} F_i, \; F = \bigcup_{i=1}^{n} F_i$

To Prove that F and F are closed sets in X.

By theorem 4, F_i is closed $\Leftrightarrow$ F_i' is open.

By theorem 8, $\bigcup_{i=1}^{\infty} F_i', \bigcap_{i=1}^{n} F_i'$ are open in X. ...(1)

By, De Morgan's law, $\bigcup_{i=1}^{\infty} F_i', = (\bigcap_{i=1}^{\infty} F_i) = F'$

and $\quad \bigcap_{i=1}^{\infty} F_i' = (\bigcup_{i=1}^{\infty} F_i) = F'$

From (1), F' , F' are open in X

i.e., F, F are closed in X. [by Theorem 4]

The union of an infinite number of closed sets may or may not be closed.

Let $F_n = \left[0, \dfrac{n}{n+1}\right]$

Then $\quad \bigcup_{i=1}^{\infty} F_n = [0, 1)$ =semi open set

$\therefore \quad \bigcup_{n=1}^{\infty} F_n$ is not closed, though each F_n is closed.

Examples:

To find the boundary of set of integers Z and set of rational Q.

$Z^o = \cup \{ G \subset R : G \text{ is open and } G \subset Z \} = \phi$

(1) For every subset of R contains fractions also.

Similarly $(R - Z)^o = \phi$

$$b(Z) = R - Z^o \cup (R - Z)^o = R - \phi \cup \phi = R$$

or, $\quad b(Z) = R$

Similarly $\quad b(Q) = R$

(2) Give two examples of limit points.

(i) If $A = \left\{1 + \dfrac{1}{n} : n \in N\right\}$,

$A = \left\{2, \dfrac{3}{2}, \dfrac{4}{3}, \dfrac{5}{4}, \dfrac{6}{5}, \ldots\right\}$, *then*

1 is limit point of A for $\lim_{n \to \infty} \left(1 + \dfrac{1}{n}\right) = 1.$

(ii) If $A = \left\{1, \frac{1}{2}, \frac{1}{3}, \frac{1}{4}, \ldots\right\} = \left\{\frac{1}{n} : n \in N\right\}$,

then 0 is the limit point of A.

For $\lim_{n\to\infty} \frac{1}{n} = 0.$

(3) Since N has no limit points and so D (N) = ϕ*. For every open sphere centred at any point of N other than that point.*

$$\overline{N} = N \cup D(N) = N \cup \phi = N$$

or $\overline{N} = N$ *(∴ N is closed.)*

so that N' is open and so $(N')^o = N'$.

$$ext(N) = (N')^o = N'$$

$$b(N) = [N^o \cup ext(N)]' = [\phi \cup ext(N)']$$

$$= (N')' = N$$

Note that N is non-dense set.

For $(\overline{N})^o = N^o = \phi$

Theorem 27:

A subset in a metric space is open iff it is a nbd of each of its point.

Proof:

Let A be a subset of a metric space (X, d).

Step I : Given A is and of each of its points.

Aim : A is an open set. Recall that a set N is called nbd of a point $x \in X$ if ∃ open set

$G \subset X$ s.t. $x \in G \subset N$.

X s.t. $p \in G_p \subset A$.

It is true $\forall\ p \in A$.

Take $A = \cup\ \{G_p :\ p \in G_p,\ G_p$ is open set, $G_p \subset A\}$

= An arbitrary union of open sets

= open set.

∴ A is an open set.

Step II : Let A be an open subset of X.

Aim : A is a nbd of each of its points. By assumption, we can write

$$p \in A \subset A\ \forall\ p \in A$$

This ⇒ A is a nbd of each of its points.

Theorem 28:

In any metric space X, each closed sphere is a closed set.

Proof:

Let $S_r[x_0]$ be a closed sphere in X. By Theorem B, it suffices to show that its complement $S_r[x_0]'$ is open. $S_r[x_0]'$ is open if it is empty, so we may assume that it is non-empty. Let a; be a point in $S_r[x_0]'$. Since $d(x, x_0) > r$, $r_1 = d(x, x_0) - r$ is a positive real number. We take r_1 as the radius of an open sphere Sr_1 (x) centered on x, and we show that $S_r[x_0]'$ is open by showing that $S_{r_1}(x) \subseteq S_r[x_0]'$. Let y be a point in $S_{r_1}(x)$, so that $d(y, x) < r_1$. On the basis of this and the fact that $d(x_0, x) < d(x + y) + d(y, x)$, we see that $d(y, xa_0) > d(x, x_0) - d(y, x) > d(x, x_0) - n = d(x, x_0) - [d(x, x_0) - r] = r$, so that y is in $S_r[x_0]'$.

The main general facts about closed sets are those given in our next theorem.

Theorem 29:

A subset A of metric subspace (Y, d) of a metric space (X, d) is open in Y iff ∃ an open set G in (X, d) s.t. A = Y ∩ G.

Proof:

Let $A \cap Y$ and let (Y, d) be a subspace of a metric space (X, d).

Then $A \subset Y \subset X$ ...(1)

Step I : Let A be open in Y. ...(2)

Aim : $A = Y \cap G$ for some open set G in X. By def. of open set, $\forall\ y \in A \Rightarrow\ \exists$ open sphere $S_{1\,(Y,\,r)}$ in s.t.

$$S_{1\,(Y,\,r)} \subset A \qquad ...(3)$$

Then A is expressible as

$$A = \cup\,\{S_{1\,(Y,\,r)} : y \in A\} \qquad ...(4)$$

Suppose S(Y, r) is an open sphere in X.

Then $S_{1\,(Y,\,r)} = Y \cap S(Y, r)$

Now (4) is expressible as

$$A = \cup\,\{S_{(Y,r)} \cap Y : y \in A\}$$
$$= Y \cup [S_{(Y,r)} : y \in A\}$$

or, $A = Y \cap G$,

where $G = \cup\,\{S_{(Y,r)} : y \in A\}$

= arbitrary union of open spheres in X

= open set in X.

$\therefore$ A = Y $\cap$ G, where G is open in X.

For each open sphere is an open set.

Step II : Let $A = Y \cap G$ ...(5)

where G is an open set in X.

Aim : A is open in Y.

By def. of open set in X,

$\forall$ x $\in$ G $\subset$ $\Rightarrow$ $\exists$ open sphere $S_{(x,\epsilon)}$ in X. s.t.

$$S_{(x,\epsilon)} \subset G \quad ...(6)$$

Write $S_{1(x,\hat{\imath})} = Y \cap S_{(x,\hat{\imath})}$ ...(7)

Then $S_{1(x,\epsilon)}$ is an open sphere in Y.

By (6), $S_{(x,\epsilon)} \cap Y \subset G \cap Y$

Using (7), we get $S_{1(x,\epsilon)} \subset G \cap Y = A$, by (5)

or, $x \in S_{1(x,\epsilon)} \subset A$

This $\Rightarrow$ A is open in Y, be def. of open set.

Theorem 30:

The necessary and sufficient condition that a subset A of a metric subspace (Y, d) of a metric space (X, d) be closed in (Y, d) is that $\exists$ a set F closed in (X, d) s.t.

$$A = Y \cap F.$$

Proof:

Firstly we shall prove a lemma :

A is open in Y $\Leftrightarrow$ A Y $\cup$ G, G is open in X.

Now we come to the proof of the actual theorem.

F is closed in X $\Rightarrow$ X – F is open in X.

A is closed in Y $\Leftrightarrow$ Y – A is open in Y.

Be the lemma,

$\therefore$ $Y - A = Y \cap (X - F)$

or, $Y - Y \cap (X - F) = A$

or, $Y - (Y \cap X - Y \cap F) = A$

or, $Y - (Y - Y \cap F) = A$

or, $Y \cap F = A$

Theorem 31:

If A_1 is an open set in (X_1, d) and A_2 is an open set in (X_2, d_2), then $A_1 \times A_2$ is open in the product metric space (X,d), where $X = X \times X_2$.

Proof:

Let $x = (x_1, x_2) \in A_1 \times A_2$ be arbitrary, then $x_1 \in A_1$, $x_2 \in A_2$. By def. of open set,

$\exists$ open spheres $S(x_1, r_1)$, $S(x_2, r_2)$ s.t.

$$\left.\begin{array}{l} S(x_1, r_1) \subset A_1 \\ S(x_2, r_2) \subset A_2 \end{array}\right\} \quad ...(1)$$

Let $r = \min. \{r_1, r_2\}$, then $r \leq r_1$, $r \leq r_2$.

$\therefore$ $S(x_1, r) \subset S(x_1, r_1) \subset A_1$

$S(x_2, r) \subset S(x_2, r_2) \subset A_2$

This $\Rightarrow S(x_1, r) \times S(x_2, r_1) \subset A_1 \times A_2$

$\Rightarrow x = (x_1, x_2) \in S(x, r)$

$\Rightarrow S(x_1, r) \times S(x_2, r_1) \subset A_1 \times A_2$

$\Rightarrow x \in S(x, r) \subset A_1 \times A_2$

It is true $\forall\ x \in A_1 \times A_2$

$\therefore$ $A_1 \times A_2$ is open subset of (X, d).

Here $S(x, r)$ is an open sphere in (X, d).

Theorem 32:

Let (X, d) be metric space and $A \subset X$. Show that

$\overline{A} = \{x \in X : d(x, A) = 0\}$.

Proof:

Let (X, d) be metric space and $A \subset X$.

We claim $\overline{A} = \{x \in X : d(x, A) = 0\}$

Suppose $d(x, A) = \in > 0$

Then (i) $x \notin A$ and

(ii) $(S_{\in/2}(x) - \{x\}) \cap A = \Phi$

Evidently (ii) $\Rightarrow x \notin D(A)$.

Thus, $d(x, A) = \notin$

$\Rightarrow x \notin A, x \notin D(A)$

$\Rightarrow x \notin A \cup D(A) = \overline{A}$

$\Rightarrow x \in \overline{A}$

Finally, $d(x, A) = \varepsilon \Rightarrow x \notin \overline{A}$

Again $d(x, A) = 0 \Rightarrow$

(iii) $x \in A$,

(iv) $(S_{r(x)} - \{x\}) \cap A \neq \Phi$

(iv) $\Rightarrow x \in D(A)$

Finally, $d(x, A) = 0 \Rightarrow x \in A$ or $x \in D(A)$

$\Rightarrow x \in A \cup D(A) = \overline{A} \Rightarrow x \in \overline{A}$

Thus, we have shown that

$d(x, A) = 0 \Rightarrow x \in \overline{A}$

and $d(x, A) = \varepsilon > 0 \Rightarrow x \notin \overline{A}$

Furthermore, $d(x, A) = \geq 0$.

Consequently the required result follows :

Theorem 33:

Let (X, d) be a metric space. A subset A of X is closed if given any $x \in X - A$, $d(x, A) \neq 0$.

Proof :

Let (X, d) be a metric space and $A \subset X$ be an arbitrary closed set.

To prove that

given any $x \in X - A$, $d(x, A) \neq 0$

A is closed $\Rightarrow X - A$ is open.

By definition of open set,

any $x \in X - A \Rightarrow \exists\, r \in R^{+}$ s.t. $S_{r(x)} \subset X - A \Rightarrow S_{r(x)} \cap A = \Phi$

$\Rightarrow d(x, A) \geq r \Rightarrow d(x, A) \neq 0$.

Conversely let A be any subset of a metric space (X, d). Let

any $x \in X - A \Rightarrow d(x, A) \neq 0$

To prove that A is closed.

Let $x \in X - A$ be arbitrary so that, by assumption,

$d(x, A) = r \neq 0 \Rightarrow S_{r(x)} \cap A = \phi$

$\Rightarrow \qquad S_{r(x)} \subset X - A$

$\therefore \qquad x \in X - A \Rightarrow \exists\, r \in R^+$ s.t. $S_{r(x)} \in X - A$.

By definition, this $\Rightarrow X - A$ is open

$\Rightarrow A$ is closed.

Theorem 34:

Let A be a subset of a metric space (X, d). Then $x_0 \in X$ is a limit point of A iff every open sphere centred at x_0 contains an infinitely many; points of A, other than x_0.

Proof:

Let a be a subset of a metric space (X, d) and let $x_0 \in X$.

Step I : Given x_0 is a limit point of A.

Aim. $S_{r\,(x0)}$ contains an infinite number of points of A other than x_0 ...(*)

By def. of limit point,

$$(S_{r\,(x0)}) - \{x_0\}) \cap A \neq \phi \qquad ...(1)$$

for every open sphere centred at x_0.

Suppose (*) is not true.

Then $S_{r\,(x0)}$ contains a finite number of points

$$x_1, x_2, \ldots, x_n, \in A \qquad ...(2)$$

Let $r = \min \{d\,(x_0, x_i) : = 1, 2. \ldots, n\}$

and $S_{r(x0)} = \{x \in X : (x, x_0) < r_0\}$

Evidently $(S_{r\,(x0)}) - \{x_0\}) \cap A = \phi$

This $\Rightarrow x_0$ is not a limit point of A.

This is contrary to what is given . This contradiction arises by assuming that $S_{r(x0)}$ contains a finite number of points of A other than x_0. Hence the required result follows.

Step II: Given $S_{r(x0)}$ contains an infinite number of points of A other than x_0.

Aim. x_0 is a limit point of A.

By assumption, $(S_{r\,(x0)}) - \{x_0\}) \cap A \neq \phi$

This $\Rightarrow x_0$ is a limit point of A.

Corollary : *If p is a limit point of A in a metric space, then every nbd N(p) contains infinitely many points of A.*

Or

Every open sphere centred at a limit point of a sat in a metric space contains infinitely many points of the set.

Hint : Open sphere $S_{r(p)}$ can be taken in place of $N_{(p)}$. Here write step I of the above previous theorem.

Theorem 35:

Let (X, ρ) be a metric space.

Define $$d(x, y) = \frac{\rho(x, y)}{1+\rho(x, y)} \; \forall \; x, y, \in X.$$

Then (X, y) is metric space.

Proof:

Let (X, ρ) be a metric space

Let $$d(x, y) = \frac{\rho(x, y)}{1+\rho(x, y)} \; \forall \; x, y, \in X.$$

To prove that (X, d) is a metric space.

Let x, y, z, ∈ X be arbitrary.

∵ (X, ρ) is a metric space.

∴ (i) ρ (x, y) ≥ 0

(ii) ρ (x, y) = ρ (y, x)

(iii) ρ (x, y) = 0 iff x = y

(iv) ρ (x, y) + ρ (y, z) ≥ ρ (x, z).

Now (i)' d (x, y) ≥ 0 in accordance with (i).

(ii)' d (x, y) = d (y, x).

For $$d(y, x) = \frac{\rho(y, x)}{1+\rho(y, x)}$$

$$= \frac{\rho(x, y)}{1+\rho(x, y)} = d(x, y)$$ in accordance with (ii)

(iii)'d (x, y) = 0, iff x = y

$$d(x, y) = 0 \Leftrightarrow \frac{\rho(x, y)}{1+\rho(x, y)} = 0$$

⇔ ρ (x, y) = 0 ⇔ x = y, according to (iii).

(iv)'d (x, y) + d (y, z) ≥ d (x, z).

For $$d(x, y) + d(y, z) = \frac{\rho(x, y)}{1+\rho(x, y)} = \frac{\rho(y, z)}{1+\rho(y, z)}$$

$$\geq \frac{\rho(x, y)}{1+\rho(x, y)+\rho(y, z)} + \frac{\rho(y, z)}{1+\rho(y, z)+\rho(x, y)}$$

$$\geq \frac{\rho(x, y)+\rho(y, z)}{1+\rho(x, y)+\rho(y, z)}$$

$$\geq \frac{\rho(x, z)}{1+\rho(x, z)}$$

$$= d(x, z)$$

i.e., d (x, y) + d (y, z) ≥ d (x, z).

The results (i)', (ii)', (iii)', (iv)' taken together prove the required result.

Theorem 36:

Let (X, y) be a metric space and let m a positive real number. Show that there exists a metric d' for X such that the metric space (X, y) is bounded with d(X) ≤ m.

Proof:

Let (X, y) be a metric space and m > 0.

we define d (x, y) = $\dfrac{md\,(x,y)}{1+d\,(x,y)}\ \forall\ x, y \in X.$

To prove that (X) (X, d) is a metric space and m > 0.

Let x, y, x ∈ X be arbitrary.

d is metric ⇒ d (x, y) ∈ R

Also m ∈ R

⇒ d' (x, y) $\dfrac{md\,(x,y)}{1+d\,(x,y)} \in R.$

(X, d) is metric space

(i) d (x, y) ≥ 0

(ii) d (x, y) = d (y, x)

(iii) d (x, y) = 0 iff x = y

(iv) d (x, y) + d (y, z) ≥ d (x, y).

(1) Evidently d' (x, y) ≥ 0, according to (i) as m > 0.

(2) d' (x, y) = d' (y, x)

For d' (y, x) $\dfrac{md\,(y,x)}{1+d\,(y,x)}$

$$= \frac{md\,(x\,y)}{1 + d\,(x,y)}\text{ , according to (ii)}$$

$= d'\,(x, y)$

(3) $d'\,(x, y) = 0$ iff $x = y$.

For $$d'\,(x, y) = 0 \Leftrightarrow \frac{md\,(x\,y)}{1 + d\,(x,y)} = 0$$

$\Leftrightarrow md\,(x\,,y) = 0$

$\Leftrightarrow d\,(x, y) = 0$

$\Leftrightarrow x = y$, according to (iii).

(4) $d'\,(x, y) + d'\,(y, z) \geq d'\,(x, z)$.

For $$d'\,(x, y) + d'\,(y, z) = \frac{md\,(x,y)}{1 + d\,(x,y)} + \frac{md\,(y,z)}{1 + d\,(y,z)}$$

$$= m\left[\frac{d\,(x,y)}{1 + d\,(x,y)} + \frac{d\,(y,z)}{1 + d\,(y,z)}\right]$$

$$\geq m\left[\frac{d\,(x,y)}{1 + d\,(x,y) + d\,(y,z)} + \frac{d\,(y,z)}{1 + d\,(x,y) + d\,(y,z)}\right]$$

$$= m\left[\frac{d\,(x,y) + d\,(y,z)}{1 + d\,(x,y) + d\,(y,z)}\right] \geq \frac{md(x,z)}{1 + d(x,z)}$$

$= d'\,(x, z)$

The results (1), (2), (3) and (4) taken together prove that (X, d') is metric space. Remains to prove that d' is bounded.

$$d'\,(x, y) = \frac{md(x,y)}{1 + d(x,y)} = \frac{m}{1 + \frac{1}{d(x,y)}}$$

$d'\,(x, y) \leq m\ \forall\ x, y, \in X$.

This $\Rightarrow$ (x, d') is a bounded metric space.

If $$d'\,(x, y) = \frac{d(x,y)}{1 + d(x,y)}$$

Then $d'\,(x, y) \leq 1$ so that (X, d') is bounded.

Example:

Define non where dense set and give an example.

(4) Consider the set R. Evidently $Q \subset R$. Every real number is a limit point so that rational limit points belong to the set Q.

It means that $$D(R) = R,\ D(Q) = R$$

$$\because \quad \overline{A} = A \cup D(A)$$

$$\therefore \quad \overline{Q} = Q \cup D(Q) = Q \cup R = R$$

This proves that the closure of the set of rational numbers w.r.t. usual metric no R is R.

Thus $Q \subset R$, Q is countable and $\overline{Q} = R$.

This proves that r with usual metric is separable.

$Q^o = \cup \{G \subset R : G \text{ is open}, G \subset Q = \phi.$

Since every subset of R contains rationals and irrationals. Hence their exists no open set.

$$G \subset R \text{ s.t. } G \subset Q.$$

Thus, $$Q^o = \phi.$$

Similarly $$\text{ext}(Q) = (R - Q)^o = \phi.$$

$$\text{int}(\overline{Q}) = \text{int}(R) = R \neq \phi.$$

$\therefore$ *Q is some where dense set in R.*

$$b(Q) = R - Q^o \cup (R - Q)^o = R - \phi \cup \phi = R$$

(5) If A = (a, b) or [a,b) or (a, b] or [a, b],

then in every case,

$\overline{A} = [a, b],\ A^o = (a, b),\ b(A) = \{a, b\}.$

(6) If A = [0, 1), then $(\overline{A})^o = [0, 1]^o = (0, 1) \neq \phi$.

and so A is somewhere dense set in R.

(7) To show that N and Z are non-dense sets in R relative to the usual metric.

$$D(N) = \phi.$$

For if a is any real number, then any nbd = $(a - \varepsilon, a + \varepsilon) \subset R$ does not contain any point of N other than a

$$Z = \{n : n \,\varepsilon\, N\} \cup \{0\} \cup \{-n : n \,\varepsilon\, N\}$$

$$D(Z) = \phi \cup \phi \cup \phi = \phi$$

or, $$D(Z) = \phi \subset Z$$

or, $$D(Z) \subset Z.$$

$\therefore$ *Z is a closed set. Similarly N is a closed set.*

For $D(A) \subset A \Leftrightarrow A$ is closed.

$\therefore \qquad \overline{Z} = Z$ and $\overline{N} = N$

$Int\ \overline{Z} = Int\ (Z) = \cup\{G \subset R : G$ is open and $G \subset Z\}' = \phi$

For any open subset $(a_1, a_2) \subset R$ *contains fractions also.*

Now $\qquad Int\ (\overline{Z}) = (\overline{Z})^o = \phi$.

Similarly $\qquad (\overline{N})^o = \phi$.

$\therefore$ *Z and N are nowhere dense subsets of R.*

Theorem 37:

In any metric space X, the empty set ϕ *and the full space X are closed sets.*

Proof:

The empty set has no limit points, so it contains them all and is therefore closed. Since the full space X contains all points, it automatically contains its own limit points and thus is closed.

The following theorem characterizes closed sets in terms of open sets. We already know a good deal about open sets, so that characterization provides us with a useful tool for establishing properties of closed sets.

Theorem 38:

Let X be a metric space. A subset F of X is closed $\Leftrightarrow$ *its complement F' is open.*

Proof:

Assume first that F is closed. We show that F' is open. If F' is empty, it is open, so we may suppose that F' is non-empty. Let x be a point in F'. Since F is closed and x is not in F, x is not a limit point of F. Since x is not in F and is not a limit point of F, there exists an open sphere $S_r(x)$ which is disjoint from F. $S_r(x)$ is an open sphere centered on x and contained in F', and since x was taken to be any point of F', F' is open.

We now assume that F' is open and show that F is closed. The only way F can fail to be closed is to have a limit point in F'. This cannot happen, for since F' is open, each of its points is the center of an open sphere disjoint from F, and no such point can be a limit point of F.

If x_0 is a point in our metric space X, and r is a non-negative real number, the closed sphere $S_r[x_0]$ with *center* x_0 and *radius* r is the subset of X defined by

$$S_r[x_0] = \{x : d(x, x_0) \leq r\}.$$

$S_r[x_0]$ contains its center, and when $r = 0$ it contains only its center. The closed spheres on the real line are precisely the closed intervals. In this connection, we observe that though open spheres on the real line are open intervals, there are open intervals which are not open spheres, *e.g.*, $(-\infty, +\infty)$.

The following theorem justifies the adjective in the phrase "closed sphere."

Theorem 39:

If A and B are subsets of a metric space (X, d), then

(i) $(A \cap B)^o = A^o \cap B^o$.

(ii) $A^o \cup B^o \subset (A \subset B)^o$.

Also show that $A^o \cup B^o \neq (A \cup B)^o$ *by means of an example.*

Proof:

(i) Any $x \in (A \cap B)^o$

$\Rightarrow$ x is an interior point of $A \cap B$

$\Rightarrow S_r(x) \subset A \cap B$

$\Rightarrow S_r(x) \subset A \cap B \subset A,\ S_r(x) \subset A \cap B \subset B$

$\Rightarrow S_r(x) \subset A,\ S_r(x) \subset B$

$\Rightarrow x \in A^o,\ x \in B^o$

$\Rightarrow x \in A^o \cap B^o$

$\therefore$ $(A \cap B^o) \subset A^o \cap B^o$.

Conversely, any $x \Rightarrow A^o \cap B^o$. ...(1)

$\Rightarrow x \in A^o$ and $x \in B^o$

$\Rightarrow \exists$ open spheres $S_{r1}(x)$, $S_{r2}(x)$, s.t.

$S_{r1}(x) \subset A,\ S_{r2}(x) \subset B$

Choose $r = \min.(r_1, r_2)$

$\Rightarrow S_r(x) \subset S_{r1}(x) \subset A,\ S_r(x) \subset S_{r2}(x) \subset B$

$\Rightarrow S_r(x) \subset A,\ S_r(x) \subset B$

$\Rightarrow S_r(x) \subset A \cap B$

$\Rightarrow x \in (A \cap B)^o$

$\therefore$ $A^o \cap B^o \subset (A \cap B)^o$ ... (2)

Combining (1) and (2), we get the result (i)

(ii) Any $x \in A^o \cup B^o$

$$\Rightarrow x \in A^o \text{ or } x \in B^o.$$

$$\Rightarrow S_{r1}(x) \subset A \text{ or } S_{r2}(x) \subset B$$

Choose the two spheres s.t. one is contained in the other so that $r_2 \geq r_1$ and $S_{r1}(x) \cup S_{r2}(x) = S_{r2}(x)$.

$$\Rightarrow S_{r2}(x) = S_{r1}(x) \cup S_{r2}(x) \subset A \cup B$$

$$\Rightarrow S_{r2}(x) \subset A \cup B$$

$$\Rightarrow x \in (A \cup B)^o$$

$\therefore \quad A^o \cup B^o \subset (A \cup B)^o$.

Theorem 40:

Let (X, d) be a metric space and let

$$d'(x, y) = \min\{1, d(x, y)\}.$$

Then d' is a metric for X.

Proof:

Let (X, d) be a metric space so that

(i) $d(x, y) \geq 0$

(ii) $d(x, y) = d(y, x)$

(iii) $d(x, y) = 0$ iff $x = y$

(iv) $d(x, y) + d(y, x) \geq d(x, z)$, where $x, y, z \in X$.

Let $d'(x, y) = \min, \{1, d(x, y)\}$.

Then $\quad d'(x,y) = \begin{cases} 1 \\ or\ d(x,y) \end{cases}$ ...(1)

(2) $d'(x, y) \geq 0$, according to (1) and (i),

(3) $d'(x, y) = d'(y, x)$.

For if $d'(x, y) = 1$, then $d'(y, x) = 1$ so that $d'(x, y) = d'(y, x)$.

Again if $d'(x, y) = d(x, y)$, then $d'(y, x) = d(y, x)$ using (ii) and combining the last two, we get

$$d'(x, y) = d'(y, x)$$

$\therefore$ In either case, $d'(x, y) = d'(y, x)$

(4) $d'(x, y) = 0$ iff $x = y$.

For $d'(x, y) = 0 \Leftrightarrow d(x, y) = 0$ according to (1)

$\Leftrightarrow x = y$ according to (iii).

(5) To prove $d'(x, y) + d'(y, z) \geq d'(x, z)$

If d' (x, y) = 1, then d' (x, z) = 1, d' (y, z) = 1 and

so d' (x, y) + d' (y, z) > d' (x, z)

If we take $\quad d'(x, y) = d(x, y)$, then

$$d'(y, z) = d(y, z),$$

$$d'(x, z) = d(x, z)$$

and so $\quad d'(x, y) + d'(y, z) = d(x, y) + d(y, z)$

$$\geq d(x, z) = d', (x, z) \text{ d by (iv)}$$

or, $d'(x, y) + d'(y, z) \geq d'(x, z)$

∴ (5) is true in both cases.

From what has been it follows that d' is a metric on X.

Example : Show that if d is a pseudo metric on X and e(x, y) = min {1, d (x, y)}, then e is also a pseudo metric.

Theorem 41:

If $1 < p < \infty$ *and* $\frac{1}{p}+\frac{1}{q}=1$, *then*

$ab \leq \frac{a^p}{p}+\frac{b^q}{q}$ *where a, b are two non-negative reals.*

Proof :

if a = 0 or b = 0 the result is obvious.

Now consider the case in which a ≠ 0, b ≠ 0.

Let $f(t) = 1 - \lambda + \lambda t - t^{\lambda}$ for $0 < \lambda < 1$.

Then $\quad f'(t) = \lambda - \lambda t^{\lambda-1}$

$$f'(t) = 0 \Rightarrow \lambda(1 - t^{\lambda-1}),$$

where $\quad \lambda, \lambda - 1 \neq 0$

$$\Rightarrow t = 1$$

$$f''(t) = -\lambda(\lambda - 1)\, t^{\lambda-2}.$$

At t = 1, $f''(t) = -\lambda(\lambda - 1) = +$ ive as $0 < \lambda < 1$.

Thus, $\quad \frac{df}{dt} = 0, \frac{d^2f}{dt^2} >$ at t = 1.

This proves that f is minimum at t = 1.

∴ $\quad f(1) \leq f(t)$

This $\quad \Rightarrow 1 - \lambda + \lambda - 1 \leq 1 - \lambda + \lambda t - t^{\lambda}$

or $\quad 1 - \lambda + \lambda t - t^{\lambda} \geq 0.$

or $\quad (1 - \lambda) + \lambda t \geq t^{\lambda}$ (Put $\lambda = 1/p$)

or $$\left(1 - \frac{1}{p}\right) + \frac{t}{p} \geq \frac{t^1}{r}$$

But $$\frac{1}{p} + \frac{1}{q} = 1$$

Hence $$\frac{1}{q} + \frac{t}{p} \geq \frac{t^1}{p}.$$ Put $t = \dfrac{a^p}{b^q}$

Then $$\frac{1}{q} + \frac{a^p}{pb^q} \geq \frac{a}{b^{q/p}}$$

or $$\frac{b^q}{q} + \frac{a^p}{p} \geq ab^{q-q/p} = ab.$$

or $$\frac{a^p}{p} + \frac{b^q}{q} \geq ab$$

as $$\frac{1}{p} + \frac{1}{q} = 1 \Rightarrow q - \frac{q}{p} = 1$$

Theorem 42:

In any metric space, prove that difference between the two sides of a triangle is less than the third side.

Or

Let (X, d) be a metric space and let x, y, z be any three points of X. Then $|d(x, z) - d(y, z)| \leq d(x, y)$

Proof :

Let x, y, z denote the co-ordinates of vertices of a triangle.

If we show that $|d(x, z) - d(y, z)| \leq d(x, y)$, the result will follow.

We know that

$$d(x, y) + d(y, z) \geq d(x, z)$$

Similarly $\quad d(y, x) + d(x, z) \geq d(y, z)$

From Which $d(x, y) \geq d(x, z) - d(y, z)$ ...(1)

and $\quad d(y, x) \geq -[d(x, z) - d(y, z)]$

The last gives $-d(x, y) \leq d(x, z) - d(y, z)$ ...(2)

For $d(x, y) = d(y, x)$.

Combining (1) and (2),

$$-d(x, y) \leq d(x, z) - d(y, z) \leq d(x, y)$$

This $\Rightarrow |d(x, z) - d(y, z)| \geq d(x, y)$.

Theorem 43:

Let (X, d) be a metric space, then $\forall\, x, y, x', y'$, in X, $|d(x, y) - d'(x' y')| \leq d(x, x') + d(y, y')$.

Proof:

For proving the required result, we make repeated use of "triangle inequality."

$$d(x, y) \leq d(x, x') + d(x', y)$$

$$\leq d(x, + d(x', y') + d(y',$$

or, $\quad d(x, y) - d(x', y') \leq d(x, + d(y', \qquad \text{...(1)}$

as $\quad d(y, = d(y',$

Again $\quad d(x', y') \leq d(x', + d(x, y')$

$$\leq d(x', + d(x, y) + d(y,$$

Using the fact that

$d(a, b) = d(b, a)$,

we get $\quad d(x', y') - d(x, y) \leq d(x, + d(y,$

or, $d(x, y) - d(x', y') \geq -[d(x, + d(y, \qquad \text{...(2)}$

Combining (1) and (2), we get,

$-[d(x, + d(y, \leq d(x, y) - d(x', y') \leq d(x, + d(y,$

or, $|d(x, y) - d(x', y')| \leq d(x, + d(y,$

Theorem 44:

Let (X, d) be a metric space. If $\langle x_n \rangle$ and $\langle y_n \rangle$ are sequences in X, such that $x_n \to x$, $y_n \to y$, then show that

$$d(x_n, y_n) \to d(x, y).$$

Proof :

Suppose $\langle x_n \rangle$ *and* $\langle y_n \rangle$ are sequences in a metric space (X, d).

Let $x_n \to x$, $y_n \to y$.

To prove that

$$d(x_n, y_n) \to d(x, y).$$

Let x, y, z $\in$ X be arbitrary.

Firstly we shall prove that

$$| d (x, z) - (y, z)| \leq d (x, y) \quad ...(3)$$

Now write the proof of previous theorem .

$\because$ $\quad x_n \to x, y_n \to y.$

$\therefore$ $\quad \varepsilon > 0, \exists\ n_0, m_0, \in N$ s.t.

$\forall\ n \geq n_0 \Rightarrow d (x_n, x) < \varepsilon/2$

$\forall\ n \geq m_0 \Rightarrow d (y_n, y) < \varepsilon/2$

Take $\quad k_0 = \max. \{m_0, n_0\}$

Then $\quad d (x_n, x) < \varepsilon/2, d (y_n, y) < \varepsilon/2\ \forall\ n \geq k_0$

Now $\quad |d (x_n, y_n) - d (x, y)|$

$= |d (x_n, y_n) - d (x_n, y) + d (x_n, y) - d (x, y)|$

$\leq |d (x_n, y_n) - d (x_n, y)| + |d (x_n, y) - d (x, y)|$

$= |d (y_n, x_n) - d (y, x_n)| + |d (x_n, y) - d (x, y)|$

$\leq d (y_n, y) + d (x_n, x)$, according to (3)

$< \varepsilon/2 + < \varepsilon/2 = \varepsilon$

$\therefore \quad |d (x_n, y_n) - d (x, y)| < \varepsilon/2\ \forall\ n \geq k_0$

This $\quad \Rightarrow d (x_n, y_n) \to d (x, y).$

PROJECTIONS

A projection on a Banach space B is an idempotent operator on B, that is, an operator P with the property that P2 = P. It was proved in that section that each projection.

P determines a pair of closed linear subspaces M and N–the range and null space of P–such that B = M $\oplus$ N, and also, conversely, that each such pair of closed linear subspaces M and N determines a projection P with range M and null space N.

In this way, there is established a one-to-one correspondence between projections on B and pairs of closed linear subspaces of B which span the whole space and have only the zero vector in common.

The context of our present work, however, is the Hilbert space H, and not a general Banach space, and the structure which H enjoys in addition to being a Banach space enables us to single out for special attention those projections whose range and null space are orthogonal. Our first theorem gives a convenient characterization of these projections.

Theorem 1:

If P is a projection on H with range M and null space N, then $M \perp N \Leftrightarrow$ P is self-adjoint; and in this case, $N = M^{\perp}$.

Proof:

Each vector 2 in H can be written uniquely in the form z = x + y with x and y in M and N. If $M \perp N$, so that $x \perp y$, then P* = P will follow (P*z, z) = (Pz, z), and this is a consequence of

(P*z, z) = (z, Pz) = (z, x) = (x, + y, x) = (x, x) = (y, x) = (x, x) and (Pz, z) = (x, z) = (x, x + y) = (x, x) + (x, y) = (x, x). If, conversely, P* = P, then the conclusion that $M \perp N$ follows from the fact that for any x and y in M and N we have

$$(x, y) = (Px, y) = (x, P^*y) = (x, Py) = (x, 0) = 0.$$

All that remains is to see that if $M \perp N$, then $N = M^{\perp}$. It is clear that $N \subseteq M^{\perp}$; and if *N* is a proper subset of $M^{\perp}$, and therefore a proper closed linear subspace of the Hilbert space $M^{\perp}$, then Theorem 2, implies that there exists a non-zero vector z_0 in $M^{\perp}$ such that $z_0 \perp N$. Since $z_o \perp$. N and $z_0 \perp N$, and since $H = M \oplus N$, it follows that $z_0 \perp H$. This is impossible, so we conclude that $N = M^{\perp}$.

A projection on H whose range and null space are orthogonal is sometimes called a perpendicular projection. The only projections considered in the theory of Hilbert spaces are those which are perpendicular, so it is customary to omit the adjective and to refer to them simply as projections. In the light of this agreement and Theorem 1, a projection on H can be defined as an operator P which satisfies the conditions $P^2 = P$ and P* = P. The operators 0 and I are projections, and they are distinct $\Leftrightarrow H \neq \{0\}$.

The great importance of the projections on H rests mainly which allows us to set up a natural one-to-one correspondence between projections and closed linear subspaces. To each projection P there corresponds its range M = $\{Px : x \varepsilon H\}$, which is a closed linear subspace; and conversely, to each closed linear subspace M there corresponds the projection P with range M defined by P (x + y) = x, where x and y are in M and $M^{\perp}$. Either way, we speak of P as the projection on M.

It is clear that P is the projection on $M \Leftrightarrow I - P$ is the projection on $M^{\perp}$. Also, if P is the projection on M, then

$$x \varepsilon M \Leftrightarrow Px = x \Leftrightarrow \|Px\| = \|x\|.$$

The first equivalence here was proved in Problem 44-11; and since for every x in H we have

$$\| x \|^2 = \| Px \mid (I - P) x \|^2 = \| Px \|^2 + \| (I - P) x \|^2. \qquad ...(1)$$

the non-trivial part of the second is given by the following chain of implications:

$$\| Px \| = \| x \| \Rightarrow \| P* \|^2 = \| * \|^2$$

$$\Rightarrow \qquad \| (I - P) x \|^2 = 0 \Rightarrow Px = x.$$

Relation (20) also shows that $\| Px \| \leq \| x \|$ for every x, so $\| P \| \leq 1$. If x is an arbitrary vector in H, it is easy to see that

$$(Px, x) = (PPx, x) = (Px, P(x) = (Px, Px) = \| px \|^2 \geq 0. \qquad ...(2)$$

so P is a positive operator ($0 \leq P$) in the sense. Since $I - P$ is also a projection, we also have $0 \leq I - P$ or $P \leq I$, so $0 \leq P \leq I$. Let T be an operator on H. A closed linear subspace M of H is said to be invariant under T if $T(M) \subseteq M$. When this happens, the restriction of T to M can be regarded as an operator on M alone, and the action of T on vectors outside of M can be ignored. If both M and Mx are invariant under T, we say that M reduces T, or that T is reduced by M. This situation is much more interesting, for it allows us to replace the study of *T* as a whole by the study of its restrictions to *M* and $M^\perp$, and it invites the hope that these restrictions will turn out to be operators of some particularly simple type. In the following four theorems, we translate these concepts into relations between T and the projection on *M*.

Theorem 2:

A closed linear subspace M of H is invariant under an operator $T \Leftrightarrow M^\perp$ *is invariant under T*.*

Proof:

Since $M^{\perp\perp} = M$ and $T^{**} = T$, it suffices by symmetry to prove that if M is invariant under T, then M1- is invariant under T*. If y is a vector in M1, our conclusion will follow from $(x, T^*y) = 0$ for all x in M. But this is an easy consequence of $(x, T^*y) = (Tx, y)$, for the invariance of M under T implies that $(Tx, y) - 0$.

Theorem 3:

A closed linear subspace M of H reduces an operator $T \Leftrightarrow M$ *is invariant under both T and T*.*

Proof:

This is obvious from the definitions and the preceding theorem.

Theorem 4:

If P is the projection on a closed linear subspace M of H, then M is invariant under an operator $T \Leftrightarrow TP = PTP$

Proof:

If M is invariant under T and x is an arbitrary vector in H, then TPx is in M, so PTPx = TPx and PTP = TP. Conversely, if TP = PTP and x is a vector in M, then Tx = TPx = PTPx is also in M, so M is invariant under T.

Theorem 5:

If P is the projection on a closed linear subspace M of H, then M reduces an operator T $\Leftrightarrow$ TP = PT.

Proof:

M reduces T $\Leftrightarrow$ M is invariant under T and T* $\Leftrightarrow$ TP = PTP and T*P = PT*P $\Leftrightarrow$ TP = PTP and PT = PTP. The last statement in this chain clearly implies that TP = PT, it also follows from it, as we see by multiplying TP = PT on the right and left by P.

Our next theorem shows how projections can be used to express the statement that two closed linear subspaces of H are orthogonal.

Theorem 6:

If P and Q are the projections on closed linear subspaces M and N of H, then $M \perp N \Leftrightarrow$ PQ = 0 $\Leftrightarrow$ QP = 0.

Proof:

We first remark that the equivalence of PQ = 0 and QP – 0 is clear by taking adjoints. If M $\perp$ N, so that N $\subseteq M^{\perp}$, then the fact that Qx is in N for every x implies that PQx – 0, so PQ = 0. If, conversely, PQ = 0, then for every x in N we have Px = PQx = 0, so N $\subseteq M^{\perp}$ N.

Motivated by this result, we say that two projections P and Q are orthogonal if PQ = 0. Our final theorem describes the circumstances under which a sum of projections is also a projection.

Theorem 7:

If $P_1, P_2, \ldots, P_n$ are the projections on closed linear sub-spaces $M_1, M_2, \ldots, M_n$ of H, then $P = P_1 + P_2 + \ldots + P$» is a projection $\Leftrightarrow$ the P_i's are pairwise orthogonal (in the sense that $P_iP_j = 0$ whenever $i \neq j$), and in this case, P is the projection on

$$M = M_1 + M_2 + \ldots + M_n.$$

Proof:

Since P is clearly self-adjoint, it is a projection $\Leftrightarrow$ it is idem-potent. If the P_i,s are pairwise orthogonal, then a simple computation shows at once that P is idempotent. To prove the converse, we assume that P is idempotent. Let a; be a vector in the range of P_i, so that $x = P_i x$. Then

$$\| x \|^2 = \| P_i x \|^2 \leq \sum_{j=1}^{n} \|P_j x\|^2 = \sum_{j=1}^{n} (P_j x, x) = (Px, x) = \| Px \|^2 \leq \| x \|^2.$$

We conclude that equality must hold all along the line here, so

$$\sum_{j=1}^{n} \|P_j x\|^2 = \| P_i x \|^2$$

and $\qquad \| P_j x \| = 0$ for $j \neq i$.

Thus the range of P_i is contained in the null space of P_j, that is, $M_i \subseteq M_j^{\perp}$L, for every $j \neq i$. This means that $M_i \perp M$, whenever $i \neq j$, and our conclusion that the P_i's are pairwise orthogonal now follows from the preceding theorem. We prove the final statement in two steps. First, we observe that since $\| Px \| = \| x \|$ for every x in M_i, each M_i is contained in the range of P, and therefore M is also contained in the range of P. Second, if x is a vector in the range of P, then

$$x = Px = P_1 x + P_2 x + \ldots + P_n x$$

is evidently in M.

There are many other ways in which the algebraic structure of the set of all projections on H can be related to the geometry of its closed linear subspaces, and several of these are given in the problems below. The significance of projections in the general theory of operators on H is the theme of the next chapter. As we shall see, the essence of the matter (the spectral theorem) is that every normal operator is made of projections in a way which clearly reveals the geometric nature of its action on the vectors in H.

SOME MISCELLANEOUS EXAMPLES

Example 1:

Show that a subset F of a metric space X is closed iff $\{x \in X : d(x, F) = 0\} \subset F$.

Solution:

Let (X, d) be a metric space.

Let $F \subset X$ be closed.

To prove that

$$\{x \in X : d(x, F) = 0\} \subset F.$$

We know that $\overline{F} = \{x \in X : d(x, F) = 0\}$.

F is closed $\Rightarrow \overline{F} = F \Rightarrow \overline{F} \subset F$

$\Rightarrow \{x \in X : d(x, F) = 0\} \subset F.$ on using (1)

Conversely suppose that F is a subset of a metric space (X, d) with the property that

$$\{x \in X : d(x, F) = 0\} \subset F \quad \text{...(2)}$$

To prove that F is closed, it suffices to show that $F = \overline{F}$.

But $\overline{F} = \{x \in X : d(x, F) = 0\}$ for any set $F \subset X$. ...(3)

Now (2) and (3) $\Rightarrow \overline{F} \subset F$ ...(4)

But $F \subset \overline{F}\ \forall\ F$ ...(5)

Combining (4) and (5), $\overline{F} = F$.

Example 2:

In every metric space, all finite sets are closed

Solution:

Let (X, d) be a metric space.

To prove that every finite subset of X is closed.

For any subset A of X, we know that

$\overline{A} = \{x \in X : d(x, A) = 0\}$

Let $p \in X$ be arbitrary.

Then $\{\overline{p}\} = \{x \in X : d(x, \{p\}) = 0)$
$= \{p\}$

$\therefore$ $d(x, \{p\}) = 0 \Rightarrow x \in \{p\} \Rightarrow x = p$

$\{p\} = \{\overline{p}\}$,

Showing thereby, {p} is closed in X.

Hence every singleton set in X is closed. Moreover a finite union of closed sets is closed. A finite set is a finite union of singleton sets in which every singl^ton set is closed. Accordingly every finite set in a metric space is closed.

Example 3:

If d is a metric on X, prove that the metric ρ defined by

$$\rho(a,b) = \frac{d(a,b)}{1 + d(a,b)}$$

is also a metric on X.

Solution:

Do yourself.

Example 4:

Let R be the set of all real numbers and let

$$d(x,y) = \frac{|x-y|}{1+|x-y|} \text{ for all } x, y, \in R.$$

Prove that d is a metric for R.

Solution:

Let x, y, z ∈ R be arbitrary.

Let $\rho(x,y) = |x-y|$.

Then (1) ⇒ ρ (x, y) ≥ 0.

For modulus of any quantity is always non-negative.

(2) ρ (x, y) = 0 iff x = y.

For ρ (x, y) = 0

$\Rightarrow |x-y| = 0 \Rightarrow x = y.$

(3) ρ (x, y) = ρ (y, x).

For |x − y| = |y − x|

(4) ρ (x, y) + ρ (y, z) ≥ ρ (x, z).

For ρ (x, y) + ρ (y, z) = |x − y| + |y − z|.

≥ |x − y + y − z| = |x − z| = ρ (x, z).

i.e., ρ (x, y) + ρ (y, z) ≥ ρ (x, z).

The above arguments prove that ρ is a metric on R.

By what is given $d(x, y) = \frac{|x-y|}{1+|x-y|}$.

This is expressible as

$$d(x, y) = \frac{\rho(x,y)}{1+\rho(x,y)}$$

where ρ is a metric on R. We have to prove that d is a metric on R.

Example 5:

Let ρ_1 and ρ_2 be metrics on the same set X, show that there exists a metric ρ on X such that

$\rho(x, y) \geq \rho_i(x, y)$ *for* $i = 1, 2$

and $\forall x, y \in X.$

Solution:

Let ρ_1 and ρ_2 be metrics on the same set X, Let x, y, z ∈ X be arbitrary.

To prove that ∃ a metric ρ on X s.t.

$$\rho(x, y) \geq \rho_i(x, y) \text{ for } i = 1, 2$$

Write $\rho(x, y) = \rho_1(x, y) + \rho_2(x, y).$ ...(1)

We claim ρ is metric on X.

(i) $\rho(x, y) \geq 0$

∵ $\rho_1(x, y) \geq 0$, $\rho_2(x, y) \geq 0.$

(ii) $\rho(x, y) = 0$ iff $x = y$

For $\rho(x, y) = 0 \Leftrightarrow \rho_1(x, y) + \rho_2(x, y) = 0.$

$$\Leftrightarrow \rho_1(x, y) = 0 = \rho_2(x, y).$$

Since ρ_1 and ρ_2 both are non-negative.

$$\Leftrightarrow x = y$$

(iii) $\rho(y, x) = \rho(x, y)$

For $\rho(y, x) = \rho_1(y, x) + \rho_2(y, x) = \rho_1(x, y) + \rho_2(x, y)$

$$\rho(x, y)$$

∵ $\rho_1(x, y) = \rho_1(y, x),\ \rho_2(x, y) = \rho_2(y, x)$

(iv) $\rho(x, y) + \rho(y, z) \geq \rho(x, z)$

For $\rho(x, y) + \rho(y, z) = \rho_1(x, y) + \rho_2(x, y) + \rho_1(y, z) + \rho_2(y, z)$

$$[\rho_1(x, y) + \rho_1(y, z)] + [\rho_2(x, y) + \rho_1(y, z)]$$

$$^3\ \rho_1(x, z) + \rho_2(x, z) = \rho(x, z).$$

From what has been done it follows that ρ is a metric on X.

Moreover (1) ⇒ $\rho(x, y) \geq \rho_1(x, y)$ for i = 1, 2.

Hence ρ is the required metric.

Example 6:

Let l_∞ *denote the set of all bounded sequences* $\langle x_n \rangle$. *If* $x = \langle x_n \rangle$, $y = \langle y_n \rangle$ *in* $l_{\infty,}$ *then we define*

$$d(x, y) = \sup\{|x_n - y_n| : n \in N\}.$$

Show that d is a metric on $l_{\infty.}$

Solution:

(1) $d(x, y) = \sup |x_n - y_n|$

$\leq \sup |x_n| + \sup |y_n| < \infty$

or d (x, y) is finite. For $<x_n>$ and $<y_n>$ both are bounded.

This $\Rightarrow$ d (x, y) is well defined map.

(2) Evidently $d(x, y) \geq 0$.

(3) $d(x, y) = 0$ iff $x = y$

For $x = y \Leftrightarrow x_n = y_n \ \forall\ n \Leftrightarrow |x_n - y_n| = 0 \ \forall\ n$

$\Leftrightarrow \sup |x_n - y_n| = 0 \Leftrightarrow d(x, y) = 0$

(4) $d(x, y) = d(y, x)$.

For $|x_n - y_n| = |y_n - x_n|$

(5) $d(x, y) \leq d(x, z) + d(z, y)$

For $d(x, y) = \sup |x_n - y_n| = \sup |x_n - z_n + z_n - y_n|$

$\leq \sup [|x_n - z_n| + |z_n - y_n|]$

$\leq \sup |x_n - z_n| + \sup |z_n - y_n|$

$= d(x, z) + d(z, y)$

$\therefore$ (l_∞, d) is a metric space.

Example 7:

Every derived set in a metric space is a closed set.

Solution:

Let (X, d) be a metric space and $A \subset X$. Let x be a limit point of D (A) . If we show that x ε D (A), the result will be proved.

x is limit point of D (A) $\Rightarrow (S_r(x) - \{x\} \cap D(A) \neq \varnothing$ ...(1)

for every open sphere $S_r(x)$.

Then $\exists\ y_r \neq x \in S_r(x) \cap D(A)$

i.e. $y_r \in D(A)$, $y_r \in S_r(x)$

$\therefore$ $d(x, y) < r$.

Take $r - d(x, y) = r_1 > 0$.

Evidently $S_{r1}(y_r) \subset S_r(x)$ ...(2)

$$\left[\begin{array}{l} \textit{For } z \in S_{r1}(y_r) \Rightarrow d(z, y_r) < r_1 = r - d(x, y_r) \\ \qquad \Rightarrow d(x, y_r) + d(y_r, z) < r \\ \qquad \Rightarrow d(x, z) < r \Rightarrow z \in S_r(x) \end{array}\right]$$

$y_r \in D(A) \Rightarrow S_{r1}(y_r) \cap A \neq \varnothing \;\forall\; S_{r1}(y_r)$

$\Rightarrow \quad S_r(x) \cap A \neq \varnothing$, by (2)

$\Rightarrow \quad \in D(A).$

Example 8:

Prove that the mapping defined by $d : R \times R \to R$ such that

$$d(x,y) = \begin{cases} o \text{ if } x = y \\ 1 \text{ if } x \neq y \end{cases}$$

is a metric on R.

Solution:

Let x, y, z ∈ R be arbitrary and let

$$d(x,y) = \begin{cases} o \text{ if } x = y \\ 1 \text{ if } x \neq y \end{cases}$$

Now (i) $d(x, y) \geq 0$

For $d(x, y) = 0$ or 1.

(ii) $d(x, y) = 0$ or iff $x = y$

(iii) $d(x, y) = d(y, x)$

For if $d(x, y) = 0$. Then $d(y, x) = 0$ so that $d(x, y) = d(y, x)$; and if $d(x, y) = 1$, then $d(y, x) = 1$ so that $d(x, y) = d(y, x)$

∴ In either case $d(y, x) = d(x, y)$

(iv) To prove $d(x, y) + d(y, z) \geq d(x, z)$.

If $x = z$, then $d(x, z) = 0$, so that

$$d(x, y) + d(y, z) \geq 0 = d(x, z).$$

If $x \neq z$, then $d(x, z) = 1$ and if y is any real number, then y is at least different from one of x and z ad hence at least one of $d(y, x)$ and $d(y, z)$ be equal to one. Consequently,

$$d(x, z) \leq d(x, y) + d(y, z)$$

The above facts prove that d is a metric on R.

Example 9:

Prove that the function $d : R^2 \times R^2 \to R$ s.t.

$$d(x, y) = [(x_1 - y_1)^2 + (x_2 - y_2)^2]^{1\,2},$$

where $x = (x_{1.}\, x_1) = (y_2 - y_2) \in R^2$

Is a metric on R.

[This is called unusual metric on R^2]

Solution:

Let $x = (x_1, x_2)$, $y = (y_1, y_2)$, $z = (z_1, z_2)$ be arbitrary elements of R^2

(i) $d(x, y) \geq 0$.

For $(x_1 - y_1)^2 \geq 0$, $(x_2 - y_2)^2 \geq 0$ and hence

$$[(x_1 - y_1)^2 + (x_2 - y_2)^2]^{1/2} \geq 0$$

(ii) $d(x, y) = 0$ iff $x = y$

$$d(x, y) = 0$$
$$\Leftrightarrow (x_1 - y_1)^2 + (x_2 - y_2)^2 = 0$$
$$\Leftrightarrow x_1 - y_1 = 0,\ x_2 - y_2 = 0$$
$$\Leftrightarrow x_1 = y_1,\ x_2 = y_2$$
$$\Leftrightarrow (x_1, x_2) = (y_1, y_2)$$
$$\Leftrightarrow x = y$$

(iii) $d(x, y) = d(y, x)$.

For $[(x_1 - y)^2 + (x_2 - y_2)^2]^{1/2} = [(y_1 - x_1)^2 + (y_2 - x_2)^2]^{1/2}$

(iv) $d(x, y) + d(y, z) \geq d(x, z)$

For $d(x, y) + d(y, z) \geq [(x_1 - y_1)^2 + (x_2 - y_2)^2]^{1/2}$

$+ [(y_1 - z_1)^2 + (y_2 - z_2)^2]^{1/2} \geq [(x_1 - y_1 + y_1 - z_1)^2$

$+ (x_2 - y_2 + y_2 - z_2)^2]^{1/2}$ be Minkowski inequality.

$= [(x_1 - z_1)^2 + (x_2 - z_2)^2]^{1/2} = d(x, z)$

$\therefore d(x, y) + d(y, z) \geq d(x, z)$.

Example 10:

Contor's set contains exactly as many points as the interval [0, 1] .

Hint : To prove $|E| = |[0,1]|$.

Also $|[0,1]| = C$

E is a closed set.

Since an infinite disjoint union of open sets is open, Hence the family of removed open intervals form an open set.

$E = [0,1]$ – family of removed open intervals.

$= [0,1]$ – open set = closed set.

For complement of open set is a closed set.

Example 11:

Given an example of two closed subset A and B of the real line r such that d (A, B) = 0, but A/B ∅.

Solution:

Let d be the usual metric on R. Write

$$N_1 = \{ 2, 3, 4, 5 \ldots \} = \{n + 1 : n \in N\}$$

$$A = \{2, 3, 4, 5 \ldots\},$$

$$B = \left\{2\frac{1}{2}, 3\frac{1}{3}, 4\frac{1}{4}, 5\frac{1}{5}\right\}$$

Then A and B are expressible as

$$A = \{n : n \in N_1\},$$

$$B = \left\{n + \frac{1}{n} : n \in N_1\right\}$$

If $n \in A$, then $n + 1/n \in B$ and the distance between $n + 1/n$ and n is 1/n, which $\to 0$ as $n \to \infty$.

$$d (A, B) = \inf \{d (x, y) : x \in A, y \in B\}$$

$$= \inf \{| x - y| : x \in A, y \in B\}$$

$$= \inf \{1/n : n \in A, n + 1/n \in B\} = 0.$$

Thus, d (A, B) = 0. Also $A \text{ ç} B = \varnothing$.

Therefore A and B are the required sets.

Example 12:

Every set in a discrete metric space is open.

Solution:

Let (X,d) be a discrete metric space. Let $x, y \in X$ be arbitrary.

By def. of discrete metric, $d (x, y) = \begin{cases} 1 \text{ if } x \neq y \\ 0 \text{ if } x = y \end{cases}$.

Let r be any positive real number s.t. $r \le 1$.

Then $Sr(x) = \{y \in X : d (y, x) < r \le 1\}$

$= \{y \in X : d (y, x) < 1\}$

$= \{y \in X : d (y, x) = 0$, by def. of d.

$= \{y \in X : y= x\} = \{0\}$.

or, $S_{r(x)} = \{0\}$.

But every open sphere is an open set.

$\therefore$ $\{x\}$ is an open set in X $\forall$ $x \in X$.

If $A = \{x_1, x_2, ..., x_n\}$ = finite set $\subset X$, then

$$A = \bigcup_{r=1}^{n} \{x_r\} = \text{finite union of open sets.}$$

$$= \text{open set.}$$

Hence every finite sub set of X is open set. ...(1)

If $B = \{x_1, x_2, x_3, ...\} \subset X$, then

B is an infinite subset of X.

Now $$B = \bigcup_{r=1}^{\infty} \{x_r\}$$

= Arbitrary union of open sets

= open set, by Theorem 8.

$\therefore$ B is an open set, ...(2)

From (1) and (2), it follows that every subset (finite or infinite) is an open set in X.

Example 13:

In a discrete metric space, every subset of X is closed.

Solution:

We know that G is open

$\Leftrightarrow X - G = G'$ is closed set.

Let A be a subset of a discrete metric space (X, d), then A is open.

Also $A' = X - A$ is again sub set of X and so A' is open. Consequently A is closed.

Thus every subset of X is open as well as closed.

Example 14:

A finite set in any metric space has no limit point.

Solution:

Let A be a finite subset of a metric space (X, d). We know that "$x \in X$ is a limit point of any set B if every open sphere $S_{r(x)}$ contains an infinite number of points of B other than x."

This condition can not be satisfied here as A is finite set.

Hence A has no limit point.

Example 15:

In any metric space, show that

$$x - \overline{A} = (X - A)^{o}$$

or, $(\overline{A})' = (A')^{o}$

Solution:

$(\overline{A})' = X - \overline{A}$

$= X -$ Intersection of all closed super sets of A.

$= X - \bigcap_i F_i$ where F_i is closed and $F_i \supset A$.

$= \bigcup_i (X, F_i)$ where $X - F_i$ is open

and $X - F_i \subset X - A$

$=$ Union of open subsets of $X - A = A'$

$= (A')^{o}$

Example 16:

Given an example of Pseudo metric which is not a metric.

Solution:

Suppose C is the family of all Riemann integral functions f; $[0, 1] \to R$.

Let f, g, h, $\in$ C be arbitrary. Define a map

$d : C \times C \to R$ s.t. $d(f, g) = \int_0^1 |f - g|(x)\, dx$.

To prove d is a Pseudo metric but; not metric on C.

$d(f, g) \geq$ For $|f - g|(x) \geq 0$ and so.

$\int_0^1 |f - g|(x)\, dx \geq 0$

(1) $d(f, g) = d(g, f)$.

This follows form the fact that $|f - g| = |g - f|$

(2) $d(f, g) + d(g, f)$.

$= \int_0^1 |f - g|(x)\, dx + \int_0^1 |g - h|(x)\, dx$

$= \int_0^1 [|f - g| + |g - h|](x)\, dx$

$\geq \int_0^1 |f - g + g - h|(x)\, dx$

$= \int_0^1 |f - h|(x)\, dx = d(f, h)$

(3) d (F, g) = 0 if f = g

If f = g, then

d (f, g) = $\int_0^1 |f - g|(x)\, dx = \int_0^1 |g - g|(x)\, dx = \int_0^1 0\, dx = 0$

(4) d (f, g) = 0 does not necessarily imply f = g.

That is to say, d (f, g) = 0

$\Rightarrow$ f $\neq$ g, in general.

Solution:

Suppose $f(x) = \begin{cases} 4\, where\, x = 0 \\ 0\, when\, 0 < x \leq 1 \end{cases}$

and $g(x) = \begin{cases} 9\, when\, x = 0 \\ 0\, when\, 0 < x \leq 1 \end{cases}$

The f(x) $\neq$ g (x), so that f $\neq$ g.

Also $|f - g|(x) = \begin{cases} 9 - 4 = \, when\, x = 0 \\ 0 \quad when\, 0 < x \leq 1 \end{cases}$

Evidently f – g is Riemann integrable,

x = o is the only point of discontinuity.

$$d(f, g) = \int_0^1 |f - g|(x)\, dx = \int_0^1 0\, dx = 0$$

Thus, d (f, g) = 0 but f $\neq$ g.

This proves that d is a Pseudo metric but metric on C.

Ex : *Let C be the class of bounded continuous functions f : [0, 1 → R. Define a map d : C × C → R such that*

$$d(f, g) = \int_0^1 |f - g|(x)\, dx \;\forall f, g \in C = \|f - g\|$$

Then d is a metric on C.

Example 17:

Describe open spheres for the usual metric on R^2.

Solution:

Let d be the usual metric on R^2 so that d is given by

$$d(x, y) = [(x_1 - x_1)^2 + (y_1 - y_2)^2]^{1/2}$$

where $x = (x_1, y_1) \in R^2$, $y = (x_2, y_2) \in R^2$.

Let $z_0 = (x_0, y_0) \in R^2$ and r > 0 be any real number

$S_r(z_0) = \{(x, y) \in R^2 : z = (x, y), d(z, z_0) < r\}$

$= \{(x, y) \in R^2 : z = \sqrt{[(x - x_0)^2 + (y - y_0)^2]} < r\}$

$= \{(x, y) \in R^2 : (x - x_0)^2 + (y - y_0)^2 < r^2\}$

This proves that open sphere $S_r(z_0)$ is the interior of the circle $(x - x_0)^2 + (y - y_0)^2 = r^2$, where $z_0 = (x_0, y_0)$.

Example 18:

Describe open sphere of unit radius having the centre at (0, 0) for each of the following metric on R.

(i) $d_1(z_1, z_2) = |x - x_2| + |y_1 - y_2|$

(ii) $d_2(z_1, z_2) = max. \{|x_1 - x_2|, |y_1 - y_2|\}$

where $z_1 = (x_1, y_1)$, $z_2 = (x_2, y_2) \in R^2$.

Solution:

Let $z_0 = (0, 0)$, $z = (x, y) \in R^2$.

To determine $S_1(z_0)$.

Step (i) $S_1(z_0) = \{(x, y) \in R^2 : d_1(z, z_0) < 1\}$

$= \{(x, y) \in R^2 : |x - 0| + |y - 0| < 1\}$

$= \{(x, y) \in R^2 : |x| + |y| < 1\}$

This shows that the open sphere $S_1(z_0)$ is the interior of the square bounded by the lines

$x + y = 1, - x - y = 1, - x + y = 1, x - y = 1.$

Step (ii) $S_1(z_0) = \{(x, y) \in R^2 : d_1(z, z_0) < 1\}$

$= \{(x, y) \in R^2 : max. \{|x - 0|, |y - 0|\} < 1\}$

$= \{(x, y) \in R^2 : max. \{|x|, |y|\} < 1\}$

This proves that the open sphere $S_1(z_0)$ is the interior of the square bounded by the lines

$x = 1, - x = 1, y = 1. - y = 1.$

Example 19:

Show that C the set of complex numbers is metric space under the metric d defined as follows:

$$d(z_1, z_2) = \frac{|z_1 - z_2|}{\left[1 + |z_1|^2\right]^{1/2} \left[1 + |z_2|^2\right]^{1/2}}$$

Solution:

Let z_1, z_2, z_3 be arbitrary complex numbers . It is given that

$$\therefore \quad d(z_1, z_2) = \frac{|z_1 - z_2|}{\left[1 + |z_1|^2\right]^{1/2}\left[1 + |z_2|^2\right]^{1/2}}$$

To prove that (C, d) is a metric space.

We know that $|z| \geq 0$ for every complex number z.

If view of this, (1) says that $d(z_1, z_2) \geq 0$. ...(1)

Since $|z_1 - z_2| = |z_2 - z_1|$. In view of this, (1) $\Rightarrow$

$$d(z_1, z_2) = d(z_2, z_1) \quad ...(2)$$

$$d(z_1, z_2) = 0 \Leftrightarrow \frac{|z_1 - z_2|}{\left[1 + |z_1|^2\right]^{1/2}\left[1 + |z_2|^2\right]^{1/2}} = 0$$

$$\Leftrightarrow |z_1 - z_2| = 0 \Leftrightarrow z_1 = z_2$$

$$\therefore \quad d(z_1, z_2) = 0 \Leftrightarrow z_1 = z_2 \quad ...(3$$

$d(z_1, z_2) + d(z_2, z_3)$

$$= \frac{|z_1 - z_2|}{\left[1 + |z_1|^2\right]^{1/2}\left[1 + |z_2|^2\right]^{1/2}} + \frac{|z_2 - z_3|}{\left[1 + |z_2|^2\right]^{1/2}\left[1 + |z_3|^2\right]^{1/2}}$$

$$\geq \frac{|z_1 - z_2|}{\left[1 + |z_1|^2\right]^{1/2}\left[1 + |z_2|^2\right]^{1/2}\left[1 + |z_3|^2\right]^{1/2}}$$

$$+ \frac{|z_2 - z_3|}{\left[1 + |z_1|^2\right]^{1/2}\left[1 + |z_2|^2\right]^{1/2}\left[1 + |z_3|^2\right]^{1/2}}$$

$$= \frac{|z_1 - z_2| + |z_2 - z_3|}{\left[1 + |z_1|^2\right]^{1/2}\left[1 + |z_2|^2\right]^{1/2}\left[1 + |z_3|^2\right]^{1/2}}$$

$$\geq \frac{|z_1 - z_3|}{\left[1 + |z_1|^2\right]^{1/2}\left[1 + |z_2|^2\right]^{1/2}\left[1 + |z_3|^2\right]^{1/2}}$$

$$\geq \frac{|z_1 - z_3|}{\left[1 + |z_1|^2\right]^{1/2}\left[1 + |z_3|^2\right]^{1/2}} = d(z_1, z_3)$$

or $d(z_1, z_2) + d(z_2, z_3) \geq d(z_1, z_3)$

From (1), (2), (3) and (4), it follows that (C, d) is a metric space.

Example 20:

Show that the function d defined by $d(p, d) = \|p - q\|$, where p, q are vectors in a normal space V, is a metric on V.

Solution:

If p, q, r ∈ V and K is any scalar, then by def. of normed vector space,

(N_1) $\|p - q\| \geq 0$, (N_2) $\|p - q\| = 0 \Leftrightarrow p = 0$

(N_3) $\|kp\| = |k|\,\|p\|$, (N_4) $\|p + q\| \leq \|p\| + \|q\|$

We have $d(p, q) = \|p - q\|$ ...(1)

Using (N_1) in (1), we find that $d(p, q) \geq 0$ ...(M_1)

Using (N_2) in (1), we see that

$$d(p, q) = 0 \Leftrightarrow \|p - q\| = 0 \Leftrightarrow p = q = 0$$

$$\Leftrightarrow p = q \quad ...(M_2)$$

or $$d(p, q) = 0 \Leftrightarrow p = q \quad ...(M_3)$$

Using (N_3) in (1), we see that

$$d(q,p) = \|q - p\| = \|(-1)(p - q)\| = \|p - q\| = d(p,q)$$

or $$d(q, p) = d(p, q) \quad ...(M_3)$$

Using (N_4) in (1), we see that

$$d(p,r) = \|p - r\| = \|(p - q) + (q, r)\| \leq \|p - q\| + \|q - r\|$$

$$d(p, q) + d(q, r)$$

or $$d(p, r) \leq d(p, q) + d(q, r)$$

From (M_1), (M_2), (M_3) and (M_4), it follows that (C, d) is a metric space.

Example 21:

In a metric space (X, d), $S = S(x_0, r)$ is an open sphere and p is any point of it. Prove that ∃ an open sphere $T = S(p, \varepsilon)$ s.t. $T \subset S$.

Solution:

$p \in S(x_0, r) \Rightarrow d(p, x_0) < r \Rightarrow r - d(p, x_0) = r_1 > 0.$

Take $\varepsilon < r_1$, then $x \in S\,p\,\varepsilon \Rightarrow d(x, p) < \varepsilon < r_1 \Rightarrow d(x, p) < r_1$

or $$d(x, p) < r - d(p, x_0)$$

or $$d(x, p) + d(p, x_0) < r$$

or $$d(x, x_0) \leq d(x, p) + d(p, x_0) < r$$

or $d(p, x_0) < r$

or $x \in S(x_0, r)$ ...(M_4)

Thus $x \in S(p, \varepsilon)$

$\Rightarrow$ $x \in S(x_0, r)$.

Hence $S(p, \varepsilon) \subset S(x_0, r)$ or $T \subset S$.

Example 22:

Let X be a non-empty set and let d be a real valued function of ordered pairs of elements of X which satisfies the following conditions:

(i) $d(x, y) = 0 \Leftrightarrow x = y$

(ii) $d(x, y) \leq d(x, z) + d(y, z)$

Then prove that d is a metric on X.

Solution:

(1) To prove $d(x, y) \geq 0$.

By (ii), $d(x, x) \leq d(x, z) + d(x, z)$, on putting $x = y$.

or, $0 \leq 2d(x, z)$ by (i)

or, $d(x, z) \geq 0$

or, $d(x, y) \geq 0$

(2) To prove $d(x, y) = d(y, x)$.

putting $z = x$ in (ii),

$$d(x, y) \leq 0 + d(y, x)$$

or, $d(x, y) \leq d(y, x)$

Interchanging x and y,

$d(y, x) \leq d(x, y)$

Combining the last two, $d(x, y) = d(y, x)$.

(3) $d(x, y) = 0 \Leftrightarrow x = y$ by (i)

(4) To prove $d(x, z) + d(z, y) \geq d(x, y)$

Using (2) in (ii), $d(x, y) \leq d(x, z) + d(z, y)$

or, $d(x, z) + d(z, y) \geq d(x, y)$

Remark : This gives an alternative definition of metric.

Example 23:

In any metric space (X, d), show that $A \subset X$ *is closed iff* $A = \overline{A}$

Solution:

Let (X, d) be metric space and $A \subset X$. Firstly we shall establish the following lemmas :

(i) $\overline{A}$ is closed and $A \subset \overline{A}$.

(ii) $F \supset A$, F is closed $\Rightarrow F \supset C$

By definition, $\overline{A} = \cap\{F \subset X : F \text{ is closed}, F \supset A\}$, ...(1)

(i) Since an arbitrary; intersection of closed sets is closed and hence $\overline{A}$ is closed.

By (1), $\overline{A} \supset A$

(ii) Suppose $F \subset X$ is closed and $F \supset A$.

$\therefore$ F is one of those members whose intersection is $\overline{A}$ and hence $F \supset \overline{A}$. .E.D. lemma (ii)

Now we come to the proof of actual theorem.

Let $A = \overline{A}$. We claim A is closed. By (i). $\overline{A}$ is closed.

$\overline{A}$ is closed, $A = \overline{A} \Rightarrow$ is closed.

Conversely let A be closed. To prove that $A = \overline{A}$.

By the lemma (i), $A \subset \overline{A}$. ...(2)

By the lemma (ii), f is closed, $F \supset A \Rightarrow F \supset \overline{A}$. ...(3)

Since $A \supset A$ for every set A.

In particular (3) gives A is closed, $A \supset A \Rightarrow A \supset \overline{A}$ ...(4)

From (2) and (4), $A = \overline{A}$.

This concludes the problem.

Example 24:

In any metric space (X, d), prove that A is open

$$\Leftrightarrow A^{o} = A.$$

Solution:

Let A be a subset of a metric space (X, d). By def. of interior,

$$A^{o} = \cup \{S_{r(x)} : S_{r(x)} \subset A\} \qquad ...(1)$$

Since every open sphere is an open set and arbitrary union of open sets is open. Consequently,

A^{o} is an open set. ...(2)

By (1), it is clear that $A^{o} \subset A$ ...(3)

and A^{o} is largest open subset of A. ...(4)

(i) Given $A = A^{o}$ ...(5)

Aim. A is an open set

(2) and (5) $\Rightarrow$ A is an open set.

(ii) Given A is an open set ...(6)

Aim, $A = A^o$

(4) and (6) $\Rightarrow = A\ A^o$

Example 25:

Let $A = [0, 1], B = [1, 2]$

Then $A^o = (0, 1), B^o = (1, 2).$

$A^o \cup B^o = (0, 1), \cup (1, 2)$

$= \{x \in R : 0 < x < 1 \text{ or } 1 < x < 2\}$

$\because$ $(A \cup B) = [0, 2]$

$\therefore$ $(A \cup B)^o = (0, 2).$

Evidently $(A \cup B)^o \neq A^o \cup B^o.$

Solution:

Do yourself.

Example 26:

If G_1, G_2 are open sets in metric space (X_1, d_1) and (X_2, d_2) then $G_1 \times G_2$ is open in the product metric space (X, d).

Solution:

Let $x = (x_1, x_2)$, $y\ (y_1, y_2) \in X = X_1 \times X_2$, then $x_1, y_1 \in X_1$ and $x_2\ y_2 \in X_2$. Let $x \in G_1 \times G_2$, then $x_1 \in G_1$, $x_2 \in G_2$.

By def. of open set,

$x \in G_1 \Rightarrow \exists\ r_1 > 0$ s.t. $S\ (x_1, r_1) \subset G_1$ ...(1)

Similarly, $S\ (x_2, r_2) \subset G_2$ ...(2)

By (1) and (2),

$S\ (x_1, r_1) \times S\ (x_2, r_2) \subset G_1 \times G_1$ ...(3)

Take $r = \min \{r_1, r_2\}$, then $r \leq r_1$, $r \leq r_2$

$\therefore$ $S\ (x_1, r) \times S\ (x_2, r) \subset S\ (x_1, r_1) \times S\ (x_2, r_2) \subset G_1 \times G_2$

or $S\ (x, r) \subset G_1 \times G_2$,

for any $x \in G_1 \times G_2$

By def., this proves that $G_1 \times G_2$ is open.

Example 27:

In a metric space, prove that $d(A) = d(\overline{A})$.

Solution:

Evidently $A \subset B \Rightarrow d(A) \leq d(B)$.

But $A \subset \overline{A}$ and so that $d(A) \leq d(\overline{A})$. ...(1)

Let $x, y \in \overline{A}$, then \$ $u, v \in A$ s.t.

$$d(x, u) < r,\ d(u, v) < r \text{ where } r > 0.$$

But $d(x, y) \leq d(x, u) + d(u, v) + d(v, y)$

$$< d(u, v) + 2r \leq d(A) + 2r$$

or $\quad d(x, y) \leq d(A) + 2r.$

For $d(A) = \sup d(x, y)$ where $x, y \in A$.

This$\Rightarrow d(\overline{A}) \leq d(A)$. ...(2)

Combining (1) and (2), $d(A) = d(\overline{A})$.

Example 28:

Let X be a non-empty set and let d be a real valued function of ordered pairs of elements of X which satisfies the following conditions:

(i) $d(x, y) = 0 \Leftrightarrow x = y$

(ii) $d(x, y) \leq d(x, z) + d(y, z)$

Then prove that d is a metric on X.

Solution:

(1) To prove $d(x, y) \geq 0$.

By (ii), $d(x, x) \leq d(x, z) + d(x, z)$, on putting $x = y$.

or, $0 \leq 2d(x, z)$ by (i)

or, $d(x, z) \geq 0$

or, $d(x, y) \geq 0$

(2) To prove $d(x, y) = d(y, x)$.

putting $z = x$ in (ii),

$$d(x, y) \leq 0 + d(y, x)$$

or, $d(x, y) \leq d(y, x)$

Interchanging x and y,

$d(y, x) \leq d(x, y)$

Combining the last two, d (x, y) = d (y, x).

(3) d (x, y) = 0 ⇔ x = y by (i)

(4) To prove d (x, z) + d (z, y) ≥ d (x, y)

Using (2) in (ii), d (x, y) ≤ d (x, z) + d (z, y)

or, d (x, z) + d (z, y) ≥ d (x, y).

Example 29:

Every set in a discrete metric space is open.

Solution:

Let (X,d) be a discrete metric space. Let x, y ∈ X be arbitrary.

By def. of discrete metric, $d(x, y) = \begin{cases} 1 \text{ if } x \neq y \\ 0 \text{ if } x = y \end{cases}$.

Let r be any positive real number s.t. r ≤ 1.

Then $Sr(x) = \{y \in X : d(y, x) < r \leq 1\}$

$= \{y \in X : d(y, x) < 1\}$

$= \{y \in X : d(y, x) = 0$, by def. of d.

$= \{y \in X : y = x\} = \{0\}$.

or, $S_{r(x)} = \{0\}$.

But every open sphere is an open set.

∴ {x} is an open set in X ∀ x ∈ X.

If $A = \{x_1, x_2, ..., x_n\}$ = finite set ⊂ X, then

$A = \bigcup_{r=1}^{n} \{x_r\}$ = finite union of open sets.

= open set.

Hence every finite sub set of X is open set. ...(1)

If $B = \{x_1, x_2, x_3, ...\} \subset X$, then

B is an infinite subset of X.

Now $B = \bigcup_{r=1}^{\infty} \{x_r\}$

= Arbitrary union of open sets

= open set, by Theorem 8.

∴ B is an open set, ...(2)

From (1) and (2), it follows that every subset (finite or infinite) is an open set in X.

Example 30:

Show that the map $d : R^2 \times R^2 \to R$

define by $d(x, y) = max. \{|x_1 - y_1|, |x_2 - y_2|\}$.

where $x = (x_1, x_2), y = (y_1, y_2) \in R_2$

is a metric on R^2.

Solution:

Let $x = (x_1, x_2)$, $y = (y_1, y_2)$, $z = (z_1, z_2)$ be arbitrary elements or R^2. Also let

$$d(x, y) = \max \{|x_1 - y_1|, |x_2 - y_2|\}$$

(i) $d(x, y) \geq 0$.

Since modulus of any quantity ;is non-negative.

(ii) $d(x, y) = 0$ iff $x = y$.

For $d(x, y) = 0 \Leftrightarrow \max.\{|x_1 - y_1|, |x_2 - y_2|\} = 0$

$$\Leftrightarrow |x_1 - y_1| = 0 = |x_2 - y_2|$$

$$\Leftrightarrow x_1 = y_1, |x_2 = y_2$$

$$\Leftrightarrow (x_1, x_2) = y_2, y_2)$$

$$\Leftrightarrow x = y$$

(iii) $d(x, y) = d(y, x)$.

For $|a - b| = |b - a|$ and so

$\max.\{|x_1 - y_1|, |x_2 - y_2|\} = \max \{|y_1 - x_1|, |y_2 - x_2|\}$

(iv) $d(x, y) + d(y, z) \geq d(x, z)$

For $d(x, y) = \max \{|x_1 - z_1|, |x_2 - z_2|\}$

$$= \max. \{|x_1 - y_1 + y_1 - z_1|, |x_2 - y_2 + y_2 - z_2|\}$$

$$\leq \max. \{|x_1 - y_1| + |y_1 - z_1|, |x_2 - y_2| + |y_2 - z_2|\}$$

$$\leq \max. \{|x_1 - y_1|, |x_2 - y_2|\} + \max. \{|y_1 - z_1|, |y_2 - z_2|\}.$$

$$= d(x, y) + d(y, z).$$

The above facts lead to the conclusion that d is a metric on R^2.

Example 31:

Let X be a non-empty set and define d by

$$d(x,y) = \begin{cases} 0 \text{ if } x = y \\ 1 \text{ if } x \neq y \end{cases}$$

Prove that d is metric on X.

Solution:

Let x, y, z ∈ X be arbitrary.

Then

(1) d (x, y) ≥ 0

This follows from the definition of d.

(2) d (x, y) = 0 iff x = y, by def. of d.

(3) d (x, y) = d (y, x) as value of d is free from order or x and y.

(4) d (x, y) + d (y, z) ≥ d (x, z)

For if x ≠ y ≠ z, then

$$d(x, y) + d(y, z) = 1 + 1 = 2 > 1 = d(x, z)$$

or, $$d(x, y) + d(y, z) > d(x, z) \quad \text{...(a)}$$

Again if x = y ≠ z, then

$$d(x, y) + d(y, z) = 0 + 1$$
$$= d = (x, z)$$

or, $$d(x, y) + d(y, z) = d(x, z) \quad \text{...(b)}$$

Combining (a) and (b), we get result (4).

∴ d is a metric on X.

Example 32:

Given two different metrics for the set R of all real numbers.

Solution:

Consider the map

$$d : R \times R \to R \text{ s. t.}$$

$$d(x,y) = \begin{cases} 0 \text{ if } x = y \\ 1 \text{ if } x \neq y \end{cases}$$

where (x, y) ∈ R^2.

Then d is metric on R.

Next consider the map

$$d' : R \times R \to R$$

$$d'(x, y) |x - y|.$$

where x, y ∈ R

Then d' is metric on R.

Thus d and d' are two different metrics on R.

Example 33:

Describe open spheres for the usual metric on R, *i.e.*, open spheres on the real line are precisely the bounded open intervals.

Solution:

Let d be the usual metric on R; then

$$d(x, y) = |x - y| \;\forall\; x, y \in R$$

Let $x_0 \in R$ and $r > 0$ be any real number.

$$\begin{aligned} S_r(x_0) &= \{x \in R : d(x, x_0) < r\} \\ &= \{x \in R : |x - x_0| < r\} \\ &= \{x \in R : -r < x - x_0 < r\} \\ &= \{x \in R : x_0 - r < x < x_0 + r\} \\ &= (x_0 - r, x_0 + r). \end{aligned}$$

This proves that the open spheres on the real line R are open intervals.

Example 34:

Show that the map $d : R^2 \times R^2 \to R$

define by $d(x, y) = max. \{|x_1 - y_1|, |x_2 - y_2|\}$.

where $x = (x_1, x_2), y = (y_1, y_2) \in R_2$

is a metric on R^2.

Solution:

Let $x = (x_1, x_2), y = (y_1, y_2), z = (z_1, z_2)$ be arbitrary elements or R^2. Also let

$d(x, y) = \max \{|x_1 - y_1|, |x_2 - y_2|\}$

(i) $d(x, y) \geq 0$.

Since modulus of any quantity ;is non-negative.

(ii) $d(x, y) = 0$ iff $x = y$.

For $d(x, y) = 0 \Leftrightarrow \max.\{|x_1 - y_1|, |x_2 - y_2|\} = 0$

$$\begin{aligned} &\Leftrightarrow |x_1 - y_1| = 0 = |x_2 - y_2| \\ &\Leftrightarrow x_1 = y_1, |x_2 = y_2 \\ &\Leftrightarrow (x_1, x_2) = y_2, y_2) \\ &\Leftrightarrow x = y \end{aligned}$$

(iii) $d(x, y) = d(y, x)$.

For $|a - b| = |b - a|$ and so

max.$\{|x_1 - y_1|, |x_2 - y_2|\}$= max $\{|y_1 - x_1|, |y_2 - x_2|\}$

(iv) $d(x, y) + d(y, z) \geq d(x, z)$

For $d(x, y) = \max\{|x_1 - z_1|, |x_2 - z_2|\}$

$$= \max.\{|x_1 - y_1 + y_1 - z_1|, |x_2 - y_2 + y_2 - z_2|\}$$

$$\leq \max.\{|x_1 - y_1| + |y_1 - z_1|, |x_2 - y_2| + |y_2 - z_2|\}$$

$$\leq \max.\{|x_1 - y_1|, |x_2 - y_2|\} + \max.\{|y_1 - z_1|, |y_2 - z_2|\}.$$

$$= d(x, y) + d(y, z).$$

Example 35:

Let ρ_1 and ρ_2 be metrics on the same set X, show that there exists a metric ρ on X such that

$$\rho(x, y) \geq \rho_i(x, y) \text{ for } i = 1, 2$$

and $\forall\, x, y \in X$.

Solution:

Let ρ_1 and ρ_2 be metrics on the same set X, Let x, y, z $\in$ X be arbitrary.

To prove that $\exists$ a metric ρ on X s.t.

$$\rho(x, y) \geq \rho_1(x, y) \text{ for } i = 1, 2$$

Write $\rho(x, y) = \rho_1(x, y) + \rho_2(x, y)$. ...(1)

We claim ρ is metric on X.

(i) $\rho(x, y) \geq 0$

$\because$ $\rho_1(x, y) \geq 0$, $\rho_2(x, y) \geq 0$.

(ii) $\rho(x, y) = 0$ iff $x = y$

For $\rho(x, y) = 0 \Leftrightarrow \rho_1(x, y) + \rho_2(x, y) = 0.$

$$\Leftrightarrow \rho_1(x, y) = 0 = \rho_2(x, y).$$

Since ρ_1 and ρ_2 both are non-negative.

$$\Leftrightarrow x = y$$

(iii) $\rho(y, x) = \rho(x, y)$

For $\rho(y, x) = \rho_1(y, x) + \rho_2(y, x) = \rho_1(x, y) + \rho_2(x, y)$

$$\rho(x, y)$$

$\because$ $\rho_1(x, y) = \rho_1(y, x)$, $\rho_2(x, y) = \rho_2(y, x)$

(iv) $\rho(x, y) + \rho(y, z) \geq \rho(x, z)$

For $\rho(x, y) + \rho(y, z) = \rho_1(x, y) + \rho_2(x, y) + \rho_1(y, z) + \rho_2(y, z)$

$$[\rho_1 (x, y) + \rho_1 (y, z)] + [\rho_2 (x, y) + \rho_1 (y, z)]$$
$$\geq \rho_1 (x, z) + \rho_2 (x, z) = \rho (x, z).$$

From what has been done it follows that ρ is a metric on X.

Moreover (1) ⇒ $\rho (x, y) \geq \rho_1 (x, y)$ for i = 1, 2.

Hence ρ is the required metric.

Example 36:

If A is any non-empty subset of a metric space (X, d), prove that

$|d (x, A) - d (y, A)| \leq d (x, y)$

for any points x, y ∈ X.

Solution:

By def. of metric,

$$d (x, a) \leq d (x, y) + d (y, a)$$

and $d (y, a) \leq d (y, x) + d (x, a)$

Taking infimum over a ∈ A, we obtain

$$d (x, A) \leq d (x, y) + d (y, A)$$

and $$d (y, A) \leq d (y, x) + d (x, A)$$

i.e., $$d (x, A) - d (y, A) \leq d (x, y) \quad ...(1)$$

$$d (y, A) - d (x, A) \leq d (x, y)$$

The last gives

$$d (x, A) - d (y, A) \geq - d (x, y)$$

From (1) and (2),

$$- d (x, y) \leq d (x, A) - d (y, A) \leq d (x, y)$$

⇒ $|d (x, A) - d (y, A)| \leq d (x, y)$.

Example 37:

Let A, B be non-empty subsets of a metric space (X, d).

Then show that

$$d (A \cap B) \leq d (A) + d (B) + d (A, B)$$

Solution:

Let x, y ∈ A ∩ B be arbitrary. Then there arise four cases.

(i) x, y ∈ A

This ⇒ $d (x, y) \leq d (A)$

For $d(A) = \sup\{d(x, y) : x, y \in A\}$

(ii) $x, y \in B$.

This $\Rightarrow d(x, y) \leq d(B)$

(iii) $x \in A$ and $y \in B$

(iv) $x \in B$ and $y \in A$

In case (iii), we have

$$d(x, y) \leq d(x, a) + d(a, y), \quad a \in A$$
$$\leq d(x, a) + d(a, b) + d(b, y), \; b \in B$$

$$\therefore \quad d(x, y) \leq d(x, a) + d(a, b) + d(b, y).$$

Similarly in case (iv), we have

$$d(x, y) \leq d(x, b) + d(b, a) + d(a, y)$$

$$\Rightarrow \quad d(A \cup B) \leq d(B) + d(B, A) + d(A)$$

or $$d(A \cup B) \leq d(A) + d(B) + d(A, B).$$

Example 38:

Describe open spheres for the discrete metric on a space X.

Solution:

Let (X,d) be a discrete metric space so that d is given by

$$d(x,y) = \begin{cases} 0 \textit{ if } x = y \\ 1 \textit{ if } x \neq y \end{cases}$$

Let $x_0 \in X$ and $r > 0$ be any real number.

$$S_r(x_0) = \{x \in X : (x, x_0) < r\}.$$

Case (i) : When $r > 1$.

By definition metric, any $x \in X$, $x \neq x_0$

$$\Rightarrow d(x, x_0) = 1 < r$$
$$\Rightarrow d(x, x_0) = < r$$
$$\Rightarrow x \in S_r(x_0)$$

Also $x_0 \in S_r(x_0)$

$$\because \quad d(x_0, x_0) = 0 < r.$$

Combining these two results, we find that

any $\quad x \in X \Rightarrow x \in S_r(x_0)$

This $\quad \Rightarrow X = S_r(x)$

Case (ii) : when $r \leq 1$.

any $x \in X, x \neq x_0 \Rightarrow d(x, x_0) = 1 < r$

$\Rightarrow d(x, x_0) < r.$

$\Rightarrow x \notin S_r(x_0)$

$x_0 \in S_r(x_0)$

Thus, any $x \in X, x \neq x_0 \Rightarrow x \notin S_r(x_0)$

and $x_0 \in S_r(x_0)$.

From this it follows that $S_r(x_0) = \{x_0\}$.

Finally, $S_r(x_o) = \begin{cases} X & \text{if } r . 1. \\ \{x_0 & \text{if } r \le 1.\} \end{cases}$.

Example 39:

Let d be a Pseudo metric on X. Define a relation on X by setting X~y iff d (x, y) = 0. Show that this relation is an equivalence relation whose corresponding equivalence class can be made into a metric space in a natural way.

Solution:

Let d be a Pseudo metric on X so that

$\forall$ x, y, z, $\in$ X we have

(1) $d(x, y)\ ^3\ 0$

(2) $d(x, y) = d(y, x)$

(3) $d(x, x) = 0$

(4) $d(x, y) + d(y, z)\ ^3\ d(x, z)$

Define a relation ~ on X s.t.

$x \sim y$ iff $d(x, y) = 0$

To prove that ~ is an equivalence relation

(i) ~ is reflexive

For $d(x, x) = 0$

$\Rightarrow$ $x \sim x$, by (3)

(ii) ~ is symmetric.

For $x \sim y$

$\Rightarrow$ $d(x, y) = 0$

$\Rightarrow$ $d(y, x) = 0$, by (2)

$\Rightarrow y \sim x$

(iii) ~ is transitive.

For $x \sim y,\ y \sim z$

$\Rightarrow$ $d(x, y) = 0 = d(y, z)$

$\Rightarrow$ $d(x, z) \le (x, y) + d(y, z) = 0 + 0 = 0$

$\Rightarrow$ $d(x, z) \le 0$

But $d(x, z) \ge 0$, by (1)

$\Rightarrow$ $d(x, z) = 0$

$\Rightarrow$ $x \sim z$

Thus we have proved that $\sim$ is an equivalence relation of X.

Hence the relation $\sim$ decomposes X into family of disjoint subsets of X, known as equivalence classes. Let [X] denote the equivalence class determined by $x \in X$ and let C be the set of all equivalence classes,

i.e., $C = \{[x] : x \in X\}$

Define a map $f : C \times C \to R$ s.t.

$f([x_1], [x_2]) = d(x_1, x_2)$

where $[x_1], [x_2] \in C$

Then f is a metric for C on account of the fact that d is a pseudometric on X. The metric f defined as above is called natural metric for C.

Example 40:

Let C be the set of all real valued bounded continuous functions defined on [0, 1].

If d is defined by

$d(f, g0 = \sup\{|f(x) - g(x)| : x \in [0, 1]\}$

then d is a metric on C,

where $f, g \in C$.

Solution:

Let $f, g, h \in C$ and $x \in [0, 1]$ be arbitrary. Then

(1) $d(f, g) \ge 0$. For mod of any thing ≥ 0.

(2) $d(f, g) = d(g, f)$. For $|f(x) - g(x)| = |g(x) - f(x)|$.

(3) $d(f, g) = 0 \Leftrightarrow f = g$

For $\sup\{|f(x) - g(x)| : x \in [0, 1]\} = 0$

$\Leftrightarrow |f(x) - g(x)| = 0$

$\Leftrightarrow f(x) - g(x)\ \forall x$

$\Leftrightarrow f = g$

(4) To prove $d(f, h) \leq d(f, g) + d(g, h)$

Since $|f - h| = |f - g + g - h|$

$\leq |f - g| + |g - h|$

This $\Rightarrow \sup |f(x) - h(x)| \leq \sup |f(x) - g(x)| + \sup |g(x) - h(x)|$

$\Rightarrow d(f, h) \leq d(f, g) + d(g, h)$

From what has been it follows that d is a metric on C.

Example 41:

Describe open spheres for the usual metric on R^2.

Solution:

Let d be the usual metric on R^2 so that d is given by

$$d(x, y) = [(x_1 - x_1)^2 + (y_1 - y_2)^2]^{1/2}$$

where $x = (x_1, y_1) \in R^2, y = (x_2, y_2) \in R^2$.

Let $z_0 = (x_0, y_0) \in R^2$ and $r > 0$ be any real number

$$S_r(z_0) = \{(x, y) \in R^2 : z = (x, y), d(z, z_0) < r\}$$

$$= \{(x, y) \in R^2 : z = \sqrt{[(x - x_0)^2 + (y - y_0)^2]} < r\}$$

$$= \{(x, y) \in R^2 : (x - x_0)^2 + (y - y_0)^2 < r^2\}$$

This proves that open sphere $S_r(z_0)$ is the interior of the circle $(x - x_0)^2 + (y - y_0)^2 = r^2$, where $z_0 = (x_0, y_0)$.

EXERCISES

1. If P and Q are the projections on closed linear subspaces M and N of H, prove that PQ is a projection $\Leftrightarrow$ PQ = QP. In this case, show that PQ is the projection on $M \cap N$.
2. Let X and Y be metric spaces and f a mapping of X into Y. Show that f is continuous $\Rightarrow f^{-1}(F)$ is closed in X whenever F is closed in Y $\Leftrightarrow f(\overline{A}) \subseteq f(\overline{A})$ for every subset A of X.
3. Show that any mapping of the metric space any other metric space is continuous.
4. Consider the real function f defined on the real line R by $f(x) = x^2$. If b is a given positive real number, show that the restriction of f to the closed interval [0, b] is uniformly continuous by starting with an $\in > 0$ and exhibiting a $\delta > 0$ which satisfies the requirement of the definition.

5. Let f be a continuous real function defined on R which satisfies the functional equation f(x + y) = f(x) + f(y). Show that this function must have the form f(x) = mx for some real number m.

6. Let X be a metric space, proving the following statements:
   (a) any point and disjoint closed set in X can be separated by open sets, in the sense that if a: is a point and F a closed set which does not contain x, then there exists a disjoint pair of open sets G_1 and G_2 such that $x \varepsilon G_1$ and $F \subseteq G_2$;
   (b) any disjoint pair of closed sets in X can be separated by open sets, in the sense that if Fi and Fi are disjoint closed sets, then there exists a disjoint pair of open sets Gi and Gi such that $F_1 \subseteq G_1$ and $F_2 \subseteq G_2$.

7. Let X and Y be metric spaces with metrics d_1 and d_2, and $\{f_n\}$ $\{f_n\}$ be a sequence of mappings of X into Y which converges pointwise to a mapping f of X into Y, in the sense that $f_n(x) \to f(x)$ for each x in X. Define what ought to be meant by the statement that f, converges uniformly to f_n and prove that under this assumption f is continuous if each f_n is continuous.

8. Let X and Y be metric spaces and f a mapping of X into Y. If f is a constant mapping, show that f is continuous. Use this to show that a continuous mapping need not have the property that the image of every open set is open.

9. Let X be a metric space with metric d, and let x_0 be a fixed point in X. Show that the real function f_{x0} defined on X by $f_{x0}(x) = d(x, x_0)$ is continuous. Is it uniformly continuous?

10. Let X and Y be metric spaces and A a non-empty subset of X. If f and g are continuous mappings of X into Y such that f(x) = g(x) for every x in A, show that f(x) = g(x) for every x in $\overline{A}$.

11. Let X be a metric space, and let A be a subset of X. If a: is a limit point of A, show that each open sphere centered on x contains an infinite number of distinct points of A. Use this result to show that a finite subset of X is closed.

12. Show that a subset of a metric space is bounded <=> it is non-empty and is contained in some closed sphere.

13. If P and Q are the projections on closed linear subspaces M and N of H, 'prove that the following statements are all equivalent to one another:

(a) $P \leq Q$,

(b) $\| Px \| \leq \| Qx \|$ for every x,

(c) $M \subseteq N$,

(d) $PQ = P$,

(e) $QP = P$.

(**Hint:** the equivalence of (a) and (b) is easy to prove, as is that of (c), (d), and (e), prove that (d) implies (a) by using $(Px, x) = \| Px \|^2 = \| PQx \|^2 \leq \| Qx \|^2 = (Qx,, x)$, and prove that (b) implies (c) by observing that if x is in M, then $\| x \| = \| Px \| \leq \| Qx \| \leq \| x \|$.)

14. Determine which of the following functions are uniformly continuous on the open unit interval (0, 1) : $1/(1 - x)$; $1/(2–x)$; sin x; sin (1/x); $x^{1/2}$, x^3. Which are uniformly continuous on the open interval $(0, +\infty)$?
15. In the proof of Theorem D we used the following fact: the image of a Cauchy sequence under a uniformly continuous mapping is again a Cauchy sequence. Give the details of the proof.
16. Show that the projections on H form a complete lattice with respect to their natural ordering as self-adjoint operators.
17. If P and Q are the projections on closed linear subspaces M and N of H, prove that Q–P is a projection $\Leftrightarrow P \leq Q$. In this case, show that Q–P is the projection on $N \cap M^{\perp}$.

3

BANACH SPACE

INTRODUCTION

Banach space is a linear space which is also, in a special way, a complete metric space. This combination of algebraic and metric structures opens up the possibility of studying linear transformations of one Banach space into another which have the additional property of being continuous.

Most of our work in this chapter centers around three fundamental theorems relating to continuous linear transformations. The *Hahn-Banach theorem* guarantees that a Banach space is richly supplied with continuous linear transformations, and makes possible an adequate theory of conjugate spaces. The *open mapping theorem* enables us to give a satisfactory description of the projections on a Banach space, and has the important closed graph theorem as one of its consequences. We use the *uniform boundedness theorem* in our discussion of the conjugate of an operator on a Banach space, and this in turn provides the setting for our treatment in the next chapter of the adjoint of an operator on a Hilbert space.

Virtually all this theory had its origins in analysis. Our present interest, however, lies in the study of form and structure, not in exploring the many applications of these ideas to specific problems. This chapter is therefore strongly oriented toward the algebraic and topological aspects of the matters at hand.

THE DEFINITION AND SOME EXAMPLES

We begin by restating the definition of a Banach space.

A normed linear space is a linear space N in which to each vector x there corresponds a real number, denoted by $\| x \|$ and called the norm of x, in such a manner that

(1) $\| x \| \geq 0$, and $\| x \| = 0 \Leftrightarrow x = 0$;

(2) $\| x + y \| \leq \| x \| + \| y \|$

(3) $\| ax \| = | a | \| x \|$.

The non-negative real number $\| x \|$ is to be thought of as the length of the vector x. If we regard $\| x \|$ as a real function defined on N, this function is called the *norm* on N. It is easy to verify that the normed linear space N is a metric space with respect to the metric d defined by $d(x, y) = \| x - y \|$. A *Banach space* is a complete normed linear space. Our main interest in this chapter is in Banach spaces, but there are several points in the body of the theory at which it is convenient to have the basic definitions and some of the simpler facts formulated in terms of normed linear spaces. For this reason, and also to emphasize the role of completeness in theorems which require this assumption, we work in the more general context whenever possible. The reader will find that the deeper theorems, in which completeness hypotheses are necessary, often make essential use of Baire's theorem.

Several simple but important facts about a normed linear space are based on the following inequality:

$$\| x \| - \| y \| I \leq \| x - y \| \qquad \text{...(1)}$$

To prove this, it suffices to prove that

$$\| x \| - \| y \| \leq \| x - y \| \qquad \text{...(2)}$$

follows from (2) that we also have

$$-(\| x \| - \| y \|) = \| y \| - \| x \| \leq \| y - xx \| = \| -(x - y) \| = \| x - y \|,$$

which together with (2) yields (1). We now prove (2) by observing that $\| x \| = \| (x - y) + y \| \leq \| x - y \| + \| y \|$. The main conclusion we draw from (1) is that the norm is a continuous function:

$$x_n \to x \Rightarrow \| x_n \| \to \| x \|.$$

This is clear from the fact that $| \| x_n \| | - \| x \| \leq \| x_n - x \|$, since $x_n \to x$ means that $\| x_n - x \| \to 0$. In the same vein, we can prove that addition and scalar multiplication are jointly continuous.

$$x_n \to x \text{ and } y_n \to y \Rightarrow x_n + y_n \to x + y$$

and

$$\alpha_n \to \alpha \text{ and } x_n \to x \Rightarrow \alpha_n x_n \to \alpha x.$$

These assertions follow from

$$\| (x_n + y_n) - (x + y) \| = \| (x_n - x) + (y_n - y) \|$$

$$\| x_n - x \| + \| y_n - y \|$$

and

$$\| \alpha_n x_n - \alpha x \| = \| \alpha_n (x_n - x) + (\alpha_n - a) x \|.$$

Our first theorem exhibits one of the most useful ways of forming new normed linear spaces out of old ones.

Theorem:

Let M be a closed linear subspace of a normed linear space N. If the norm of a coset x + M in the quotient space N/M is defined by

$$|| x + M || = inf\{|| x + m || : m \varepsilon M\}, \quad ...(3)$$

then N/M is a normed linear space. Further, if N is a Branch space, then so is N/M.

Proof:

We first verify that (3) defines a norm in the required sense. It is obvious that $\| x + M \| \geq 0$, and since M is closed, it is easy to see that $\| x + M \| = 0 \Leftrightarrow$ there exists a sequence $\{m_k\}$ in M such that $\| x + m_k \| \to 0 \Leftrightarrow$ x is in M $\Leftrightarrow x + M = M$ = the zero element of N/M. Next, we have $\| x + M) + (y + M) \| = \| (x + y) + M \| = \inf\{\| x + y + m \| : m \varepsilon M\} = \inf\{ \| x + y + m + m' \| : m \text{ and } m' \leq M\} = \inf\{\| (x + m) + (y + m') \| : m \text{ and } m' \varepsilon M\} \leq \inf\{ \|x + m \| + \| y + m' \| : m \text{ and } m' \varepsilon M] = \inf\{\| y + m \| : m \varepsilon M\} + \inf\| y + m' \| : m' \varepsilon M\} = \| x + M \| + \| y + M \|$. The proof of $\| \alpha (x + M) \| = | a |x + M \|$ is similar.

Finally, we assume that N is complete, and we show that N/M is also complete. If we start with a Cauchy sequence in N/M, then it suffices to show that this sequence has a convergent subsequence. It is clearly possible to find a subsequence $\{x_n + M \}$ of the original Cauchy sequence such that $\| (x_1 + M) - (x_1 + M\} \| \leq 1/2$, $\| (x_2 + M) - (x_3 + M) \| < 1/4$ and, in general, $\| (x_1 + M) - (x_{n+1} + M) \| < 1/2$.

We prove that this sequence is convergent in N/M. We begin by choosing any vector y_1 in $x_1 + M$, and we select y_2 in $x_2 + M$ such that $\| y_1 \ y_2 \| < 1/4$. We next select a vector y_3 in $x_3 + M$ such that $\| y_2 - y_3 \| < 1/4$. Continuing in this way, we obtain a sequence {yn} in N such that $\| y_n - y_{n+1} \| < 1/2^n$. If $m < n$, then

$$\| y_m - y_n \| = \| y_m - y_{m+1} (+ (y_{m+1} - y_{m+2}) + ...$$

so $\{y_n\}$ is a Cauchy sequence in N. Since W is complete, there exists a vector y in N such that $y_n \to y$. It now follows from $\| (x_n + M) - (y + M) \| \leq \| y_n - y \|$ that $x_{n'} + M \to y + M$, so N/M is complete. In the following sections and chapters, we shall often have occasion to consider the quotient space of a normed linear space with respect to a closed linear subspace. In accordance with our theorem, a quotient space of this kind can always be regarded as a normed linear space in its own right.

We now describe some of the main examples of Banach spaces. In each of these, the linear operations are understood to be defined either coordinate wise or point wise, whichever is appropriate in the circumstances.

Example 1:

The spaces R and C– the real numbers and the complex numbers–are the simplest of all normed linear spaces. The norm of a number x is of course defined by $\| x \| = | x |$, *and each space is a Banach space.*

Example 2:

The linear spaces R^n *and* C^n *of all n-tuples* $x = (x_1, x_2, \ldots, x_n)$ *of real and complex numbers can be made into normed linear spaces in an infinite variety of ways, as we shall see below. If the norm is defined by*

$$\| x \| = \left(\sum_{i=1}^{n} |x_i|^2 \right)^{1/2} \qquad \ldots(4)$$

then we get the n-dimensional Euclidean and unitary spaces familiar to us from our earlier work. We denoted these spaces by R^n and C^n in Part 1 of this book, and we know by the theorems that both are Banach spaces.

Each of the following examples consists of n-tuples of scalars, sequences of scalars, or scalar-valued functions defined on some none empty set, where the scalars are the real numbers or the complex numbers. We do not normally specify which system of scalars is to be used, and it should be emphasized that both possibilities are allowed unless the contrary is clearly stated. Also, we make no distinction in notation between the real case and the complex case. When it turns out to be necessary to distinguish these two cases, we do so verbally, by referring, for instance, to "the complex space –." These conventions are in accord with the standard usage preferred by most mathematicians, and they enable us to avoid a good deal of cumbersome notation and many unnecessary case distinctions.

Example 3:

Let p be a real number such that $1 \le p < \infty$. *We denote by* l_p^n *the space oj all n-tuples* $x = (x_1, x_2, \ldots, x_n)$ *of scalars, with the norm defined by*

$$\| x \|p = \left(\sum_{i=1}^{n} |x_i|p \right)^{1/p} \qquad \ldots(5)$$

Obviously the special case of (5) which corresponds to $p = 2$, so the real and complex spaces l_2^n are the n-dimensional Euclidean and unitary spaces R^n and C^n. It is easy to see that (5) satisfies conditions (1) and (3) required by the definition of a norm. In Problem 4 we outline a proof of the fact that (5) also satisfies condition (2), that is, that $\| x + y \|_p \le \| x \|_p + \| y \|_p$. The completeness of Inf follows from substantially the same reasoning as that used in the proof of so l_p^n is a Banach space.

Example 4:

We again consider a real number p *with the property that* $1 < p < \infty$ *(and we denote by lp the space of all sequences*

$$x = \{x_1, x_2, \ldots, x_n, \ldots\}$$

of scalars such that $\Sigma_{n=1}^{\infty} |x_n|^p < \infty$, with the norm defined by

$$\|x\|_p = \left(\sum_{i=1}^{n} |x_n|p\right)^{1/p} \qquad \ldots(6)$$

The reader will observe that the real and complex spaces It are precisely the infinite-dimensional Euclidean and unitary spaces R^{∞} and C^{∞} defined. The proof of the fact that I_p actually is a Banach space requires arguments similar to those used.

The Banach spaces discussed in these examples are all special cases of the important L_p spaces studied in the theory of measure and integration. A detailed treatment of these spaces is outside the scope of this book, but we can describe them loosely as follows. An L_p space essentially consists of all measurable functions defined on a measure space X with measure m which are such that $|f(x)|^p$ is integrable, with

$$\|f\|^p = (\int |f(x)|^p \, dm(x))^{1/p} \qquad \ldots(7)$$

taken as the norm. In order to include the l_p^n spaces and l_p within the theory of L_p spaces, we have only to consider the sets $\{1, 2, \ldots, n\}$ and $\{1, 2, \ldots, n, ..\}$ as measure spaces in which each point has measure 1, and to regard n-tuples and sequences of scalars as functions defined on these sets. Since integration is a generalized type of summation.

Example 5:

Just we start with the linear space of all n-tuples $x = (x_1, x_2, \ldots, x_n)$ *of scalars, but this time we define the norm by*

$$\|x\| = \max\{|X_1, ||x_2|, \ldots, |x_n|\}. \qquad \ldots(8)$$

This Banach space is commonly denoted by l_{∞}^n and the symbol $\|x\|_{\infty}$ is occasionally used for the norm given by (8). The reason for this practice lies in the interesting fact that

$$\|x\|_{\infty} = \lim \|x\|_p \text{ as } p \to \infty$$

that is, that

$$\max\{|x_i|\} = \lim \left(\sum_{i=1}^{n} |x_i|p\right)^{1/p} \quad \text{as } p \to \infty \qquad \ldots(9)$$

We briefly inspect the case n = 2 to see why this is true. Let $x = (x_1, x_2)$ be an ordered pair of real numbers with x_1 and $x_2 \geq 0$. It is clear that $\| x \|_\infty = \max \{x_1, x_2\} \leq (x_1{}^p + x_2{}^p)^{1/p} = \| x \|_p$. If $x_1 = x_2$, then $\lim \| x \|_p = \lim (x_1{}^p + x_2{}^p)^{1/p} = \lim (2x_2p^p)^{1/p} = x_2 = \| x |_\infty$. And if $x_1 < x_2$, then

$$\lim \| x \|_p = \lim (x_1{}^p + x_2{}^p)^{1/p} = \lim ([(x_1/x_2)^{1/p} + 1]x_2{}^p)^{1/p}$$
$$= \lim [(x_1/x_2)^p + 1]^{1/p} x_2 = x_2 = || x ||_\infty.$$

Example 6:

Consider the linear space of all bounded sequences $x = | x |, x_l, x_2 \ldots, x_n, \ldots\}$ of scalars. By analogy with Example 5, we define the norm by

$$|| x || = sup \mid x_n \mid, \qquad ...(10)$$

and we denote the resulting Banach space by l_∞. The set *c* of all convergent sequences is easily seen to be a closed linear subspace of l_∞ and is therefore itself a Banach space. Another Banach space in this family is the subset c_o of *c* which consists of all convergent sequences with limit 0.

Example 7:

The Banach space of primary interest to us is the space $\mathcal{C}(X)$ of all bounded continuous scalar-valued functions defined on a topological space X, with the norm given by

$$\| f \| = \sup | f(x) |. \qquad ...(11)$$

This norm is sometimes called the uniform norm, because the statement that f_n converges to f with respect to this norm means that f_n converges to f uniformly on X. The fact that this space is complete amounts to the fact that if f is the uniform limit of a sequence of bounded continuous functions, then f itself is bounded and continuous. If, as above, we consider n-tuples and sequences as functions defined on {1, 2, . . . , n} and (1, 2, . . . , n, . . .}, then the spaces l_∞^n and l_∞ are the special cases of $\mathcal{C}$(X) which correspond to choosing X to be the sets just mentioned, each with the discrete topology.

Many important properties of a Banach space are closely linked to the shape of its *closed unit sphere*, that is, the set $S = \{x : \| x \| \leq 1\}$. One basic property of S is that it is always convex, in the sense that if x and y are any two vectors in S, then the vector $z = \alpha x + \beta y$ is also in S, where α and β are non-negative real numbers such that $\alpha + \beta = 1$; for $\| z \|| = \| \alpha x + \beta y \| \leq \alpha \| x \| + \beta \| y \| \leq \alpha + \beta = 1$. In this connection, it is illuminating to consider the shape of S for certain simple examples. Let our underlying linear space be the real linear space R^2 of all ordered pairs $x = (x_1, x_2)$ of real numbers. As we have seen, there are many different norms which can be defined on R^2, among which are the following: $\| x \|_1 = | x_1 + | x_2 |$; $\| x \|_2 = (| x_1 |^2 + | x_2 |^2)^{1/2}$ and $\| x \|_\infty = \max \{| x_1 |, | x_2 |\}$. Fig. 3.1

illustrates the closed unit sphere which corresponds to each of these norms. In the first case, S is the square with vertices (1, 0), (0, 1), (–1, 0), (0, – 1); in the second, it is the circular disc of radius 1; and in the third, it is the square with vertices (1, 1), (– 1, 1), (–1, –1), (1, –1). If we consider the norm defined by

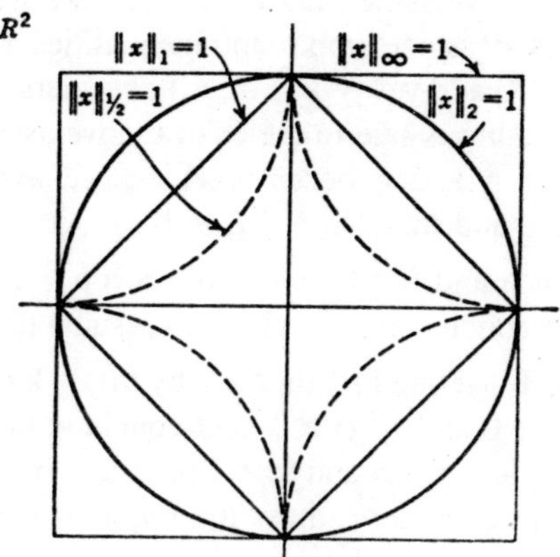

Fig. 3.1 : Some closed unit spheres.

$$\| x \|_p = (|x_1|^p + | x_2 |^p)^{1/p} \qquad ...(12)$$

where $1 \leq p \leq \infty$, and if we allow p to increase from 1 to ∞, then the corresponding S's swell continuously from the first square mentioned to the second. We note that S is truly "spherical" $\Leftrightarrow$ p = 2. These considerations also show quite clearly why we always assume that $p \geq 1$; for if we were to define $\| x \|_p$ with $p < 1$, then $S = \{x : \| x \|_p \leq 1 \}$ would not be convex. For $p < 1$, therefore, formula (12) does not yield a norm. In the above examples, we have exhibited several different types of Banach spaces, and there are yet others which we have not mentioned. Amid this diversity of possibilities, it is well to realize that any Banach space can be regarded–from the point of view of its linear and norm structures alone–as a closed linear subspace of $\mathcal{C}(X)$ for a suitable compact Hausdorff space *X*. We prove this below, in our discussion of the natural embedding of a Banach space in its second conjugate space.

EXERCISES

1. Let N be a non-zero normed linear space, and prove that N is a Banach space $\Leftrightarrow \{x : \| x \| = 1\}$ is complete.

2. Let a Banach space B be the direct sum of the linear subspaces M and N, so that $B = M \oplus N$. If $z = x + y$ is the unique expression of a vector z in B as the sum of vectors x and y in M and N, then a new norm can be defined on the linear space B by $\| z \|' = \| x \| + \| y \|$. Prove that this actually is a norm. If B' symbolizes the linear space

B equipped with this new norm, prove that B' is a Banach space if M and N are closed in B.

3. Prove for the case of an arbitrary positive integer n.

4. In this problem we sketch the proofs–and we ask the reader to fill in the details–of some important inequalities relating to n-tuples $x = (x_1, x_2, ..., x)$ and $y = (y_1, y_2, ..., y_n)$ of scalars. Whenever p occurs alone, and nothing is said to the contrary, we assume that $1 \leq p < \infty$, and whenever p and q occur together, we assume that both are greater than 1 and that $1/p + 1/q = 1$.

   (a) Show that o and $b \geq 0 \Rightarrow a^{1/p}\, b^{1/q} \leq a/p + 1/q = 1$. (If a = 0 or b = 0, the conclusion m is clear, so assume that both are positive. If $k \in (0, 1)$, define f (t) for $t \geq 1$ by $f(t) = k(t-1) - t^k + 1$. Note that f (1) = 0 and f′ (t) ≥ 0, and conclude that $tk \leq k^t + (1 - k)$. If $a \geq b$, put t = a/b and k = 1/p; if a < b, put t = n/a and k = 1/q; and in each case, draw the required conclusion.)

   (b) Prove Holder's inequality: $\Sigma_{i=1}^{n} |x, y_i| \leq \|x\|_p \|y\|_q$. (If x = 0 or y = 0, the inequality is obvious, so assume that both are ≠ 0. Put $a_i = (|x_i|/\|x\|_p)^p$ and $b_i = (|y_i - |/\|y\|_q{}^q$ and use part (a) to obtain $(x, y_i |/\|x\|_p \|y\|_q$ £ $a_i/p + b_i/q$. Add these inequalities for i = 1, 2, ..., n, and conclude that

$$\left(\sum_{i=1}^{n} |x, y_i|\right) / \|x\|_p \|y\|_q \leq 1/p + 1/q = 1.)$$

   (c) Prove Minkowski's inequality: $\|x + y\|_p$ £ $\|x\|_p + \|y\|_p$. (The inequality is evident when p – 1, so assume that p > 1. Use Holder's inequality to obtain

$$\|x+y\|_p{}^p = \sum_{i=1}^{n} |x_i + y_i|^p = \sum_{i=1}^{n} |x_i + y_i||x_i + y_i|^{p-1}$$

$$\leq \sum_{i=1}^{n} |x_i||x_i + y_i|^{p-1} + \sum_{i=1}^{n} |y_i||x_i + y_i|^{p-1}$$

$$\leq (\|x\|p + \|y\|p)\|x+y\|_p{}^{p/q}.)$$

When p = q = 2, Holder's inequality becomes Cauchy's inequality as stated and proved. The Holder and Minkowski inequalities can easily be extended from finite sums to series. For readers with some knowledge of the theory of measure and integration, we remark that these inequalities can also be stated in the following much more general forms : if f is in L_p and g is in L_q then their point wise product *fg* is in L_1 and

$$\|fg\|_1 \leq \|f\|_p \|q\|_q,$$

and if f and g are both in L_p, then f + g is also in L_p and

$$\| f+g \|_p \leq \| f \|_p + \| g \|_p.$$

It is to be understood that f and g are measurable functions defined on an arbitrary measure space and that the norms occurring in these inequalities are those defined by formula.

CONTINUOUS LINEAR TRANSFORMATIONS

Let *N* and *N'* be normed linear spaces with the same scalars, and let *T* be a linear transformation of *N* into *N'*.*1* When we say that *T* is continuous, we mean that it is continuous as a mapping of the metric space *N* into the metric space *N'*. This amounts to the condition that $x_n \to x$ in $N \Rightarrow T(x_n) \to T(x)$ in *N'*. Our main purpose in this section is to convert the requirement of continuity into several more useful equivalent forms and to show that the set of all continuous linear transformations of *N* into *N'* can itself be made into a normed linear space in a natural way.

Theorem 1:

Let N and N' be normed linear spaces and T a linear transformation of N into N'. Then the following conditions on T are all equivalent to one another:

(1) T is continuous;

(2) T is continuous at the origin, in the sense that $x_n \to 0 \Rightarrow T(x_n) \to 0$;

(3) *there exists a real number* $K \geq 0$ *with the property that* $\| T(x) \| \leq K \| x \|$ *for every* $x \varepsilon N$;

(4) if $S = (a : \| x \| \leq 1)$ *is the closed unit sphere in N, then its image T (S) is a bounded set in N'.*

Proof:

(1) ⇒ (2). If T is continuous, then since T (0) = 0 it is certainly continuous at the origin. On the other hand, if T is continuous at the origin, then $x_n \to x \Leftrightarrow x_n - x \to 0 \Rightarrow T(xn - x) \to 0 \Leftrightarrow T(x_n) - T(x) \to \Leftrightarrow T(x_n) \to T(x)$, so T is continuous.

(2) ⇔ (3). It is obvious that (3) ⇒ (2), for if such a K exists, then $x_n \to 0$ clearly implies that $T(x_n) \to 0$. To show that (2) ⇒ (3), we assume that there is no such K. It follows from this that for each positive integer n we can find a vector xn such that $\| T(x_n) \| > n \| x_n \|$, or equivalently, such that $\| T(x_n/n \| x_n \|) \| > 1$. If we now Put

$$y_n = x_n/n \| x_n \|,$$

then it is easy to see that $y_n \to 0$ but $T(y_n) + \to 0$, so T is not continuous at the origin.

(3) $\Leftrightarrow$ (4). Since a non-empty subset of a normed linear space is bounded $\Leftrightarrow$ it is contained in a closed sphere centered on the origin, it is evident that (3) $\Rightarrow$ (4); for if $\| x \| \leq 1$, then $\| T(x) \| \leq K$. To show that (4) $\Rightarrow$ (3), we assume that T (S) is contained in a closed sphere of radius K centered on the origin. If $x = 0$, then $T(x) = 0$, and clearly $\| T(x/\| x \|) \| \leq K \| x \|$, and if $x \neq 0$, then $x/\| x \| \in S$, and therefore $\| T(x/\| x \|) \| \leq K$, so again we have $\| T(x) \| \leq K \| x \|$.

If the linear transformation T in this theorem satisfies condition (3), so that there exists a real number $K \geq 0$ with the property that

$$\| T(x) \| \leq K \| x \|$$

for every x, then K is called a *bound* for T, and such a T is often referred to as a *bounded linear* transformation.

According to our theorem, T is bounded $\Leftrightarrow$ it is continuous, so these two adjectives can be used interchangeably. We now assume that T is continuous, so that it satisfies condition (4), and we define its norm by

$$\| T \| = \sup \{ \| T(x) \| : \| x \| \leq 1 \}. \quad \text{...(13)}$$

When $N \neq \{ 0 \}$, this formula can clearly be written in the equivalent form

$$\| T \| = \sup \{ \| T(x) \| : \| x \| = 1 \}. \quad \text{...(14)}$$

It is apparent from the proof of Theorem A that the set of all bounds for T equals the set of all radii of closed spheres centered on the origin which contain T (S). This, yields yet another expression for the norm of T, namely,

$$\| T \| = \inf \{ K : K \geq 0 \text{ and } \| T(x) \| \leq K \| x \| \text{ for all } x \}; \quad \text{...(15)}$$

and from this we see at once that

$$\| T(x) \| \leq \| T \| \| x \| \quad \text{...(16)}$$

for all x.

We now denote the set of all continuous (or bounded) linear transformations of N into N′ by $\mathcal{B}(N,N')$, where the letter "$\mathcal{B}$" is intended to suggest the adjective "bounded." It is a routine matter to verify that this set is a linear space with respect to the point wise linear operations defined by Eqns. (13) and (14) and to show that actually does define a norm on this linear space. We summarize and extend these remarks in

Theorem 2:

If N and N' are normed linear spaces, then the set $\mathcal{B}(N, N')$ of all continuous linear transformations of N into N' is itself a normed linear space with respect to the point wise linear operations and the norm defined by (1). Further, if N' is a Banach space, then $\mathcal{B}(N,N')$ is also a Banach space.

Proof:

We leave to the reader the simple task of showing that $\mathcal{B}(N, N')$ is a normed linear space, and we prove that this space is complete when N' is.

Let $\{T_n\}$ be a Cauchy sequence in $\mathcal{B}(N, N')$. If x is an arbitrary vector in N, then $\| T_m(x) - T_n(x) \| = \| (T_m - T_n)(x) \| \leq \| T_m - T_n \| \| x \|$ shows that $\{T_n(x)\}$ is a Cauchy sequence in N'; and since N' is complete, there exists a vector in N'– we denote it by T (x) – such that $T_n(x) \to T(x)$. This defines a mapping T of N into N', and by the joint continuity of addition and scalar multiplication, T is easily seen to be a linear transformation. To conclude the proof, we have only to show that T is continuous and that $T_n \to T$ with respect to the norm on $\mathcal{B}(N, N')$. By the inequality, the norms of the terms of a Cauchy sequence in a normed linear space form a bounded set of numbers, so

$$\| T(x) \| = \| \lim T_n(x) \| = \lim \| T_n(x) \| \leq \sup (\| T_n \| \| x| |) = (\sup \| T_n \|) \| x \|$$

shows that T has a bound and is therefore continuous. It remains to be proved that $\| T_n - T \| \to 0$. Let $\varepsilon > 0$ be given, and let n_0 be a positive integer such that $m, n \geq n_0 = \| T_m - T_n \| < \varepsilon$. If $\| z \| \leq 1$ and $m, n \geq n_0$, then

$$\| T_m(x) - T_N(x) \| = \| (T_{m'} - T_n)(x) \| \leq \| T_{m'} - T_n \| \| x \|$$
$$\leq \| T_m - T_n \| \leq \varepsilon.$$

We now hold m fixed and allow n to approach ∞,, and we see that $\| T_m(x) - T_n(x) \| \to \| T_m(x) - T(x) \|$, from which we conclude that $\| T_m(x) - T(x) \| \leq \varepsilon$ for all $m \geq n_0$ and all x such that $\| x \| \leq 1$. This shows that $\| T_m - T \| \leq \varepsilon$ for all TO $\geq n_0$, and the proof is complete.

Let N be a normed linear space. We call a continuous linear transformation of N into itself an *operator* on N, and we denote the normed linear space of all operators on N by $\mathcal{B}(N)$ instead of $\mathcal{B}(N, N)$. Theorem B shows that $\mathcal{B}(N)$ is a Banach space when A⊥ is. Furthermore, if operators are multiplied in accordance then $\mathcal{B}(N)$ is an algebra in which multiplication is related to the norm by

$$\| TT' \| \leq \| T \| \| T' \| \qquad ...(17)$$

This relation is proved by the following computation:

$$\| TT \| = \sup \{\| (TT')(x) \| : \| x \| \leq 1\} = \sup \{\| T(T'(x)) \| : \| x \| \leq 1\}$$
$$\leq \sup \{\| T \| \| T'(x) \| \ : \ \| x \| \leq 1\} = | \ T \| \sup \{\| T'(x) \| -. \| x \| \leq 1\}$$

We know from the previous section that addition and scalar multiplication in $\mathcal{B}(N)$ are jointly continuous, as they are in any normed linear space. Permits us to conclude that multiplication is also jointly continuous:

$T_n \to T$ and T¢ and $T'_n \Rightarrow T_n T'_n \to TT'$.

This follows at once from

$$\| T_n T'_n - TT' \| = \| T_n (T'_n - T') + (T_n - T) T' \|$$
$$\leq \| T_n \| \|T'_n - T' \| + \| T_n - T \| \| T' \|.$$

We also remark that when $N \neq \{0\}$, then the identity transformation I ie an identity for the algebra $\mathcal{B}(N)$. In this case, we clearly have

$$\| I \| = 1 \qquad ...(18)$$

for $\quad \| I \| = \sup \{\| I(x) \| : \| x \| \leq 1\} = \sup \{\| x \| : \| x \| \leq 1\} = 1.$

We complete this section with some definitions which will often be useful in our later work. Let N and N′ be normed linear spaces. An *isometric isomorphism* of N into N′ is a one-to-one linear transformation T of .N into N' such that $\| T(x) \| = \| x \|$ for every x in N, and N is said to be isometrically isomorphic to W if there exists an isometric isomorphism of N onto N′. This terminology enables us to give precise meaning to the statement that one normed linear space is essentially the same as another.

EXERCISES

1. If M is a closed linear subspace of a normed linear space N, and if T is the natural mapping of N onto N/M defined by $T(x) = x + M$, show that T is a. continuous linear transformation for which $\| T \| \leq 1$.
2. If T is a continuous linear transformation of a normed linear space N into a normed linear space N′, and if M is its null space, show that T induces a natural linear transformation T′ of N/M into N′ and that $\| T \| = \| T' \|$.
3. Let N and N′ be normed linear spaces with the same scalars. If N is infinite-dimensional and $N' \neq \{0\}$, show that there exists a linear transformation of N into N' which is not continuous. (We shall see in that if N is finite-dimensional, then every linear transformation of N into N' is automatically continuous.)
4. Let a linear space L be made into a normed linear space in two ways, and let the two norms of a vector x be denoted by $\| x \|$ and $\| x \|'$. These norms are said to be equivalent if they generate the same topology on L. Show that this is the case ⇔ there exist two positive real numbers K_1 and K_2 such that $K_1 \| x \| \leq \| x \|' \leq K_2 \| x \|$ for all x. (If L is finite-dimensional, then any two norms defined on it are equivalent.
5. If n is a fixed positive integer, the spaces l_p^n $(1 \leq p \leq \infty)$ consist of a single underlying linear space with different norms defined on it. Show that these norms are all equivalent to one another. (Hint: show that convergence with respect to each norm amounts to coordinatewise convergence.)

6. If N is an arbitrary normed linear space, show that any linear transformation T of l_p^n ($1 \le p \le \infty$) into N is continuous. (Hint: if $\{e_1, e_2, ..., e_n\}$ is the natural basis for l_p^n, where e_i is the n-tuple with 1 in the ith place and 0's elsewhere, then an arbitrary vector x in l_p^n can be written uniquely in the form

$$x = \alpha_1 e_1 + \alpha_2 e_2 + \ldots + \alpha_n e_n,$$

and from this we get $T(x) = a_1 T(e_1) + a_2 T(e_2) + \ldots + a_n T(e_n)$, now apply the hint given.

7. Let N be a finite-dimensional normed linear space with dimension n > 0, and let $\{e_1, e_2, ..., e_n\}$ be a basis for N.

Each vector x in N can be written uniquely in the form

$$x = \alpha_1 e_2 + \alpha_2 e_2 + \ldots + \alpha_n e_n.$$

If T is the one-to-one linear transformation of N onto l_1^n defined by $T(x) = (\alpha_1, \alpha_2, \ldots, \alpha_n)$, then T^{-1} is continuous.

(a) Prove that T is continuous. (Hint: if T is not continuous, then for some $\varepsilon > 0$ there exists a sequence $\{y_n\}$ in N such that $y_n \to 0$ and $\| T(y_n) \| \ge \in$, if $z_1 = y_n / \| T(y_n) \|$, then $z_n \to 0$ and $\|T(z_n)\| = 1$, the subset of I" consisting of all vectors of norm 1 is compact, so $\{T(z_n)\}$ has a subsequence which converges to a vector with norm 1; now use the continuity of T^{-1}.)

(b) Show that every linear transformation of N into an arbitrary normed linear space N' is continuous.

(c) Show that any other norm defined on N is equivalent to the given norm.

(d) Show that N is complete, and infer from this that every finite-dimensional linear subspace of an arbitrary normed linear space is closed.

8. It is a simple consequence of Problem 7 that every finite-dimensional normed linear space is locally compact. Prove the converse, that is, that a locally compact normed linear space N is finite-dimensional. Hint: the closed unit sphere S of N is compact, so there is a finite subset of S, say $\{x_1, x_2, ... x_n\}$, with the property that each point of S is distant by less than 1/2 from one of the x_i's let M be the linear subspace of N spanned by the x_i's, and show that M = N (to do this, assume that there exists a vector y not in M, use the fact that M is closed to infer that $d = d(y, M) > 0$, find m_0 in M such that $d \le \| y - m_0 \| \le 3d/2$, and deduce the contradiction that the vector y_0

in S defined by $y_0 = (y - m_o)/\| y - m_o \|$ is distant from M by at least 2/3).

THE HAHN-BANACH THEOREM

One of the basic principles of strategy in the study of an abstract mathematical system can be stated as follows: consider the set of all structure-preserving mappings of that system into the simplest system of the same type. This principle is richly fruitful in the structure theory (or representation theory) of groups, rings, and algebras, and we shall see in the next section how it works for normed linear spaces.

We have remarked that the spaces R and C are the simplest of all normed linear spaces. If N is an arbitrary normed linear space, the above principle leads us to form the set of all continuous linear transformations of N into R or C, according as N' is real or complex. This set – it is $\mathcal{C}$(N, R) or $\mathcal{B}$(N, C)–is denoted by N* and is called the conjugate space of N. The elements of N* are called *continuous linear functional*, or more briefly, *functional*. It follows from our work in the previous section that if these transformations are added and multiplied by scalars point wise, and if the norm of a functional f is defined by

$$\| f \| = \sup \{| f(x) | : \| x \| \le 1\}|$$
$$= \inf \{K : K \ge 0 \text{ and} | f(x) | \le K \| x \| \text{ for all } x \},$$

then N* is a Banach space.

When we consider various specific Banach spaces, the problem arises of determining the concrete nature of the transformations associated with these spaces. It is not our aim in this section to explore the ample body of theory which centers around this problem, and in any case, the machinery necessary for such an enterprise (mostly the theory of measure and integration) is not available to us.

Nevertheless, for the reader who may have the required background, we mention some of the main facts without proof.

Let X be a measure space with measure m, and let p be a given real number such that $1 < p \le \infty$. Consider the Banach space L_p of all measurable functions f defined on X for which $|f(x)|^p$ is integrable. If g is a function in L_g, where $1/p + 1/q = 1$, we define a function F_g on L_p by

$$F_g(f) = \int f(x)\ f(x)\, dm(x).$$

The Hölder inequality for integrals mentioned at the end.

$$\| F_q(f) | = | \int f(x)\ | g(x)\, dm(x) |$$
$$\le \int |f(x) g(x)|\ dm(x)$$

We conclude from this that Fa is a well-defined scalar-valued linear function on Lp with the property that $\| F_q \| \leq \| ga \|_q$, and is therefore a functional on L_p. It can be shown that equality holds here, so that

$$\| F_a \| = \| g \|_{q.}$$

It can also be shown that every functional on L_p arises in this way, so the mapping $g \rightarrow F_g$ (which is clearly linear) is an isometric isomorphism of L_q onto L_p. This statement is usually expressed by writing

$$L_p = L_q \qquad ...(19)$$

where the equality sign is to be interpreted in the sense just explained. If we specialize these considerations to n-tuples of scalars, we see that (1) becomes

$$(l_p^n)^* = (l_q^n) \qquad ...(20)$$

Further, it can be shown that

$$(l_1^n)^* = l_\infty^n \qquad ...(21)$$

and that $$(l_\infty^n)^* = l_1^n \qquad ...(22)$$

We sketch proofs of (20), (21), and (22) in the problems. When we consider sequences of scalars, then for $1 < p < \infty$ we have the following special i case of:

$$l_p^* = l_q. \qquad ...(23)$$

If p = 1, we obtain a natural extension of (22):

$$Ji^* = L. \qquad ...(24)$$

The corresponding extension of (4) is another matter, for it is false that $l_\infty = l_1$. Instead, we have

$$c_0^* = l_1. \qquad ...(25)$$

What is l_∞ *?. We saw that l_∞, is a special case of $\mathcal{C}$(X), so this question leads naturally to the problem of determining the nature of the conjugate space $\mathcal{C}$*(X). The classic solution of this problem for a space X which is compact Hausdorff (or even normal) is known as the *Riesz representation theorem*, and it depends on some of the deeper parts of the theory of measure and integration.

The situation is somewhat simpler for the case in which X is an interval [a, b] on the real line, but even here an adequate treatment requires a knowledge of Stieltjes integrals.

Most of the theory of conjugate spaces rests on the Hahn-Banach theorem, which asserts that any functional defined on a linear subspace of a normed linear space can be extended linearly and continuously to the whole space without increasing its norm. The proof is rather complicated, so we begin with a lemma which serves to isolate its most difficult parts.

Lemma:

Let M be a linear subspace of a normed linear space N, and let f be a functional defined on M. If x_0, is a vector not in M, and if

$$M_0 = M + [x_0]$$

is the linear subspace spanned by M and x_0, then f can be extended to a functional fa defined on M_0, such that $\|f_0\| = \|f\|$.

Proof:

We first prove the lemma under the assumption that N is a real normed linear space. We may assume, without loss of generality, that $\| f \| = 1$. Since x_0 is not in M, each vector y in M_0 is uniquely expressible in the form $y = x + \alpha x_0$ with x in M. It is clear that the definition $f_0(x + \alpha x_0) = f_0(x) + \alpha f_0(x_0) = f(x) + \alpha r_o$ extends f linearly to M_0 for every choice of the real number $r_0 = f_0(x_0)$. Since we are trying to arrange matters so that $\| f_0 \| = 1$, our problem is to show that r_0 can be chosen in such a way that $| f_0(x + \alpha x_0)| \leq | x + \alpha x_0 \|$ for every x in M and every $\alpha \neq 0$. Since $f_0(x + \alpha r_0) = f(x) + \alpha r_0$, this inequality can be written as

$$-\| x + \alpha x_0 \| \leq f(x) + \alpha r_0 \leq \| x + \alpha x_0 \|$$

or
$$-f(x) - \| x + \alpha x_0 \| \leq ar0 \leq \alpha r_0 \leq -f(x) + \| x + \alpha x_0 \|,$$

which in turn is equivalent to

$$-f\left(\frac{x}{\alpha}\right) - \left\| \frac{x}{\alpha} + x_0 \right\| \leq r_0 \leq -f\left(\frac{x}{\alpha}\right) + \left\| \frac{x}{\alpha} + x_0 \right\|. \qquad ...(26)$$

We now observe that for any two vectors x_1 and x_2 in M we have

$$\begin{aligned} f(x_2) - f(x_1) &= f(x_2 - x_1) \leq | f(x_2 - x_1) | \leq \| f \| \| x_2 - x_1 \| \\ &= \| x_2 - x_1 \| = \| (x_2 + x_0) - (x_1 + x_0) \| \\ &\leq \| x_2 + x_0 \| + \|x_1 + x_0 \|. \end{aligned}$$

so $\quad -f(x_1) - \| x_1 + x_0 \| \leq -f(x_2) + \| x_2 + x_0 \|.$

then shows that $a \leq b$. If we now choose r_o to be any real number such that $a \leq R_0 \leq b$, then the required inequality (26) is satisfied and this part of the proof is complete.

We next use the result of the above paragraph to prove the lemma for the case in which N is complex. Here f is a complex-valued functional defined on M for which $\| f \| = 1$. We begin by remarking that a complex linear space can be regarded as a real linear space by simply restricting the scalars to be real numbers. If g and h are the real and imaginary parts of f, so that $f(x) = g(x) + ih(x)$ for every x in M , then both g and h are easily seen to be real-valued functionals on the real space M; and since $\| f \| = 1$, we have $\| g \| \leq 1$. The equation

$$f(ix) = if(x),$$

together with $f(ix) = g(ix) + ih(ix)$ and

$$if(x) = i(g(x) + ih(x)) = ig(x) - h(x),$$

Shows that $h(x) = -g(ix)$, so we can write $f(x) = g(x) - ig(ix)$. By the above paragraph, we can extend g to a real-valued functional g_0 on the real space M_0 in such a way that $\| g_0 \| = \| g \|$, and we define f_0 for x in M_0 by $f0(x) = g_0(x) - ig_0(ix)$. It is easy to see that f_0 is an extension of from M to M_0, that $f_0(x + y) = f_0(x) + f_0(y)$, and that $f_0(\alpha x) = \alpha f_0(x)$ for all real α's. The fact that the property last stated is also valid for all complex a's is a direct consequence of

$$f_0(ix) = g_0(ix) - ig_0(i^2x) = g_0(ix) + ig_0(x) = i(g_0(x) - igo(ix)) = if_0(x),$$

so f_0 is linear as a complex-valued function defined on the complex space M_0. All that remains to be proved is that $\| f_0 \| = 1$, and we dispose of this by showing that if x is a vector in M_0 for which $\| x \| = 1$, then $| f_0(x) | \leq 1$. If $f_0(x)$ is real, this follows from $f_0(x) = g_0(x)$ and $\| g_0 \| \leq 1$. If $f_0(x)$ is complex, then we can write $f_0(x) = re^{i\theta}$ with $r > 0$, so

$$| f_0(x | = r = e^{-i\theta} f_0(x) = f_0(e^{-i\theta}x),$$

and our conclusion now follows from $\| e^{-i\theta}x \| = \| x \| = 1$ and the fact that $f_0(e^{-i\theta}x)$ is real.

Theorem 1:

(The Hahn-Banach Theorem) : Let M be a linear subspace of a normed linear space N, and let f be a functional defined on M. Then f can be extended to a functional ft, defined on the whole space N such that $\| f_0 \| = \| f \|$.

Proof:

The set of all extensions of f to functionals g with the same norm defined on subspaces which contain M is clearly a partially ordered set with respect to the following relation: $g_1 \leq g_2$ means that the domain of g_1 is contained in the domain of g_2, and $g_2(x) = g_1(x)$ for all x in the domain of g_1. It is easy to see that the union of any chain of extensions is also an extension and is therefore an upper bound for the chain. Zorn's lemma now implies that there exists a maximal extension f_0. We complete the proof by observing that the domain of f_0 must be the entire space N, for otherwise it could be extended further by our lemma and would not be maximal.

As we stated in the introduction to this chapter, the main force of the Hahn-Banach theorem lies in the guarantee it provides that any Banach space (or normed linear space) has a rich supply of functionals. This property is to be understood in the sense of the following two theorems, on which most of its applications depend.

Theorem 2:

If N is a normed linear space and x_0 is a non-zero vector in N, then there exists a functional f_0 in N such that $f_0(x_0) = \|x_0\|$ and $\|f_0\| = 1$.*

Proof:

Let $M = \{\alpha x_0\}$ be the linear subspace of N spanned by x_0, and define f on M by $f(\alpha x_0) = \alpha\|x_0\|$. It is clear that f is a functional on M such that $f(x_0) = \|x_0\|$ and $\|f\| = 1$. By the Hahn-Banach theorem, f can be extended to a functional f_0 in N* with the required properties.

Among other things, this result shows that N* separates the vectors in N, for if x and y are any two distinct vectors, so that $x - y \neq 0$, then there exists a functional f in N* such that $f(x - y) \neq 0$, or equivalently, $f(x) \neq f(y)$

Theorem 3:

If M is a closed linear subspace of a normed linear space N and x_0 is a vector not in M, then there exists a functional f_0 in N such that $f_0(M) = 0$ and $f_0(x_0) \neq 0$.*

Proof:

The natural mapping *T* of *N* onto *N/M* is a continuous linear transformation such that *T(M)* = 0 and

$$T(x_0) = x_0 + M \neq 0.$$

By Theorem B, there exists a functional f in (N/M)* such that

$$f(x_0 + M)* 0.$$

If we now define f_0 by $f_0(x) = f(T(x))$, then f_0 is easily seen to have the desired properties.

These theorems play a critical role in the ideas developed in the following sections, and their significance will emerge quite clearly in the proper context.

EXERCISES

1. Let M be a closed linear subspace of a normed linear space N, and let x_0 be a vector not in M. If d is the distance from x_0 to M, show that there exists a functional f_0 in N* such that $f_0(M) = 0$, $f_0(x_0) = 1$, and $\|f_0\| = 1/d$.
2. Prove that a normed linear space N is separable if its conjugate space N* is. (Hint: let $\{f_n\}$ be a countable dense set in N* and $\{x_n\}$ a corresponding set in N such that $\|x_n\| \leq 1$ and $|f_n(x_n)| \geq \|f_n\|/2$, let M be the set of all linear combinations of the x_n's whose coefficients are rational or–if N is complex–have rational real and

imaginary parts; and use Theorem C to show that M = N.) We remark that N* need not be separable when N is, for l_1 is easily proved to be separable, $l_1^* = l_\infty$, and l_∞ is not separable.

3. In this problem we ask the reader to convince himself of the validity of Eqs. (2) to (4). Let L be the linear space of all n-tuples

$$x = (x_1, x_2, ..., x_n)$$

of scalars. If $\{e_1, e_2, ..., e_n\}$ is the natural basis described in $x = x_1e + x_2e_2 + ...+ x_ne_n$, and if f is any scalar-valued linear function defined on L, then the equation $f(x) = x_1f(e_1) + x_2f(f_2) + ... + x_nf(e_n)$ shows that f determines, and is determined by, the n scalars $y_i = f(e_i)$. The mapping

$$y = (y_1.y_2, ... ,y_n) \to f,$$

where $f(x) = \Sigma_{i=1}^n x_i, y_i$ is clearly an isomorphism of L onto the linear space L′ of all f¢s. When the space L of all x's is made into l_p^n $(1 \le p \le \infty)$ by suitably defining its norm, then the space L' of all f's equals its conjugate space $(l_p^n)^*$, where the norm of f is understood to be given by

$$\|f\| = \inf\{K : K \ge 0 \text{ and } |f(x)| \le K\|x\| \text{ for all } x\}.$$

All that remains is to see what norm for the y′s makes the mapping $y \to f$ an isometric isomorphism.

(a) If $1 < p < \infty$, then $(l_p^n)^* = l_q^n$. The norm in this case is defined by $\|x\| = (\Sigma_{i=1}^n |x_i|^p)^{1/p}$, and it follows from

$$|f(x)= \left|\sum_{i=1}^n x_i, y_i\right| \le \sum_{i=1}^n |x_i y_i| \le (\sum_{i=1}^n |x_i|^p)^{1/p} (\sum_{i=1}^n |y_i|^q)^{1/q}$$

that $\|f\| \le (\Sigma_{i=1}^n |y_i|^q)^{1/q}$. Show that $\|f\| = (\Sigma_{i=1}^n |y_i|^q)^{1/q}$ by considering the vector x defined by $x_i = 0$ if $y_i = 0$ and

$$x_i = |y_i|^q/y_i$$

otherwise.

(b) $(l_1^n)^* = l_\infty^n$. Here we have $\|x\| = \Sigma_{i=1}^n |x_i|$, and it follows from

$$|f(x)| = ||\Sigma_{i=1}^n x_i y_i| \le \Sigma_{i=1}^n |x_i||y_i| \le \max\{|y_i|\} (\Sigma_{i=1}^n |x_i|)$$

that $\|f\| \le \max\{|y_i|\}$. Show that $\|f\| = \max\{|y_i\|$ by considering the vector x defined by $x_i = |y_i|/y_i$ if $|y_i| = \max\{|y_i|\}$ and $x_i = 0$ otherwise.

(c) $(l^n_\infty) = l^n_1$. In this case, the norm is defined by

$\| x \| = \max \{| x_i |\}$,

and it follows from

$$| f(x) | = \left| \sum_{i=1}^{n} x_i y_i \right| \leq \sum_{i=1}^{n} |x_i||y_i| \leq \max \{| x_i |\} \left(\sum_{i=1}^{n} |y_i|\right)$$

that $\| f \| \leq \Sigma_{i=1}^n |y_i|$. Show that $\| f \| = \Sigma_{i=1}^n |y_i|$ by considering the vector x, defined by $x_i = 0$ if $y_i = 0$ and $x_i = | y_i |/y_i$ otherwise.

4. The following generalized form of part of the Hahn-Banach theorem is useful in certain problems of measure theory. Prove it by suitably modifying the arguments given in the text. If p is a real function defined on a real linear space L such that $p(\alpha x) = \alpha p(x)$ for $\alpha \geq 0$ and $p(x + y) \leq p(x) + p(y)$, and if f is a real linear function defined on a linear subspace M such that $(x) \leq p(x)$ for all x in M, then f can be extended to a real linear function f_0 defined on L such that $f_0(x) \leq p(x)$ for all x in L.

THE NATURAL EMBEDDING OF N IN N**

Since the conjugate space N* of a normed linear space N is itself a normed linear space, it is possible to form the conjugate space (N*)* of N*. We denote this space by N**, and we call it the second conjugate space of N.

The importance of N** rests on the fact that each vector x in N gives rise to a functional Ft in N**. If we denote a typical element of N* by f, then F_x is defined by

$$F_x(f) = f(x).$$

In other words, we invert the usual practice by regarding the symbol f(x) as specifying a function of f for each fixed x, and we emphasize this point of view by writing f(x) in the form $F_x(f)$. A simple manipulation of the definition shows that Fx is linear:

$$\begin{aligned} F_x(\alpha f + \beta g) &= (\alpha f + \beta g)(x). \\ &= \alpha f(x) + \beta g(x) \\ &= \alpha F_x(f) + \beta F_x(g) \end{aligned}$$

If we now compute the norm of F_x, we see that

$$\begin{aligned} \| Fx \| &= \sup \{| Fx(f) | : \| f \| \leq 1\} \\ &= \sup \{| f(x) | : \| f \| \leq 1\} \end{aligned}$$

$$\leq \sup \{\| f \| \| x \| : \| f \| \leq 1\}$$

Is exactly what is needed to guarantee that equality holds here, so for each x in N we have

$$\|Fx\| = \|x\|.$$

It follows from these observations that $x \to F_x$ is a norm-preserving mapping of N into N**. F_x is called the functional on N* induced by the vector x, and we refer to functionals of this kind as induced functionals. We next point out that the mapping $x \to F_x$ is linear and is therefore an isometric isomorphism of N into N**. To verify this, we must show that $F_{x+y}(f) = (F_x | F_y)(f)$ and $F_{\alpha x}(f) = (\alpha F_x)(f)$ for every f in N*. The first of these relations follows from

$$\begin{aligned} Fx + y(f) &= f(x+y) \\ &= f(x) + f(y) \\ &= F_x(f) + F_y(f) \\ &= (F_x + F_y)(f), \end{aligned}$$

and the second is proved similarly. The isometric isomorphism $x \to F_x$ is called the natural embedding of N in N**, for it allows us to regard N as part of N** without altering any of its structure as a normed linear space. We write

$$N \subseteq N^{**}$$

where this set inclusion is to be understood in the sense just explained.

A normed linear space N is said to be reflexive if N = N**. The spaces l_p (and L_p) for $1 < p < \infty$ are reflexive, for $l_p^* = l_{p'}$ and

$$l_p^{**} = l_q^* = l_{p'}.$$

It follows from that the spaces l_p^n for $1 \le p \le < \infty$ are also reflexive. Since N** is complete, N is necessarily complete if it is reflexive. If N is complete, however, it is not necessarily reflexive, as we see from $c_0^* = l_1$ and $c_0^{**} = l_1^* = l_\infty$. If X is a compact Hausdorff space, it can be shown that $\mathcal{C}(X)$ is reflexive $\Leftrightarrow$ X is a finite set.

There is an interesting criterion for reflexivity, which depends on the concept of the *weak topology* on a normed linear space N. This is defined to be the weak topology on N generated by the functions in N* in the sense that is, it is the weakest topology on N with respect to which all the functions in N* remain continuous. The criterion referred to is the following: if B is a Banach space, and if $S = \{x : \|x\| \le 1\}$ is its closed unit sphere, then B is reflexive $\Leftrightarrow$ S is compact in the weak topology. This fact is something one should know about Banach spaces, but we shall have no need for it ourselves, so we state it without proof.

Far more important for our purposes is the weak* topology on N*, which is defined to be the weak topology on N* generated by all the induced functionals F_x in N**. This situation is rather complicated, so we shall try to make clear just what is going on.

First of all, N* (like N) is a normed linear space, and it therefore has a topology derived from its character as a metric space. This is called the *strong topology*. N** is the set of all scalar-valued linear functions defined on N* which are continuous with respect to its strong topology. The *weak topology* on N* (like the weak topology on N) is the weakest topology on N* with respect to which all the functions in N** are continuous, and clearly this is weaker than its strong topology.

We have indicated, these concepts apply equally to N and N*. However, since N* is the conjugate space of N, the natural embedding enables us to consider N as part of N**. We now form the weakest topology on N* with respect to which all the functions in N–regarded as a subset of N**–remain continuous. This is the *weak* topology*, and it is evidently weaker than the weak topology. The weak* topology can be given a more explicit description, in which its defining sub basic open sets are displayed. Consider a vector x in N and its induced functional F_x in N** The weak* topology on N* is the weakest topology under which all such F_x's are continuous. If f_0 is an arbitrary element in N*, and if $\varepsilon > 0$ is given, then the set

$$\begin{aligned} S(x, f_0, \varepsilon) &= \{f : f \varepsilon N^* \text{ and } | F_x(f) - F_x(f_0) | < \varepsilon\} \\ &= \{f : f \varepsilon N^* \text{ and } | f(x) - f_0(x) < \varepsilon\} \end{aligned}$$

is an open set (in fact, a neighbourhood of f_0) in the weak* topology. Furthermore, the class of all sets of this kind, for all x's, f_0's, and ε's, is the defining open subbase for the weak* topology. All finite intersections of these sets constitute an open base for this topology, and the open sets themselves are all unions of these finite intersections.

We remark at this point that N* is a Hausdorff space with respect to its weak* topology. This follows at once from the fact that if f and g are distinct functionals in N*, then there must exist a vector x in N such that $f(x) \neq g(x)$, for if we put $\varepsilon = | f(x) - g(x) |/3$, then S (x, f, ε) and S (x, g, ε) are disjoint neighbourhoods of f and g in the weak* topology.

Let us now consider the closed unit sphere S* in N*, that is, the set S* = $\{f : f \varepsilon N^* \text{ and } \|f\| \leq 1)$. It is an easy consequence of Problem 2 that S* is compact in the strong topology ⇔ N is finite-dimensional, so the strong compactness of S* is a very stringent condition. If N is complete, it follows from Problem 3 and our unproved criterion for reflexivity that S* is compact in the weak topology ⇔ N is reflexive, so the weak compactness of S* is still a fairly substantial restriction. We state these facts to emphasize that the situation is quite different with the weak* topology, for here S* is always compact.

Theorem 1:

If N is a normed linear space, then the closed unit sphere S^* in N^* is a compact Hausdorff space in the weak topology.*

Proof:

We already know that S^*– is a Hausdorff space in this topology, so we confine our attention to proving compactness. .With each vector x in N we associate a compact space $C_{x'}$, where C_x is the closed interval $[-\|x\|, \|x\|]$ or the closed disc $\{z : |z| \leq \|x\|\}$, according as N is real or complex. By Tychonoff's theorem, the product C of all the C_x's is also a compact space. For each x, the values f (x) of all f's in S^* lie in $C_{x'}$. This enables us to imbed S^* in C by regarding each f in S^* as identical with the array of all its values at the vectors x in N. It is clear from the definitions of the topologies concerned that the weak* topology on S^* equals its topology as a subspace of C; and since C is compact, it suffices to show that S^* is closed as a subspace of C. We show that if g is in $\overline{S^*}$, then g is in S^*.

If we consider g to be a function defined on the index set N, then since g is in C we have $|g(x)| \leq \|x\|$ for every x in N. It therefore suffices to show that g is linear as a function defined on N. Let $\varepsilon > 0$ be given, and let x and y be any two vectors in N. Every basic neighbourhood of g intersects S^*, so there exists an f in S^* such that $|g(x) - f(x)| < \varepsilon/3$, $|g(y) - f(y)\| < \varepsilon/3$, and $|g(x+y) - f(x+y)| < \varepsilon/3$. Since f is linear, $f(x+y) - f(x) - f(y) = 0$, and we therefore have

$$|[g(x+y) - g(x) - g(y)| = |g(x+y) - f(x+y)] - [g(x) - f(x)])$$
$$- [g(y) - f(y)]| \leq |g(x+y) - f(x+y)| + |g(x) - f(x)|$$
$$+ |g(y) - f(y)| < \varepsilon/3 + \varepsilon/3 + \varepsilon/3 = \varepsilon.$$

The fact that this inequality is true for every $\varepsilon > 0$ now implies that $g(x+y) = g(x) + g(y)$. We can show in the same way that

$$g(\alpha x) = \alpha g(x)$$

for every scalar α, so g is linear and the theorem is proved.

We are now in a position to keep the promise made in the last paragraph for the following result is an obvious consequence if our preceding work.

Theorem 2:

Let N be a normed linear space, and let S^* be the compact Hausdorff space obtained by imposing the weak topology on the closed unit sphere in N^*. Then the mapping $x \to F_x$, where $F_x(f) = f(x)$ for each f in S^*, is an isometric isomorphism of N into $\mathcal{C}(S^*)$. If N is a Banach space, his mapping is an isometric isomorphism of N onto a closed linear subspace of $\mathcal{C}(S^*)$.*

This theorem shows, in effect, that the most general Banach space is essentially a closed linear subspace of $\mathcal{C}(X)$, where X is a compact Hausdorff space. The purpose of representation theorems in abstract mathematics is to reveal the structures of complex systems in terms of simpler ones, and from this point of view, Theorem B is satisfying to a degree. It must be pointed out, however. that we know next to nothing bout the closed linear subspaces of $\mathcal{C}(X)$, though we know a good deal bout $\mathcal{C}(X)$ itself. Theorem B is therefore somewhat less revealing than about at first glance. We shall the corresponding representation theorem for Banach algebras is much more significant and useful.

EXERCISES

1. Let X be a compact Hausdorff space, and justify the assertion that $\mathcal{C}(X)$ is reflexive if X is finite.
2. If N is a finite-dimensional normed linear space of dimension n, show that N* also has dimension n. Use this to prove that N is reflexive.
3. If ≤ is a Banach space, prove that B is reflexive ⇔ B* is reflexive.
4. Prove that if B is a reflexive Banach space, then its closed unit sphere S is weakly compact.
5. Show that a linear subspace of a normed linear space is closed Û it is weakly closed.

THE OPEN MAPPING THEOREM

In this section we have our first encounter with basic theorems which require that the spaces concerned be complete. The following rather technical lemma is the key to these theorems.

Lemma:

If B and B′ are Banach spaces, and if T is a continuous linear transformation of B onto B′, then the image of each open sphere centered on the origin in B contains an open sphere centered on the origin in B′.

Proof:

We denote by S_r and S'_r the open spheres with radius r centered on the origin in B and B′. It is easy to see that

$$T(S_r) = r(rS_1) = rT(S_1),$$

so it suffices to show that $T(S_1)$ contains some S'_r.

We begin by proving that $\overline{T(S_n)}$ contains some S'_r. Since T is onto, we see that $B' - U_{n=1}^{\infty} T(S_n)$. B′ is complete, so Baire's theorem implies that some

$\overline{T(S_{n_0})}$ has an interior point y_0, which may be assumed to lie in T (S_{n_0}). The mapping $y \to y - y_0$ is a homeomorphism of B' onto itself, so $\overline{T(S_{n_0})} - y_0$ has the origin as an interior point. Since y_0 is in T (S_{n_0}), we have T (S_{n_0}) $-y_0$ T $\subseteq$ $T(S_{2_{n_0}})$ and from this we obtain $\overline{T(S_{2_0})} - y_0 = \overline{T(S_{n_0}) - y_0} \subseteq \overline{T(S_{2n_0})}$ which shows that the origin is an interior point of $\overline{T(S_{2n_0})}$. Multiplication by any non-zero scalar is a homeomorphism of B' onto itself, so $\overline{T(S_{2n_0})}$ = $\overline{2n_0 T(S_1)} = 2n_o \overline{T(S_1)}$, and it follows from this that the origin is also an interior point of $\overline{T(S_1)}$, so $S'_{\varepsilon'} \subseteq \overline{T(S_1)})$ for some positive number ε.

We conclude the proof by showing that $S'_{\varepsilon} \subseteq$ T (S_3), which is clearly equivalent to $S'_{\varepsilon/3} \subseteq$ T (S_1). Let y be a vector in B' such that $\| yT\| < \varepsilon$. Since y is in $\overline{T(S_1)}$, there exists a vector x_1 in B such that $\| x_1 \| < 1$ and $\| y - y_1 \| < \varepsilon/2$, where y_1= T (x_1). We next observe that $S'_{\varepsilon/2} \subseteq \overline{T(S_{1/2})}$, so there exists a vector x_2 in B such that $\| x_2 \| < 1/2$ and $\| (y - y_1) - y_2 \| < \varepsilon/4$, where $y/2 = T(x_2)$. Continuing in this way, we obtain a sequence $\{x_n\}$ in B such that $\| x_n \| < 1/2^{n-1}$ and $\| y - (y_1 + y_2 + \ldots + y_n) \| < \varepsilon/2^n$, where $y_n = T(x_n)$. If we put

$$s_n = x_1 + x_2 + \ldots + x_n,$$

then it follows from $\| x_n \| < 1/2^{n-1}$ that $\{s_n\}$ is a Cauchy sequence in B for which

$$\| s_n \| \leq \| x_1 \| + \| x_2 \| + \ldots + \| x_n \| < 1 + 1/2 + \ldots + 1/2^{n-1} < 2.$$

B is complete, so there exists a vector x in B such that $s_n \to x$ and $\| x \| = \| \lim s_n \| = \lim \| s_n\| \leq 2 < 3$ shows that x is in S_3. All that remains is to notice that the continuity of T yields

$T(x) = f(\lim s_n) = \lim T(s_n) = \lim (y_1 + y_2 + \ldots + y_n) = y$, from which we see that y is in T (S_3).

This makes our main theorem easy to prove.

Theorem 1:

(The Open Mapping Theorem) : If B and B' are Banach spaces, and if T is a continuous linear transformation of B onto B', then T is an open mapping.

Proof:

We must show that if G is an open set in B, then T (G) is also an open set in B'. If y is a point in T (G), it suffices to produce an open sphere centered on y and contained in T (G). Let x be a point in G such that T (x) = y. Since G is open, x is the center of an open sphere—which can be written in the form x + S_r–

contained in G. Our lemma now implies that $T(S_r)$ contains some S'n. It is clear that $y + S'_{r_1}$ is an open sphere centered on y, and the fact that it is contained in T (G) follows at once from $y + S'_{r_1} \subseteq y + T(S_r) = T(x) + T(S_r) = T(x + S_r) \subseteq T(G)$.

Most of the applications of the open mapping theorem depend more directly on the following special case, which we state separately for the sake of emphasis.

Theorem 2:

A one-to-one continuous linear transformation of one Banach space onto another is a homeomorphism. In particular, if a one-to-one linear transformation T of a Banach space onto itself is continuous, then its inverse T^{-1} is automatically continuous.

As our first application of Theorem B, we give a geometric characterization of the projections on a Banach space. The reader will recall from Sec. 44 that a projection *E* on a linear space *L* is simply an idem potent ($E^2 = E$) linear transformation of L into itself.

He will also recall that projections on *L* can be described geometrically as follows:

(1) a projection E determines a pair of linear subspaces M and N such that $L = M \oplus N$, where $M = \{E(x) : x \varepsilon L\}$ and $N = \{x : E(x) = 0\}$ are the range and null space of E;

(2) a pair of linear subspaces M and N such that $L = M \oplus N$ determines a projection E whose range and null space are M and N (if $z = x + y$ is the unique representation of a vector in L as a sum of vectors in M and N, then E is defined by $E(z) = x$).

These facts show that the study of projections on L is equivalent to the study of pairs of linear' subspaces which are disjoint and span L.

In the theory of Banach spaces, however, more is required of a projection than mere linearity and idempotence. A projection on a Banach space B is an idempotent operator on B; that is, it is a projection on B in the algebraic sense which is also continuous.

Our present task is to assess the effect of the additional requirement of continuity on the geometric descriptions given in (1) and (2) above. The analogue of (1) is easy.

Theorem 3:

If P is a projection on a Banach space B, and if M and N are its range and null space, then M and N are closed linear subspaces of B such that $B = M \oplus N$.

Proof:

P is an algebraic projection, so (1) gives everything except the fact that M and N are closed. The null space of any continuous linear transformation is closed, so N is obviously closed; and the fact that M is also closed is a consequence of

$M = \{P(x) : x \varepsilon B\} = \{x : P(x) = x\} = \{x : (I - P)(x) = 0\}$, which exhibits M as the null space of the operator I – P.

The analogue of (2) is more difficult, for Theorem B is needed in its proof.

Theorem 4:

Let B be a Banach space, and let M and N be closed linear subspaces of B such that $B = M \oplus N$. If $z = x + y$ is the unique representation of a vector in B as a sum of vectors in M and N, then the mapping P defined by $P(z) = x$ is a projection on B whose range and null space are M and N.

Proof:

Everything stated is clear from (2) except the fact that P is continuous, and this we prove as follows. B' denotes the linear space B equipped with the norm defined by

$$\| z \|' = \| x \| + \| y \|,$$

then B' is a Banach space, and since $\| P(z) \| = \| x \| \leq \| x \| + \| y \| = \| z \|'$, P is clearly continuous as a mapping of B' into B. It therefore suffices to prove that B' and B have the same topology. If T denotes the identity mapping of B' onto B, then

$$\| T(z) \| = \| z \| = \| x + y \| \leq \| x \| + \| y \| = \| z \|'$$

shows that T is continuous as a one-to one linear transformation of B' onto B. Theorem now implies that T is a homeomorphism, and the proof is complete.

This theorem raises some interesting and significant question. Let M be a closed linear subspace of a Banach space B. As we remarked at the end there is always at least one algebraic projection defined on B whose range is M, and there may be great many. However, it night well happen that none of these are continuous, and that consequently none are projections in our present sense. In the light of our theorems, this is equivalent to saying that there might not exists any closed linear subspace N such that $B = M \oplus N$. What sorts of Banach spaces have the property that this awkward situation cannot occur? We shall see in the next chapter that a Hilbert space—which is a special type of Banach space—has this property. We shall also see that this property is closely linked to the satisfying geometric structure which sets Hilbert spaces apart from general Banach spaces.

We now turn to the closed graph theorem. Let B and B¢ be Banach spaces. It we define a metric on the product B × B¢ by

$$d((x_1, y_1), (x_2, y_2)) = \max\{((x_1 - x_2 + \| y_1 - y_2 \|\},$$

then the resulting topology is easily seen to be the same as the product topology, and convergence with respect to this metric is equivalent to coordinates wise convergence. Now let T be a linear transformation of B into B¢. We recall that the graph of T is that subset of B × B′ which consists of all ordered pairs of the form (x, T (x)). Shows that if T is continuous, then its graph is closed as a subset of B × B′. In the present context, the converse is also true.

Theorem 5:

(The Closed Graph Theorem) : If B and B′ are Banach spaces, and if T is a linear transformation of B into B¢, then T is continuous ⇔ its graph is closed.

Proof:

In view of the above remarks, we may confine our attention to proving that T is continuous if its graph is closed. We denote by B1 the linear space B renormed by

$$\| x \|_1 \doteq \| x \| + \| T(x) \| = \| x \|_1,$$

T is continuous as a mapping of B_1 into B′. It therefore suffices to show that B and B_1 have the same topology. The identity mapping of BI onto B is clearly continuous, for $\| x \| \leq \| x \| + \| T(x) \| = \| x \|_1$. If we can show that B_1 is complete, then Theorem B will guarantee that this mapping is a homeomorphism, and this will conclude the proof. Let $\{x_n\}$ be a Cauchy sequence in B_1. It follows that $\{x_n\}$ and $\{T(x_n)\}$ are also Cauchy sequences in B and B′, and since both of these spaces are complete, there exist vectors x and y in B and .B′ such that $\| x_n - x \| \to 0$ and $\| T(x_n) - y \| \to 0$. Our assumption that the graph of T is closed in B y. B′ implies that (x, y) lies on this graph, so T (x) = y. The completeness of B_1 now follows from

$$\| x_n - x \|_1 = \| x_n - x \| + \| T(x_n - x) \| = \| x_n - x \| + \| (x_n) - T(x) \|$$
$$= \| x_n - x \| + \| T(x_n) - y \| \to 0.$$

The closed graph theorem has a number of interesting applications to problems in analysis, but since our concern here is mainly with matters of algebra and topology, we do not pause to illustrate its uses in this direction.

EXERCISES

1. Let a Banach space B be made into a Banach space B′ by means of a new norm, and show that the topologies generated by these norms are the same if either is stronger than the other.

2. In the text, we used Theorem B to prove the closed graph theorem. Show that Theorem B is a consequence of the closed graph theorem.
3. Let T be a linear transformation of a Banach space B into a Banach space B′. If $\{f_i\}$ is a set of functionals in B′* which separates the vectors in B′, and if $f_i T$ is continuous for each f_i, prove that T is continuous.

THE CONJUGATE OF AN OPERATOR

.We shall see in this section that each operator T on a normed linear space N induces a corresponding operator, denoted by T* and called the conjugate of T, on the conjugate space N*. Our first task is to define T*, and our second is to investigate the properties of the mapping $T \to T^*$. We base our discussion on the following theorem.

Theorem 1:

(The Uniform Boundedness Theorem) : Let B be a Banach space and N a normed linear space. If $\{T_i\}$ is a non-empty set of continuous linear transformations of B into N with the property that $\{T_i(x)\}$ is a bounded subset of N for each vector x in B; then $\{\| T_i \|\}$ is a bounded set of numbers; that is, $\{T_i\}$ is bounded as a subset of $\mathcal{B}(B, N)$.

Proof:

For each positive integer re, the set

$F_n = [x : x \varepsilon B \text{ and } \| T_i(x) \| \leq n \text{ for all } i\}$

is clearly a closed subset of B, and by our assumption we have

$$B = \cup_{n=1}^{\infty} F_n .$$

Since B is complete, Baire's theorem shows that one of the F_n's, say F_{n_0}," has non-empty interior, and thus contains a closed sphere So with center xo and radius $r_0 > 0$. This says, in effect, that each vector in every set $T_i(S_0)$ has norm less than or equal to n_0; and for the sake of brevity, we express this fact by writing $\| T_i(S_0) \| \leq n_0$.

It is clear that $S_0 - x_0$ is the closed sphere with radius r_0 centered on the origin, so $(S_0 - x_0)/r_0$ is the closed unit sphere S. Since x_0, is in S_0, it is evident that $\| T_i(S_0 - x_0) \| \leq 2n_0$. This yields $\| T_i(S_i) \| \leq 2n_0/r_0$, so $\| T_i \| \leq 2n_0/r_0$ for every i, and the proof is complete.

This theorem is often called the *Banach-Steinhaus theorem,* and it has several significant applications to analysis. See, for example, the purposes we have in view, our main interest is in the following simple consequence of it.

Theorem 2:

If N is a normal operator on H, then $\| N^2 \| = \| N \|^2$.

Proof:

The preceding theorem shows that

$$\| N^2x \| = \| NNx \| = \| N^*Nx \|$$

for every *x*, and this implies that $\| N^2 \| = \|N^*N \|$. By theorem A, we have $\| N^*N \| = \| N \|^2$, so the proof is complete.

We know that any complex number z can be expressed uniquely in the form z = a + ib where a and b are real numbers, and that these real numbers are called the real and imaginary parts of z and are given by $a = (z + \bar{z})/2$ and $b = (z - \bar{z})/2i$. The analogy between general operators and complex numbers, and between self-adjoint operators and real numbers, suggests that for an arbitrary operator T on H we form $A_1 = (T + T^*)/2$ and $A_2 = (T - T^*)/2i$. A_1 and A_2 are clearly self-adjoint, and they have the property that $T - A_1 + i A_2$. The uniqueness of this expression for T follows at once from the fact that

$$T^* = A_1 - i A_2.$$

The self-adjoint operators A_1 and A_2 are called the real part and the imaginary part of T.

We emphasized earlier that the complicated structure of $\mathcal{B}(H)$ is due in large part to the fact that operator multiplication is in general non-commutative. Since our future work will be focused mainly on normal operators, it is of interest to see–as the following theorem shows–that the existence of non-normal operators can be traced directly to the non-commutativity of self-adjoint operators.

Theorem 3:

If T is an operator on H, then T is normal $\Leftrightarrow$ *its real and imaginary parts commute.*

Proof:

If A_1 and A_2 are the real and imaginary parts of *T*, so that $T = A_1 + iA_2$ and $T^* = A_1 - iA_2$, then

$$TT^* = (A_1 + iA_2)(All - iA_2) = A_1^2 + A_2^2 + i(A_2A_1 - A_1A_2)$$

and $T^*T = (A_1 - iA_2)(A_1 + iA_2) = A_1^2 + A_2^2 + i(A_1A_2 - A_2A_1).) =$

It is clear that if $A_1A_2 = A_2A_1$, then $TT^* = T^*T$, Conversely, if $TT^* = T^*T$, then $A_1A_2 - A_2A_1 - A_1A_2$, so $2A_1A_2 = 2A_2A_1$ and $A_1A_2 = A_2A_1$.

Perhaps the most important subsystem of the complex plane after the real line is the unit circle, which is characterized by either of the equivalent identities $|z| - 1$ or $z\bar{z} = \bar{z}z = 1$. An operator U on H which satisfies the equation UU* = U*U = I is said to be unitary. Unitary operators–which are obviously normal– are thus the natural analogues of complex numbers of absolute value 1. It is clear from the definition that the unitary operators on H are precisely the non-singular operators whose inverses equal their adjoints. The geometric significance of these operators is best understood in the light of our next theorem.

Theorem 4:

If T is an operator on H, then the following conditions are all equivalent to one another:

(1) $T^*T = I$,

(2) $(Tx, Ty) = (x, y)$ *for all x and y,*

(3) $\|Tx\| = \|x\|$ *for all x.*

Proof:

If (1) is true, then $(T^*Tx, y) = (x, y)$ or $(Tx, Ty) = (x, y)$ for all x and y, so (2) is true; and if (2) is true, then by taking y = x we obtain $(Tx, Tx) = (x, x)$ or $\|Tx\|^2 = \|x\|^2$ for all x, so (3) is true.

The fact that (3) implies (1) is a consequence of Theorem 3 and the following chain of implications:

$$\|Tx\| = \|x\| \Rightarrow \|Tx\|^2 = \|x\|^2 \Rightarrow (Tx, Tx) = (x, x) \Rightarrow \Rightarrow (T^*Tx, x)$$
$$= (x, x) \Rightarrow ([T^*T - I]x, x) = 0.$$

An operator on H with this theorem is simply an isometric isomorphism of H into itself. That an operator of this kind need not be unitary is easily seen by considering the operator on h defined by

$$T\{x_1, x_2, \ldots\} = \{0, x_1, x_2, \ldots\},$$

which preserves norms but has no inverse. These ideas lead at once to

Theorem 5:

A non-empty subset X of a normed linear space N is bounded $\Leftrightarrow$ *f (X) is a bounded set of numbers for each f in N*.*

Proof:

Since $|f(x)| \le \|f\| \|x\|$, it is obvious that if X is bounded, then f (X) is also bounded for each f.

In order to prove the other half of the theorem, it is convenient to exhibit the vectors in X by writing $X = \{x_i\}$. We now use the natural imbedding to pass from X to the corresponding subset $\{Fx_i\}$ of N**.

Our assumption that $f(X) = \{f(x_i)\}$ is bounded for each f is clearly equivalent to the assumption that $\{F_{xi} f)\}$ is bounded for each f, and since N* is complete,

Shows that $\{Fx_i\}$ is a bounded subset of N**. We know that the natural imbedding preserves norms, so X is evidently a bounded subset of N.

We now turn to the problem of defining the conjugate of an operator on a normed linear space N.

Let L be the linear space of all scalar-valued linear functions defined on N. The conjugate space N* is clearly a linear subspace of L. Let T be a linear transformation of N into itself which is not necessarily continuous. We use T to define a linear transformation T' of L into itself, as follows. If f is in L, then T' (f) is defined by

$$[T'(f)](x) = T(T(x)). \qquad ...(27)$$

We leave it to the reader to verify that T' (f) actually is linear as a function defined on N, and also that T' is linear as a mapping of L into itself.

The following natural question now presents itself. Under what circumstances does T' map N* into N*? This question has a simple and elegant answer: $T'(N^*) \subseteq N^* \Leftrightarrow T$ is continuous.

If we keep Theorem B in mind, the proof of this statement is very easy; for if S is the closed unit sphere in N, then T is continuous $\Leftrightarrow$ T (S) is bounded $\Leftrightarrow$ f(T (S)) is bounded for each f in N* $\Leftrightarrow$ [T' (f)] (S) is bounded for each f in N* $\Leftrightarrow$ T' (f) is in N* for each f in N*.

We now assume that the linear transformation T is continuous and is therefore an operator on N. The preceding developments allow us to consider the restriction of T' to a mapping of N* into itself. We denote this restriction by T*, and we call it the conjugate of T. The action of T* is given by

$$[T^*(f)](x) = f(T(x)), \qquad ...(28)$$

in which – in contrast to (1)– f is understood to be a functional on N, and not merely a scalar-valued linear function. T* is clearly linear, and the following computation shows that it is continuous:

$$\begin{aligned} \| T^* \| &= \sup \{ \| T^*(f) \| : \| f \| \le 1 \} \\ &= \sup \{ |\{T^*(f)](x)| : \| f \| \text{ and } \| x \| \le 1 \} \\ &= \sup \{ | f(T(x)) | : \| f \| \text{ and } \| x \| \le 1 \} \\ &\le \sup \{ \| f \| \| T \| \| x \| : \| f \| \text{ and } \| x \| \le 1 \} \end{aligned}$$

$$\leq \| T \|.$$

Since $\| T \| = \sup \{ \| T(x) \| : \| x \| \leq 1 \}$, we see at once from Theorem 2 that equality holds here, that is, that

$$\| T^* \| = \| T \|. \qquad ...(29)$$

The mapping $T \to T^*$ is thus a norm-preserving mapping of $\mathcal{B}(N)$ into $\mathcal{B}(N^*)$.

We continue in this vein by observing that the mapping $T \to T^*$ also has the following pleasant algebraic properties:

$$(\alpha T_1 + \beta T_2)^* = \alpha T_1^* + \beta T_2^*, \qquad ...(30)$$

$$(T_1 T_2)^* = T_2^* T_1^*, \qquad ...(31)$$

and

$$I^* = I. \qquad ...(32)$$

The proofs of these facts are easy consequences of the definitions. We illustrate the principles involved by proving (31). It must be shown that $(T_1 T_2)^* (f) = (T_2^* T_1^*) (f)$ for each f in N^*, and this means that

$$[(T_1 T_2)^* (f)] (x) = [(T_2^* T_1^*) (f)] (x)$$

for each f in N^* and each x in N. A simple computation now shows that

$$[(T_1 T_2)^* (f)] (x) = f((T_1 T_2) (x)) = f(T_1 (T_2 (x)) = [T_1^* (f)] (T_2 (x))$$
$$= [T_2^* (T_1^* (f))] (f)] (x) = [(T_2^* T_1^*) (f)] (x)$$

It may be helpful to the reader to have the following summary of the results of this discussion.

Theorem 6:

The set of all normal operators on H is a closed subset of (&(H) which contains the set of all self-adjoint operators and is closed under scalar multiplication.

It is natural to wonder whether the sum and product of two normal operators afe necessarily normal. They are not, but nevertheless, we can say a little in this direction.

Theorem 7:

If N_1 and N_2 are normal operators on H with the property that either commutes with the adjoint of the other, then $N_1 + N_2$ and $N_1 + N_2$ are normal.

Proof:

It is clear by taking adjoints that

$$N_1 N_2^* = N_2^* N_1 \Leftrightarrow N_2 N_1^* = N_1^* N_2$$

so the assumption implies that each commutes with the adjoint of the other. To show that $N_1 + N_2$ is normal under the stated conditions, we have only to compare the results of the following computations:

$$(N_1 + N_2,)(N_1 + N_1)^* = (N_1 + N_2)(N_1^* + N_1^*)$$

$$= N_1N_1^* + N_1N_1^* + N_2N_2^*$$

and $$(N_1 + N_2)^*(N_1 + N = (N_1^* + N_2^*)(N_1 + N_2)$$

$$= N_1^*N_1 + N_1^*N_2^* + N_2^* N_{1\,+} N_2^* N_2$$

The fact that N_1N_2 is normal follows similarly from

By definition, a self-ad joint operator A is one which satisfies the identity $A^*x = Ax$. Many properties of self -adjoint operators do not depend on this, but only on the weaker identity $\| A^*x \| = \| Ax \|$. Our next theorem shows that all such properties are shared by normal operators.

Theorem 8:

If T is an operator on a normed linear space N, then its conjugate T defined by Eq. (2) is an operator on N*, and the mapping $T \to T^*$ is an isometric isomorphism of $\mathcal{C}(N)$ into $\mathcal{C}(N^*)$ which reverses products and preserves the identity transformation.*

The general significance of the ideas developed here can be understood only in the light of the theory of operators on Hilbert spaces. Some preliminary comments on these matters are given in the introduction to the next chapter.

EXERCISES

1. Let B be a Banach space and N a normed linear space. If $\{T_n\}$ is a sequence in $\mathcal{B}(B,N)$ such that $T(x) = \lim T_n(x)$ exists for each x in B, prove that T is a continuous linear transformation.
2. Let T be an operator on a normed linear space N. If N is considered to be part of N** by means of the natural imbedding, show that T** is an extension of T. Observe that if N is reflexive, then $T^{**} = T$.
3. Let T be an operator on a Banach space B. Show that T has an inverse $T^{-1} \Leftrightarrow T^*$ has an inverse $(T^*)^{-1}$, and that in this case $(T^*)^{-1} = (T^{-1})^*$.

4

THE STRUCTURE OF COMMUTATIVE BANACH ALGEBRAS

INTRODUCTION

The other hands were essentially algebraic in nature. The function spaces we encountered earlier led us to begin a study of Banach spaces for their own sake, and as we proceeded, we found our attention focusing more and more closely on the properties of their operators. Except for a few elementary notions about metric spaces, we used very little genuine topology. As a matter of fact, our treatment of spectral theory was completely independent of topology, for in the finite-dimensional case, all linear transformations are continuous. The primarily concerned with topological spaces and the continuous functions carried by them.

In the following three chapters, these two apparently diverse trains of thought-the topological and the algebraic—are united by a single elegant concept : that of a Banach algebra. Our remarks in the last section of the previous chapter suggested that there may be important links between algebras of operators on Hilbert spaces and algebras of the type $\mathcal{C}(X)$, where X is a compact Hausdorff space. Banach algebras are the systems which enable us to establish these connections on a firm footing. They are also interesting in that they constitute a field of study in which a wide variety of mathematical ideas meet in significant contact.

Our main task in the present chapter is to provide a number of miscellaneous tools which are necessary for the structure theory developed letter.

THE DEFINITION AND SOME EXAMPLES

A *Banach algebra* is a complex Banach space which is also an algebra with identity 1, and in which the multiplicative structure is related to the norm by the following requirements:

(1) $\|xy\| \leq \|x\| \ \|y\|$;

(2) $\|1\| = 1$.

It follows from (1) that multiplication is jointly continuous in any Banach algebra, that is, that if $x_n \to x$ and $y_n \to y$, then $x_n y_n \to xy$ (proof:

$$\|x_n y_n - xy\| = \|x_n (y_n - y) + (x_n - x)y\|$$

$$\leq \|x_n\| \ \|y_n - y\| + \|x_n - x\| \ \|y\|)$$

A *Banach subalgebra* of a Banach algebra A is a closed subalgebra of A which contains 1. The Banach subalgebras of A are precisely those subsets of A which are themselves Banach algebras with respect to the same algebraic operations, the same identity, and the same norm.

The definition of a Banach algebra is sometimes given without the restriction that the scalars are the complex numbers. The complex case, however, is the only one that concerns us, and by framing the definition as we do, we avoid the necessity of treating the additional complications which arise in the real case. We have further assumed, for the sake of simplicity, that every Banach algebra has an identity. It is possible, at a considerable sacrifice of clarity, to develop ;most of the important ideas without this assumption, and this is done whenever the primary purpose of the theory is the study of group algebras of locally compact but not discrete groups. Since our attention will be directed chiefly to the structure of operator algebras, there is no need for us to strain for the added generality obtained by not requiring the presence of an identity.

The Banach algebras of principal interest to us ;are described in the following examples. The reader will notice that they all consist of functions or operators and that the linear operations in all of them are defined pointwise. They can be classified in a general way into *function algebras, operator algebras*, or *group algebras*, according as multiplication is defined pointwise, by composition, or by convolution.

Example 1:

(a) One of the most important Banach algebras is the set $\mathcal{C}(X)$ of all bounded continuous complex functions defined on a topological space X. The case in which X is a compact Hausdorff space will have particular significance for our later work. If X has only one point, then $\mathcal{C}(X)$ can be identified with the simplest of all Banach algebras, the algebra of complex numbers.

(b) Consider the closed unit disc $D = \{z : |z| \leq 1\}$ in the complex plane. The subset of $\mathcal{C}(D)$ which consists of all functions analytic in the interior of D is obviously a subalgebra which contains the identity. A simple application of Morera's theorem from complex analysis shows that it is closed and is therefore a Banach subalgebra of $\mathcal{C}(D)$. This Banach algebra is called the disc algebra. It has a number of interesting properties, which are, of course, intimately related to the special character of its functions.

Example 2:

(a) If B is a non-trivial complex Banach space, then the set $\mathcal{B}(B)$ of all operators on B is a Banach algebra. We assume that B is non-trivial in order to guarantee that the identity operator is an identity in the algebraic sense.

(b) If we consider a non-trivial Hilbert space H, then $\mathcal{B}(H)$, is a Banach algebra. This is a special case of $\mathcal{B}(B)$, and it is important to observe that additional structure is present here, namely, the adjoint operation $T \to T^*$.

(c) A subalgebra of $\mathcal{B}(H)$ is said to be self-adjoint if it contains the adjoint of each of its operators. Banach subalgebras of $\mathcal{B}(H)$'s which are self-adjoint are called C*-algebras. We shall return to the subject of commutative C*-algebras.

(d) The weak operator topology on $\mathcal{B}(H)$ is the weak topology generated by all functions of the form $T \to (Tx, y)$; that is, it is the weakest topology with respect to which all these functions are continuous. It is easy to see from the inequality $|(Tx,y)-(T_0x,y)| \leq \|T - T_0\| \, \|x\| \, \|y\|$ that this topology is weaker than the usual norm topology, so that its closed sets are also closed in the usual sense. A C*-algebra with the further property of being closed in the weak operator topology is called a W*-algebra. Algebras of this kind are also called *rings of operators*, or *von Neumann algebras*. They are among the most interesting of all Banach algebras, but their theory is quite beyond the scope of this book.

Example 3:

(a) If $G = \{g_1, g_2, \ldots, g_n\}$ is a finite group, then its group algebra $L_1(G)$ is the set of all complex functions defined on G. Addition and scalar multiplication are defined pointwise, and the norm by $\|f\| = \sum_{i=1}^{n} |f(g_i)|$. In order to see what underlies the definition of multiplication, it is convenient to regard a typical element f of $L_1(G)$ as a formal sum $\sum_{i=1}^{n} \alpha_i g_i$, where α_i is the value of f at g_i With this interpretation, we use the given multiplication in G to define multiplication, in $L_1(G)$, as follows:

$$\left(\sum_{i=1}^{n} \alpha_i g_i\right)\left(\sum_{j=1}^{n} \beta_j g_j\right) = \sum_{k=1}^{n} \gamma_k g_k, \tag{1}$$

where $\gamma_k = \sum g_i g_j = g_k \alpha_i \beta_j.$ (2)

The meaning of the sum is (2) is that the summation is to be extended over all subscripts i and j such that $g_i g_j = g_k$. In effect, therefore, we formally multiply out the sums on the left of (1), and we then gather together all the resulting terms which contain the same element of G. With these ideas as an intuitive guide, we revert to out first point of view, in which the elements of $L_1(G)$ are functions, and we see that our definition of multiplication can be expressed in the following way. If two functions f and g in $L_1(G)$ are given, then their product, which is denoted by f * g and called their *convolution*, is that function whose value at g_k is

$$(f*g)(g_k) = \sum_{n} g_i g_j \; f(g_i)g(g_j)$$

$$= \sum_{j=1}^{n} f\left(g_k g_j^{-1}\right) g\left(g_j\right). \tag{3}$$

We note that if each element of G is identified with the function whose value is 1 at that element and 0 elsewhere, then G becomes a subset of $L_1(G)$. Further, multiplication in G agrees with convolution in $L_1(G)$, and the element of $L_1(G)$ which corresponds to the identity in G is an identity for $L_1(G)$. We conclude this description by observing that every element of G has norm 1, so that $\|1\| = 1$, and that the basic norm inequality for a Banach algebra is satisfied:

$$\|f * g\| = \sum_{k=1}^{n} \left|(f*g)(g_k)\right|$$

$$= \sum_{k=1}^{n} \left|\sum_{j=1}^{n} f\left(g_k g_j^{-1}\right) g\left(g_j\right)\right|$$

$$\le \sum_{k=1}^{n} \sum_{j=1}^{n} \left|f\left(g_k g_j^{-1}\right)\right| \left|g\left(g_j\right)\right|$$

$$= \sum_{j=1}^{n} \sum_{k=1}^{n} \left|f\left(g_k g_j^{-1}\right)\right| \left|g\left(g_j\right)\right|$$

$$= \sum_{j=1}^{n} \left|g\left(g_j\right)\right| \sum_{k=1}^{n} \left|f\left(g_k g_j^{-1}\right)\right|$$

$$= \sum_{j=1}^{n} \left|g\left(g_j\right)\right| \; \|f\|$$

$$= \|f\| \sum_{j=1}^{n} \left|g(g_j)\right|$$

$$= \|f\| \, \|g\|.$$

(b) Let (G) = {. . . , – 2, – 1, 0, 1, 2, . . .} be the additive group of integers. Its group algebra $L_1(G)$ is the set of all complex function f defined on G for which converges. The linear operations are defined pointwise, the norm by $\|f\| = \sum_{m=-\infty}^{\infty} |f(n)|$, and the convolution of f and g.

$$(f*g)(n) = \sum_{m=-\infty}^{\infty} f(n-m)g(m).$$

Just as in (a), G is contained in $L_1(G)$ in a natural way, and $L_1(G)$ is a Banach algebra. Any attempt to discuss the group algebra of a non-discrete topological group like the real line must clearly be based on an adequate theory of integration. It should also have available a theory of Banach algebras in which no identity is assumed to be present. These ideas constitute a rich and beautiful field of modern analysis. They are, however, outside the scope of this work. [1]

The Banach algebras described above are many and diverse, and there are yet others which we have not mentioned. Our attention in the following chapters will be centered on $\mathcal{C}(X)$'s and commutative C*-algebras, but the general theory we develop is equally applicable to all. It is worthy of notice that an arbitrary Banach algebra A can be regarded as a Banach subalgebra of $\mathcal{B}(A)$. In a sense, therefore, $\propto$ and its Banach subalgebras include all possible Banach algebras. To see this, we recall that a $\to M_\propto$, where $M_\propto(x) = \propto x$ is an isomorphism of A into $\mathcal{B}(A)$. It is easy to see that M_1 is the identity operator on A, so all that remains is to observe that $\|\alpha\| = \|M_\alpha\|$ for every a (proof: $\|M_\alpha(x)\| = \|\alpha x\| \le \|\alpha\| \, \|x\|$ shows that $\|M_\alpha\| \le \|\alpha\|$, and the fact that $\|\alpha\| \le \|M_\alpha\|$ follows from

$$\|M_\alpha\| = \sup\left\{\|M_\alpha(x)\| : \|x\| \le 1\right\} \ge \|M_\alpha(1)\| = \|\alpha\|).$$

The mapping $\propto \to M\propto$ is thus an isometric isomorphism of A onto a Banach subalgebra of $\mathcal{B}(A)$, and it allows us to identify the abstract Banach algebra A with a concrete Banach algebra of operators on A.

REGULAR AND SINGULAR ELEMENTS

Let A be a Banach algebra. We denote the set of regular elements in A by G, and its complement, the set of singular elements, by S. It is clear that G contains 1 and is a group, and that S contains 0. Several important issues depend on the character of G and S. Our first result along these lines is

Theorem 1:

Every element x for which $\|x-1\| < 1$ *is regular, and the inverse of such an element is given by the formula* $x–1 = 1 + \sum_{n=1}^{\infty} (1-x)^n$.

Proof:

If we put $r = \|x-1\|$, so that $r < 1$, then

$$\left\|(1-x)^n\right\| \le \|1-x\|^n = r^n$$

shows that the partial sums of the series $\sum_{n=1}^{\infty} (1-x)^n$ form a Cauchy sequence in A. Since A is complete, these partial sums converge to an element of A, which we denote by $\sum_{n=1}^{\infty} (1-x)^n$. If we define y by $y = 1+\sum_{n=1}^{\infty} (1-x)^n$, then the joint continuity of multiplication in A implies that

$$y-xy=(1-x)y=(1-x)+\sum_{n=2}^{\infty}(1-x)^n = \sum_{n=2}^{\infty}(1-x)^n = y-1,$$

so $xy = 1$. Similarly, $yx = 1$.

We now use this as a tool to prove

Theorem 2:

G is an open set, and therefore S is a closed set.

Proof:

Let x_0 be an element in G, and let x by any element in A such that It is clear that $\|x - x_0\| < 1 / \|x_0^{-1}\|$. It is clear that

$$\left\|x_0^{-1}x - 1\right\| = \left\|x_0^{-1}(x-x_0)\right\| \le \left\|x_0^{-1}\right\| \|x - x_0\| < 1,$$

so we see by Theorem 1 that $x_0^{-1}x$ is in G. Since $x = x_0 (x_0^{-1}x)$, it follows that x is also in G, so G is open.

It was the every Banach space is locally connected, so A is also locally connected. A direct application of theorem yields the fact that the components of G are themselves open sets.

As our final result, we have

Theorem 3:

The mapping $x \to x^{-1}$ *of G into G is continuous and is therefore a homeomorphism of G onto itself.*

Proof:

Let x_0 be an element of G, and x another element of G such that $\|x - x_0\| < 1 / \left(2\|x_0^{-1}\|\right)$. Since

$$\left\|x_0^{-1}x - 1\right\| = \left\|x_0^{-1}(x - x_0)\right\| \le \left\|x_0^{-1}\right\| \|x - x_0\| < \frac{1}{2},$$

we see by Theorem A that $x_0^{-1}x$ is in G and

$$x^{-1}x_0 = \left(x_0^{-1}x\right)^{-1} = 1 + \sum_{n=1}^{\infty}\left(1 - x_0^{-1}x\right)^n.$$

Our conclusion now follows from

$$\left\|x^{-1} - x_0^{-1}\right\| = \left\|\left(x^{-1}x_0 - 1\right)x_0^{-1}\right\| \le \left\|x_0^{-1}\right\| \left\|x^{-1}x_0 - 1\right\| = \left\|x_0^{-1}\right\|$$

$$\left\|\sum_{n=1}^{\infty}\left(1 - x_0^{-1}x\right)^n\right\| \le \left\|x_0^{-1}\right\| \sum_{n=1}^{\infty}\left\|1 - x_0^{-1}x\right\|^n$$

$$= \left\|x_0^{-1}\right\| \left\|1 - x_0^{-1}x\right\| \sum_{n=0}^{\infty}\left\|1 - x_0^{-1}x\right\|^n$$

$$= \frac{\left\|x_0^{-1}\right\| \left\|1 - x_0^{-1}x\right\|}{1 - \left\|1 - x_0^{-1}x\right\|}$$

$$< 2\left\|x_0^{-1}\right\| \left\|1 - x_0^{-1}x\right\| \le 2\left\|x_0^{-1}\right\|^2 \|x - x_0\|.$$

If x is an element in A, it should always be kept in mind that the regularity or singularity of x depends on A as well as on x itself. If x is regular in A, and if we pass to a Banach subalgebra A' of a which also contains x, then x may lose its inverse and become singular in A'. By the same token, if x is singular in A, and if A is regarded as a Banach subalgebra of a larger Banach algebra A", then x may acquire an inverse and become regular in A". In the next section, we study certain element in A which are singular and remain singular with respect to all possible enlargements of A.

TOPOLOGICAL DIVISORS OF ZERO

An element z in our Banach algebra A is called topological divisor of zero if there exists a sequence $\{z_n\}$ in A such that $\|z_n\| = 1$ and either $zz_n \to 0$ or $z_n z \to 0$. It is clear that every divisor or zero is also a topological divisor of zero. We denote the set of all topological divisors of zero by z.

Theorem 1:

Z is a subset of S.

Proof:

Let z be an element of Z and $\{z_n\}$a sequence such that $\|z_n\| = 1$ and (say) $zz_n \to 0$. If z were in G, then by the joint continuity of multiplication we would have $z^{-1}(zz_n) = z_n \to 0$, contrary to $\|z_n\| = 1$

Our next theorem relates to the manner in which Z is distributed within S.

Theorem 2:

The boundary of S is a subset of Z.

Proof:

Since S is closed, its boundary consists of all points in S which are limits of convergent sequences in G. We show that if z is such a point, that is, if z is in S and there exists a sequence $\{r_n\}$ in G such that $r_n \to z$, then z is in Z. First, we see from $r_n^{-1} z - 1 = r_n^{-1}(z - r_n)$ that the sequence $\{r_n^{-1}\}$ is unbounded; for otherwise, we would have

$$\left\|r_n^{-1} z-1\right\| < 1$$

for some n, so that $r_n^{-1}z$, and therefore $z = r_n(r_n^{-1}z)$, would be regular. Since $\{r_n^{-1}\}$ is unbounded, we may assume that $\left\|r_n^{-1}\right\| \to \infty$. If z_n is now defined by $z_n = r_n^{-1}/\left\|r_n^{-1}\right\|$, then our conclusion follows from the observations that $\|z_n\| = 1$ and

$$zz_n = \frac{zr_n^{-1}}{\left\|r_n^{-1}\right\|} = \frac{1+(z-r_n)r_n^{-1}}{\left\|r_n^{-1}\right\|} = \frac{1}{\left\|r_n^{-1}\right\|} + (z-r_n)z_n \to 0.$$

In order to understand the significance of these facts, let us suppose that A is imbedded as Banach subalgebra in a larger Banach algebra A'. As we remarked in the previous section, an element which is singular in A may cease to be so in A'. However, if it is a topological divisor of zero in A, then it is in A' as well, so it is singular in A'. The topological divisors of zero in A are thus "permanently singular," in the sense that they are singular and remain so with respect to every possible enlargement of the containing Banach algebra. Theorem B tells us that no matter what happens to S as a whole in such a process, its boundary is "permanent" in this sense.

THE SPECTRUM

Let T be an operator on a non-trivial Hilbert space. In the previous chapter, we defined the spectrum of T to be the set

$$\sigma(T) = \{\lambda : T - \lambda I \text{ is singular}\},$$

and we devoted a good deal of attention to the geometric ideas leading to this concept. We found— at least in the finite-dimensional case-that a number in σ (T) is a *value assumed* by T, in the sense that t acts on some non-zero vector as if it were scalar multiplication by that number. We shall see later that this formulation of the meaning of the spectrum has a much wider significancy than we might at first suspect.

Let us now consider an element x in our general Banach algebra A. By analogy with the above, we define the spectrum of x to be the following subset of the complex plane:

$$\sigma(x) = \{\lambda : x - \lambda l \text{ is singular}\}.$$

Wherever it is desirable to express the fact that the spectrum of x depends on A as well as x, we use the notation $\sigma_A(x)$. It is easy to see that $x - \lambda l$ is a continuous function of λ with values in A; and since the set of singular elements in A is closed, it follows at one that σ (x) is closed. We further observe that σ (x) is a subset of the closed disc $\{z:|z| \le \|x\|\}$, for it λ is a complex number such that $|\lambda| > \|x\|$, then $\|x/\lambda\| < 1$, $\|1-(1-x/\lambda)\| < 1$, $1 - x/\lambda$ is regular, and therefore $x - \lambda l$ is regular.

Our first task is to establish the fact that σ (x) is always non-empty, and for this we need a few preliminary notions. The resolvent set of x, denoted by ρ (x), is the complement of σ (x); it is clearly an open subset of the complex plane which contains $\{z:|z| > \|x\|\}$. The resolvent of x is the function with values in A defined on ρ (x); by

$$x(\lambda) = (x - \lambda l)^{-1}.$$

Tells us that $x(\lambda)$ is a continuous function of λ; and the fact that $x(\lambda) = \lambda^{-1}(x/\lambda - 1)^{-1}$ for $\lambda \neq 0$ implies that $x(\lambda) \to 0$ as $\lambda \to \infty$. If λ and μ are both in ρ (x) , then

$$\begin{aligned} x(\lambda) &= x(\lambda)\,[x - \mu l]\,x(\mu) \\ &= x(\lambda)\,[x - \lambda l + (\lambda - \mu)l]\,x(\mu) \\ &= [1 + (\lambda - \mu)\,x(\lambda)]\,x(\mu) \\ &= x(\mu) + (\lambda - \mu)x(\lambda)\,x(\mu), \end{aligned}$$

so
$$x(\lambda) - x(\mu) = (\lambda - \mu)x(\lambda)\,x(\mu)$$

This relation is called the resolvent equation.

Theorem 1:

σ (x) is non-empty.

Proof:

Let f be a functional on A—that is, an element of the conjugate space A*–and define $f(\lambda)$ by $f(\lambda) = f(x(\lambda))$. It is clear that $f(\lambda)$ is a complex function which is defined and continuous on the resolvent set ρ (x). The resolvent equation shows that

$$\frac{f(\lambda)-f(\mu)}{\lambda-\mu}=f(x(\lambda)x(\mu),$$

and it follows from this that

$$\lim_{\lambda-\mu}\frac{f(\lambda)-f(\mu)}{\lambda-\mu}=f\left(x(\mu)^2\right),$$

so f(λ) has a derivative at each point of ρ (x). Further,

$$|f(\lambda)|\leq\|f\|\,\|x(\lambda)\|,$$

so f(λ) → 0 as λ → ∞. We now assume that σ (x) is empty, so that ρ (x) is the entire complex plane. Liouville's theorem from complex analysis allows us to conclude that f(λ) = 0 for all λ. Since f is an arbitrary functional on A, x(λ) = 0 for all λ. This is impossible, for no inverse can equal 0, and therefore it cannot be true that σ (x) is empty.

If the reader is surprised by the appearance of Liouville's theorem in such a context, he should recall two facts. A which is a special case of the above result, required the use of the fundamental theorem of algebra. And second, the fundamental theorem of algebra is most commonly proved as a simple consequence of Liouville's theorem. It is therefore only to be expected that some tool from analysis comparable in depth with Liouville's theorem should be necessary for the proof of Theorem A. Now that we know that σ (x) is non-empty, we also know that it is a compact subspace of the complex plane. The number r(x) defined by

$$r(x) = \sup\ \{|\lambda| : \lambda\ \varepsilon\ \sigma\ (x)\}$$

is called the spectral radius of x. It is clear that $0 \leq r(x) \leq \|x\|$. The concept of the spectral radius will be useful in certain parts of our later work.

We recall that a division algebra is an algebra with identity in which each non-zero element is regular. The most important single consequence of Theorem A is

Theorem 2:

If A is a division algebra, then it equals the set of all scalar multiples of the identity.

Proof:

We must show that if x is an element of A, then x equals λ1 for some scalar λ. Suppose, on the contrary, that x ≠ λ1 for every λ. Then x – λ1 ≠0 for every λ, x – λ1 is regular for every λ, and therefore σ(x) is empty. This contradicts Theorem 1 and completes the proof.

The mapping $\lambda 1 \to \lambda$ is clearly an isometric isomorphism of the set of all scalar multiples of the identity onto the Banach algebra C of all complex numbers. We may therefore identify this set with C; and in terms of this identification, Says that *any Banach algebra which is a division algebra equals C.* This fact is the foundation on which we build the structure theory presented in the next.

It is obvious that C itself which is the simplest of all Banach algebras, is a division algebra, so Theorem B characterizes C as the only Banach algebra with this property. In the next two theorems, we give some other interesting characterizations of C among all possible Banach algebras.

Since 0 is a divisor of zero, it is a topological divisor of zero in every Banach algebra. In the Banach algebra C, 0 is plainly the only topological divisor of zero. Conversely, we have

Theorem 3:

If 0 is the only topological divisor of zero in A, then A = C.

Proof:

Let x be an element of A. Its spectrum $\sigma(x)$ is non-empty, so it has a boundary point λ; and $x - \lambda 1$ is easily seen to be a boundary point of the set S of all singular elements. By Theorem 2, $x - \lambda 1$ is a topological divisor of zero, so it follows from our hypothesis that $x - \lambda 1 = 0$ or $x = \lambda 1$.

The basic link between multiplication in A and the norm is given by the inequality $\|xy\| \leq \|x\|\|y\|$, and when A = C, this inequality; can be reversed. The following result shows to what extent this reversibility is true in general.

Theorem 4:

If the norm in A satisfies the inequality $\|xy\| \leq K\|x\|\|y\|$ *for some positive constant K, then A = C.*

Proof:

In the light of Theorem 3, it suffices to observe that the hypothesis here implies that 0 is the only topological divisor of zero.

We next look into the question of what happens to the spectrum of an element x in A when A is enlarged.

Theorem 5:

If A is a Banach subalgebra of a Banach algebra A', then the spectra of an element x in A with respect of A and A' are related as follows : (1) $\sigma_{A'}(x) \subseteq \sigma_A(x)$*; (2) each boundary point of* $\sigma_A(x)$ *is also a boundary point of* $\sigma_{A'}(x)$.

Proof:

If $x-\lambda 1$ is singular in A', then it is certainly singular in A, so (1) is clear. To prove (2), we let λ be a boundary point $\sigma_A(x)$. It is easy to see that $x-\lambda 1$ is a boundary point of the set of singular elements in A, so by Theorem 66 - B, it is a topological divisor of zero in A. It is therefore a topological divisor of zero in A' as well, so it is singular in A' and λ is in $\sigma_{A'}(x)$. The fact that λ is actually a boundary point of $\sigma_{A'}(x)$ is immediate from (1), so the proof of (2) is complete.

This result shows that in general the spectrum of an element shrinks when its containing Banach algebra is enlarged, and further, that since its boundary points cannot be lost in this process, it must shrink by "hollowing out." An illuminating example of this phenomenon is provided by the disc algebra A of all complex functions which the defined and continuous on D $= \{z : |z| \le 1\}$ and analytic in the interior of this set. If f is a function in A, then the maximum modulus theorem from complex analysis implies that

$$\|f\| = \sup\{|f(z)| : |z| \le 1\}$$

$$= \sup\{|f(z)| : |z| = 1\}$$

This allows us to identify A with the Banach algebra of all the restrictions of its functions to the boundary of D, which is a Banach subalgebra of $A' = \mathcal{C}(\{z : |z| = 1\})$. If we now consider the element f in A defined by f(z) = z, then it is easy to see that $\sigma_A(f)$ equals D and that $\sigma_{A'}(f)$ equals the boundary of D.

THE FORMULA FOR THE SPECTRAL RADIUS

Let x be an element in our general Banach algebra A, and consider its spectral radius r(x), which is defined by

$$r(x) = \sup\{|\lambda| : \lambda \varepsilon \sigma_A(x)\}.$$

Now let A' be the Banach subalgebra of A generated by x, that is, the closure of the set of all polynomials in x. E shows that r(x) has the same value if it is computed with respect to A':

$$r(x) = \sup\{|\lambda| : \lambda \varepsilon \sigma_{A'}(x)\}$$

This suggests quite strongly that r(x) depends only on the sequence of powers of x. The formula for r(x) is given in Theorem A below, and our purpose in this section to prove it. It is convenient to begin with the following preliminary result.

Lemma:

$$\sigma(x^n) = \sigma(x)^n.$$

Proof:

Let λ be a non-zero complex number and $\lambda_1, \lambda_2, \ldots, \lambda_n$ its distinct nth roots, so that

$$x^n - \lambda 1 = (x - \lambda_1 1)\,(x - \lambda_2 1)\ldots(x - \lambda_n 1).$$

The statement of the lemma follows easily from the fact that $x^n - \lambda 1$ is singular $\Leftrightarrow$ $x - \lambda_i 1$ is singular for at least one i.

Theorem:

$$r(x) = \lim \left\|x^n\right\|^{1/n}.$$

Proof:

Our lemma shows that $r(x^n) = r(x)^n$, and since $r(x^n) \leq \left\|x^n\right\|$, we have $r(x)^n \leq \left\|x^n\right\|$ or $r(x) \leq \left\|x^n\right\|^{1/n}$ for every n. To conclude the proof, it suffices to show that if a is any real number such that $r(x) < \propto$, then $\left\|x^n\right\|^{1/n} \leq \propto$ for all but a finite number of n's, and this we now do.

A and our work in Sec. 67 that if $|\lambda| > \|x\|$, then

$$x(\lambda) = (x - \lambda 1)^{-1} = \lambda^{-1}\left(\frac{x}{\lambda} - 1\right)^{-1}$$

$$= -\lambda^{-1}\left(1 - \frac{x}{\lambda}\right)^{-1}$$

$$= -\lambda^{-1}\left[1 + \sum_{n=1}^{\infty} \frac{x^n}{\lambda^n}\right]. \tag{1}$$

If f is any functional of A, then (1) yields

$$f(x(\lambda)) = -\lambda^{-1}\left[f(1) + \sum_{n=1}^{\infty} f\left(\frac{x^n}{\lambda^n}\right)\right]$$

$$= -\lambda^{-1}\left[f(1) + \sum_{n=1}^{\infty} f(x^n)\lambda^{-n}\right] \tag{2}$$

for all $|\lambda| > \|x\|$. We saw in the proof of Theorem 67 - A that $f(x(\lambda)$ is an analytic function in the region $|\lambda| > r(x)$; and since (2) is its Laurent expansion for $|\lambda| > \|x\|$. We know from complex analysis that this expansion is valid for $|\lambda| > r(x)$. If we now let α be any real number such that r(x) $\alpha < \alpha$, then it follows from the preceding remark that the series $\sum_{n=1}^{\infty} f(x^n/\alpha^n)$ converges, so its terms form a bounded sequence. Since this is true for every f in A*, an application of Theorem 51 - B shows that the elements x^n/α^n form a bounded sequence in A. Thus

$$\left\|\frac{x^n}{\sigma^n}\right\| \leq K$$

or $\|x^n\|^{1/n} \leq K^{1/n\alpha}$ for some positive constant K and every n. Since $K^{1/n}$ $x \leq x$ for every sufficiently large n, we have $\|x^n\|^{1/n} \leq \alpha$ for all but a finite number of n's, and the proof is complete.

The applications we make of this formula will appear in the next

THE RADICAL AND SEMI-SIMPLICITY

Our final preliminary task is to reach a clear understanding of what is meant by the statement that our Banach algebra A is semi-simple. For this, it is necessary to give an adequate definition of the radical of A, and this in turn depends on a detailed analysis of its ideals.

We recall that an ideal in A was defined in Sec. 45 to be a subset I with the following three properties :]

(1) I is a linear subspace of A;

(2) i ε I ⇒ xi ε I for every element x ε A;

(3) i ε I ⇒ ix ε I for every element x ε A.

If I is assumed only to satisfy conditions it is called a left ideal (or a right ideal). For the sake of clarity and emphasis, an ideal in our previous sense—one which satisfies all three of these conditions—is often called a two-sided ideal. In the commutative case, of course, these three concepts coincide with one another.

The properties of the ideals in A are closely related to the properties of its regular and singular elements. In our work so far, the statement that an element x in A is regular has meant that there exists an element y such that xy = yx = 1. For our present purposes, it is useful to refine this notion slightly, as follows. We say that x is left regular if there exists an element y such that yx = 1; and if x is not left regular, it is called left singular. The terms right regular and right singular are defined similarly. If x is both left regular and right regular, so that there exist elements y and z such that yx = 1 and xz = 1, then the relation

$$y = y1 = y(xz) = (yx)z = 1z = z$$

shows that x is regular in the ordinary sense and that $x^{-1} = y = z$.

The concept of maximality for two-sided ideals was introduced by analogy, we define a maximal left ideal in A to be a proper left ideal which is not properly contained in any other proper left ideal. A straightforward application of Zorn's lemma shows a that any proper left ideal can be imbedded in a maximal left ideal; and since the zero ideal {0} is a proper left ideal, maximal left ideals certainly exist. We now define the radical R of A to be the intersection of all its maximal left ideals. It will be convenient to abbreviate this definition by writing $R = \cap$ MLI. R is clearly a proper left ideal.

These ideas can be formulated just as easily for right ideals as for left ideals, and there is no reason for giving preference to either side over the other. The purpose of the following chain of lemmas is to show that R is also the intersection of all the maximal right ideals in A, that is, that $R = \cap$ MRI.

Lemma 1:

If r is an element of R, then 1 – r is left regular.

Proof:

We assume that 1 – r is left singular, so that

$$L = A\,(1 - r) = \{x - xr : x \varepsilon A\}$$

is a proper left ideal which contains 1 – r. We next imbed L in a maximal left ideal M, which of course also contains 1 – r. Since r is in R, it is also in M, and therefore 1 = (1 – r) + r is in M. This implies that M = A, which is a contradiction.

Lemma 2:

If r is element of R, then 1 – r is regular.

Proof:

By the lemma just proved, there exists an element s such that s(1 – r)=1, so s is right regular and s = 1 – (–s)r. The fact that R is a left ideal implies that (–s) r is in R along with r, and another application of the preceding lemma shows that 1 – (–s)r = s is left regular. Since s is both left regular and right regular, it is regular with inverse 1 – r, so 1 – r is also regular.

Lemma 3:

If r is an element of R, then 1 – xr is regular for every x.

Proof:

R is a left ideal, so xr is in R and the statement follows from the lemma just proved.

Lemma 4:

If r is an element of A with the property that 1 – xr is regular for every x, then r is in R.

Proof:

We assume that r is not in R, so that r is not in some maximal left ideal M. It is easy to see that the set

$$M + Ar = \{m + xr : m \in M \text{ and } x \in A\}$$

is a left ideal which contains both M and r, so M + Ar = a and

$$m + xr = 1$$

for some m and x. It now follows that 1 – xr = m is a regular element in M, and this is impossible, for no proper ideal can contain any regular element.

The effect of these lemmas is to establish the equality of two sets :

$$\cap \text{ MLI} = \{ r : 1 - xr \text{ is regular for every } x \}. \qquad ...(1)$$

Precisely the same arguments, when applied to maximal right ideals, show that

$$\cap \text{ MRI} = \{ r : 1 - xr \text{ is regular for every } x \}. \qquad ...(2)$$

We now prove that all four of these sets are the same by showing that the two sets on the right of (1) and (2) are equal to one another. By symmetry, it evidently suffices to prove the

Lemma 5:

If 1 – xr is regular, then 1 – rx is also regular.

Proof:

We assume that 1 – xr is regular with inverse

$$s = (1 - xr)^{-1}.$$

This means, of course, that (1 – xr)s = s(1 – xr) = 1. We leave it to the reader to show, by a simple computation, that

$$(1 - rx) + (1 - rsx)(1 - rx) = 1,$$

so that 1 – rx is regular, with inverse 1 + rsx. (The formula for $(1 - rx)^{-1}$ is less mysterious than it looks, as the reader can see by inspecting the meaningless but suggestive expressions.

$$s = (1 - xr)^{-1} = 1 + xr + (xr)^2 \ldots$$

and $(1 - rx)^{-1} = 1 + rx + (rx)^2 + (rx)^3 + \ldots = 1 + rx + rxrx + rxrxrx + \ldots = 1 + r(1 + xr + xrxr + \ldots) x = 1 + rsx.)$

We summarize our results in

Theorem 1:

The radical R of A equals each of the four sets in (1) and (2) and is therefore a proper two-sided ideal.

A is said to be semi-simple if its radical equals the zero ideal $\{0\}$, that is, if each non-zero element of A is outside of some maximal left ideal.

It will be observed that the ideas discussed above are purely algebraic in nature. They can be applied not only to our Banach algebra A, but also to any algebra or ring with identity. Our interest, however, is in A, and we now bring to bear upon these notions the results, notably, the fact that the set S of all singular elements in A is closed.

We begin by noting that if I is any ideal in A (left, right, or two-sided), then by the joint continuity of the algebraic operations, its closure $\bar{I}$ is an ideal of the same king. Next, since any proper ideal is contained in the proper closed set S, the closure of any proper ideal is a proper ideal of the same kind. It is an easy step from these facts to

Theorem 2:

Every maximal left ideal in A is closed.

Proof:

If any maximal left ideal L is not closed, then L is a proper subset of the proper left ideal $\bar{L}$; and this cannot happen, for it contradicts the maximality of L.

Taken together, the above two theorems yield

Theorem 3:

The radical R of A is a proper closed two-sided ideal. We shall also need

Theorem 4:

If I is a proper closed two-sided ideal in A, then the quotient algebra A/I is a Banach algebra.

Proof:

A tells us that A/I is a non-trivial complex Banach space with respect to the norm defined by

$$\|x+I\| = \inf\{\|x+i\| : i \,\varepsilon\, I\}.$$

Further, A/I is clearly an algebra with identity 1 + I, and

$$\|1+I\| = \inf\{\|1+i\| : i \,\varepsilon\, I\} \le \|1\| = 1.$$

The multiplicative inequality for the norm is easily proved as follows:

$$\|(x+I)(y+I)\| = \|xy+I\| = \inf\{\|xy+i\| : i \,\varepsilon\, I\}$$

$$\le \inf\{\|(x+i_1)(y+i_2)\| : i_1, i_2 \,\varepsilon\, I\}$$

$$\le \inf\{\|(x+i_1)\| \, \|(y+i_2)\| : i_1, i_2 \,\varepsilon\, I\}$$

$$= \left[\inf\{\|x+i_1\| : i_1 \,\varepsilon\, I\}\right]\left[\inf\{\|(y+i_2)\| : i_2 \,\varepsilon\, I\}\right]$$

$$= \|x+I\| \, \|y+I\|.$$

All that remains is to show that $\|1+I\| = 1$; and since we already have $\|1+I\| \le 1$, this is an immediate consequence of the fact that

$$\|1+I\| = \left\|(1+I)^2\right\| \le \|1+I\|^2 \text{ implies } 1 \le \|1+I\|.$$

As a final result, we state

Theorem 5:

A/R is a semi-simple Banach algebra.

Proof:

It suffices to observe that the natural homeomorphism $x \to x + R$ of A onto A/R induces a one-to-one correspondence between the maximal left ideals in A and those in A/R.

In the following chapters, we shall be concerned almost exclusively with commutative Banach algebras. An algebra of this kind is of course much easier to handle than one which is not commutative, for all its ideals are two-sided and its radical is simply the intersection of its maximal ideals. Our reason for studying the general case here is that when it becomes necessary to assume commutativity, as it will in the next section, we want the force of this assumption, and the issues that depend on it, to be quite clear.

THE STRUCTURE OF COMMUTATIVE BANACH ALGEBRAS

The set $\mathcal{C}(X)$ of all bounded continuous complex functions defined on a topological space X is the simplest of the really interesting Banach algebras. Our purpose in this chapter is to prove the famous *Gelfand-Neumark theorem,* which says that every commutative Banach algebra A of a certain type is essentially identical with $\mathscr{C}(X)$ for a suitable compact Hausdroff space X. More precisely, we shall prove that a compact Hausdorff space X can be built out of the inner structure of A, that X is accompanied by a natural mapping of A into $\mathscr{C}(X)$, and that this mapping is one-to-one onto and preserves all the structure assumed to be present in A.

THE GELFAND MAPPING

Let A be an arbitrary commutative Banach algebra. Our first theorem below is the principal source of the source of the structure theory of A, and the remainder of the chapter will be devoted entirely to shaping its consequences into the elegant form of the Gelfand-Neumark theorem.

Theorem 1:

If M is a maximal ideal in A, then the Banach algebra A/M is a division algebra, and therefore equals the Banach algebra C of complex numbers. The natural homeomorphism $x \to x + M$ of A onto A/M = C assigns to each element x in A $\propto$ complex number x(M) defined by

$$x(M) = x + M,$$

and the mapping $x \to x(M)$ has the following properties :

(1) $(x + y)(M) = x(M) + y(M);$

(2) $(\propto x)(M) = \propto x(M);$

(3) $(xy)(M) = x(M)y(M);$

(4) $x(M) = 0 \Leftrightarrow x \varepsilon M;$

(5) $1(M) = 1;$

(6) $|x(M)| \leq \|x\|.$

Proof:

That A/M is indeed a Banach algebra. Since A contains an identity, M is maximal as a ring ideal and therefore, A/M is a division algebra. We now appeal to conclude that A/M equals C. (Actually, of course, A/M equals the set of all scalar multiples of its own identity, but we identify this set with C in accordance with the remarks following. Finally, properties (1) to (5) are obvious consequences of the nature of the homeomorphism under discussion, and (6) follows from

$$|x(M)| = |x + M| = \|x + M\| = \inf\{\|x + m\| : m \varepsilon M\} \leq \|x\|.$$

If is interesting to observe that proof depends, either directly or indirectly, on virtually every major theorem in the previous chapter. We also note that the ultimate reason for assuming that A is commutative lies in which is definitely not true in the non-commutative case.

The language of Theorem 1 is oriented toward the idea that x(M) is a function of x for each fixed M. The notation, however, suggests that we reverse this point of view and that for each fixed x we regard x(M) as a complex function defined on the set $\mathcal{M}$ of all maximal ideals in A. This is the direction in which we now proceed.

If x is a given element of A, we denote by $\hat{x}$ the function defined on $\mathcal{M}$ by $\hat{x}(M) = x(M)$, and we put $\hat{A} = \{\hat{x} : x \varepsilon A\}$. Our next step is to defined a topology for $\mathcal{M}$ in such a manner that every function in $\hat{A}$ is continuous. The most natural way of doing this is to introduce the weak topology generated by $\hat{A}$. It will be recalled that this is the weakest topology on $\mathcal{M}$ relative to with every function $\hat{x}$ is continuous and that a typical subbasic open set has the form

$$S(\hat{x}, M_0, \in) = \{M : M \varepsilon \mathcal{M} \text{ and } |\hat{x}(M) - \hat{x}(M_0)| < \in\}.$$

We call the topological space $\mathcal{M}$ the space of maximal ideals, or the maximal ideal space, and the mapping $x \to \hat{x}$ of A onto $\hat{A}$ will be referred to as the Gelfand mapping.

We are now in a position to reformulate to extend it, in such a way that the Galfand mapping is displayed as the object of central importance.

Theorem 2:

The Gelfand mapping $x \to \hat{x}$ is a norm-decreasing (and therefore continuous) homeomorphism of A into $\mathcal{C}(\mathcal{M})$ with the following properties:

(1) the image $\hat{A}$ of A is a subalgebra of $\mathcal{C}(\mathcal{M})$ which separates the points of @ and contains the identity of $\mathcal{C}(\mathcal{M})$;

(2) the radical R of A equals the sat of all elements x for which $\hat{x}$ = 0, so $x \to \hat{x}$ is an isomorphism $\Leftrightarrow$ A is semi-simple;

(3) an element x in A is regular $\Leftrightarrow$ it does not belong to any maximal ideal $\Leftrightarrow$ $\hat{x}(M) \neq 0$ for every M;

(4) if x is an element of A, then its spectrum equals the range of the function $\hat{x}$ and its spectral radius equals the norm of $\hat{x}$, that is, $\sigma(x) = \hat{x}(\mathcal{M})$ and $r(x) = \sup |\hat{x}(M)| = \|\hat{x}\|$.

Proof:

The definition of the topology on $\mathcal{M}$ guarantees that each function $\hat{x}$ is continuous, and part theorem 1 shows that $\hat{x}$ is bounded and that

$\|\hat{x}\| = \sup|\hat{x}(M)| \le \|x\|$, so $x \to \hat{x}$ is a norm-decreasing mapping of A into $\mathcal{C}(\mathcal{M})$. The fact that this mapping is a homeomorphism is immediate from parts (1), (2), and (3) of Theorem 1.

Since $x \to \hat{x}$ is a homeomorphism, $\hat{A}$ is obviously a subalgebra of $\mathcal{C}(\mathcal{M})$. The stated properties of $\hat{A}$ follow readily from parts (4) and (5) of Theorem 1: if $M_1 \neq M_2$, and if (say) x is in M_1 but not in M_2, then $\hat{x}(M_1) = 0$ and $\hat{x}(M_2) \neq 0$; and $\hat{1}(M) = 1$ for every M.

If we recall that R is the intersection of all the M's, then the proof of (2) is easy : we have only to notice that part (4) of Theorem A tells us that $\hat{x}(M) = 0$ for every M $\Leftrightarrow$ x is in every M.

To prove (3), it suffices - in view of part (4) of Theorem A - to show that x is regular $\Leftrightarrow$ it does not belong to any M. It is elementary that a regular element cannot lie in any proper ideal, so we confine our attention to showing that if x is singular, then it does belong to some M. We prove this by observing that the singularity of x implies that $Ax = \{yx : y \,\varepsilon\, A\}$ is a proper ideal which contains x and can therefore be imbedded be in a maximal ideal M which also contains x.

Finally, we use (3) to prove (4). By the definition of the spectrum of x, we have $\lambda \,\varepsilon \sigma(x) \Leftrightarrow x - \lambda 1$ is singular $\Leftrightarrow (x - \lambda 1)(M) = 0$ for at least one M $\Leftrightarrow (\hat{x} - \lambda 1 (M) = 0$ for at least one M $\Leftrightarrow \hat{x}(M) = \lambda$ for at least one M, so $\sigma(x)$ equals the range of $\hat{x}$ .The rest of (4) follows from this statement and the definition of the spectral radius.

We add the final touch to this portion of the theory by showing that @@ is a compact Hausdorff space. The reader will recall that if A* is the conjugate space of A, then its closed unit sphere

$$S^* = \{f : f \,\varepsilon\, A^* \text{ and } \|f\| \le 1\}$$

is a compact Hausdorff space in the weak* topology.

Our strategy is to identify $\mathcal{M}$, both as a set and as a topological space, with a closed subspace of S*.

A *multiplicative functional* on A is a functional f in the ordinary sense-that is, an element of the conjugate space A*-which is non-zero and satisfies the additional condition f(xy) = f(x)f(y). Theorem A shows that to each M in $\mathcal{M}$ there corresponds a multiplicative functional f_M defined by $f_M(x) = x(M)$. It is important for us to know that $M \to f_M$ is a one-to-one mapping of $\mathcal{M}$ onto the set of all multiplicative functional. It will facilitate our work if we begin by proving the

Lemma:

If f_1 and f_2 are multiplicative functionals on A with the same null space M, then $f_1 = f_2$.

Proof:

We first show that $f_1 = \alpha f_2$ for some scalar α. Let x_0 be an element of A which is not in M. If x is an arbitrary element of A, it is easy to see that x can be expressed uniquely in form $x = m + \beta x_0$ with m in M (set $\beta = f_2(x)/f_2(x_0)$, put $m = x - \beta x_0$, and observe that $f_2(m) = 0$). It now follows that

$$f_1(x) = f_1(m) + \beta f_1(x_0) = \beta f_1(x_0) = [f_1(x_0)/f_2(x_0)]\ f_2(x),$$

So $f_1 = \alpha f_2$ with $\alpha = f_1(x_0)/f_2(x_0)$. We complete the proof by showing that α equals 1. Let x be an element not in M, so that $f_2(x) \neq 0$. Then $\alpha f_1(x)^2 = \alpha f_2(x^2) = f_1(x^2) = f_1(x^2) = [\alpha^2 f_2(x^2) = \alpha^2 f_2(x^2)$ implies that

$$\alpha^2 = \alpha$$

so $\alpha = 0$ or $\alpha = 1$. Since $f_1 \neq 0$, we conclude that $\alpha = 1$.

We now use this to prove

Theorem 3:

$M \to f_M$ is a one-to-one mapping of the set $\mathcal{M}$ of all maximal ideals in A onto the set of all its multiplicative functionals.

Proof:

The mapping is easily seen to be one-to-one, for if $M_1 \neq M_2$, and if (say) x is in M_1 and not in M_2, then $f_{M1}(x) = 0$ and $f_{M2}(x) \neq 0$. To prove that it is onto, let f be an arbitrary multiplicative functional, and consider its null space $M = \{x : f(x) = 0\}$. It is clear by the assumed properties of that M is a proper closed ideal in A. Furthermore, M is maximal, for if it were properly contained in a proper ideal I, then f(I) would be a non-trivial ideal in C, contrary. Since f and f_M are multiplicative functionals with the same null space, the lemma just proved implies that $f = f_M$, and our proof is complete.

In some of its more concrete applications, this theorem is used to replace the algebraic problem of determining the maximal ideals in A by the analytic problem of finding its multiplicative functionals. Its importance for our current task of showing that $\mathcal{M}$ is a compact Hausdorff space is that it enables us to regard $\mathcal{M}$ as a subset of A*. We can say even more than this, for parts (5) and (6) of Theorem 1 tell us that every multiplicative functional f_M has norm 1, so $\mathcal{M}$ is a subset of the closed unit sphere S*. We recalled earlier that S* is a compact Hausdorff space with respect to the weak* topology, which is the weak topology generated by all the functions f_x defined on S* by $F_x(f) = f(x)$. We now observe that when F_x is restricted to $\mathcal{M}$, it is precisely $\hat{x}$, for

$$F_x(f_M) = f_M(x) = x(M) = \hat{x}(M).$$

Therefore, the topology which $\mathcal{M}$ has as a subspace of S* is exactly its topology as the space of maximal ideals. These considerations permit us to regard $\mathcal{M}$ as a subspace of S*.

Theorem 4:

The maximal ideal space $\mathcal{M}$ is a compact Hausdorff space.

Proof:

In view of the above discussion, it suffices to show that $\mathcal{M}$ is a closed subspace of S*. We accomplish this by forming the subspace X of S* defined by

$$X = \cap_{x,y \varepsilon A} \{f : f \varepsilon S^* \text{ and } f(xy) = f(x)\, f(y)\}.$$

It is evident that X is simply Q@ together with the zero functional; and since we have

$$\begin{aligned} X &= \cap_{x,y \varepsilon A} \{f : f \varepsilon S^* \text{ and } f(xy) - f(x)\, f(y) = 0\} \\ &= \cap_{x,y \varepsilon A} \{f : f \varepsilon S^* \text{ and } F_{xy}(f) - F_x(f)\, F_y(f) = 0\} \\ &= \cap_{x,y \varepsilon A} \{f : f \varepsilon S^* \text{ and } (F_{xy} - F_x F_y)\,(f) = 0\}, \end{aligned}$$

it is easy to see that X is closed in S* (note that each of the sets last written has this property). We next remark that F_1 is continuous on X and equals 1 on $\mathcal{M}$ and 0 at the zero functional. It follows from this that $\mathcal{M}$ is closed in X and is therefore closed in S*.

It is worthy of notice that the topology we imposed on $\mathcal{M}$ is the only one which makes it into a compact Hausdorff space on which all the functions $\hat{x}$ are continuous, for any stronger compact Hausdorff must equal the given one.

When Theorems 2 and 4 are taken together, the result is often called the Gelfand representation theorem. In essence, this tells us that every commutative semi-simple Banach algebra is isomorphic to an algebra of continuous complex functions on a suitable compact Hausdorff space.

In general, the norm is not preserved by this isomorphism and the representing algebra does not exhaust the continuous functions on the underlying space. We shall remove these deficiencies in the following sections by assuming that additional structure is present in the Banach algebra under discussion.

APPLICATIONS OF THE FORMULA R(X) = LIM $\|x^n\|^{1/n}$

We continue our study of an arbitrary commutative Banach algebra A and of the Gelfand mapping X → $\hat{x}$ of A onto the subalgebra $\hat{A}$ of $\mathcal{C}(\mathcal{M})$. Our first theorem provides a simple way of guaranteeing that this mapping preserves norms.

Theorem 1:

The following conditions on A are all equivalent to one another:

(1) $\|x^2\| = \|x\|^2$ for every x ;

(2) r(x) = $\|x\|$ for every x ;

(3) $\|\hat{x}\|$ $\|x\|$ for every x .

Proof:

It follows from condition (1) that

$$\|x^4\| = \|(x^2)^2\| = \|x^2\|^2 = \|x\|^4$$

and, in general, that $\|x^{2k}\| = \|x\|^{2k}$ for every positive integer k. The formula for the spectral radius now yields

$$r(x) = \lim \|x^n\|^{1/n} = \lim \|x^{2k}\|^{1/2k} = \lim \|x\| = \|x\|,$$

so (1) implies (2). The fact that (2) implies (1) is immediate from

$\|x^2\|$ = r(x²) = r(x)² = $\|x\|^2$. In view of the equation r(x) = $\|\hat{x}\|$

Our next problem is to devise a way of making sure that the representing algebra $\hat{A}$ comes as close as it can to exhausting $\mathcal{C}(\mathcal{M})$, and we accomplish this by introducing the following property. A is said to be self-adjoint if for each x in A there exists an element y in A such that $\hat{y}(M) = \overline{\hat{x}(M)}$ for every M.

Theorem 2:

If A is self-adjoint, then $\hat{A}$ is dense in $\mathcal{C}(\mathcal{M})$.

Proof:

By part (1) of T know that $\hat{A}$ is a subalgebra of $\mathcal{C}(\mathcal{M})$ which separates the points of $\mathcal{C}$ and contains the identity function. Our hypothesis now tell us that the closure of $\hat{A}$ is a closed subalgebra of $\mathcal{C}(\mathcal{M})$ which separates points, contains the identity function, and contains the conjugate of each of its functions. Theorem 2 (the complex Stone-Weierstrass theorem) shows that this closure equals $\mathcal{C}(\mathcal{M})$. so $\hat{A}$ itself is dense in $\mathcal{C}(\mathcal{M})$.

If we put together the results obtained in the above two theorems, we have,

Theorem 3:

If A is self-adjoint, and if $\|x^2\| = \|x\|^2$ *for every x, then the Gelfand mapping* $x \to \hat{x}$ *is an isometric isomorphism of A onto* $\mathcal{C}(\mathcal{M})$.

Proof:

By Theorem A, the mapping $x \to \hat{x}$ preserves norms. It is therefore an isometric isomorphism of A onto $\hat{A}$, and we see from this that $\hat{A}$ is closed in $\mathcal{C}(\mathcal{M})$. Since $\hat{A}$ is dense in $\mathcal{C}(\mathcal{M})$ by Theorem B, it follows that $\hat{A}$ equals $\mathcal{C}(\mathcal{M})$, and the proof is complete.

This theorem lacks a certain simplicity which it ought to have, for the condition of self-adjointness is rather far removed from the intrinsic structure of A. Our work in the next two sections will remedy this defect and at the same time will establish closer connections with the operator algebras to which we apply our final result.

INVOLUTIONS IN BANACH ALGEBARS

A Banach algebra A is called a Banach *-algebra if it has an involution, that is, if there exists a mapping $x \to x^*$ of A into itself with the following properties :

(1) $(x + y)^* = x^* + y^*$;

(2) $(\alpha x)^* = \bar{\alpha} x^*$;

(3) $(xy)^* = y^*x^*$;

(4) $x^{**} = x$.

It is an easy consequence of (4) that the involution $x \to x^*$ is actually a one-to-one mapping of A onto itself. We also note that $0^* = 0$ and $1^* = 1$, as we see from $0 + x^* = x^*$ $(0 + x)^* = 0^* + x^*$ and $1^* = 11^* = 1^{**}1^* = (11^*)^* = (1^*)^* = 1^{**} = 1$. The element x^* is called the adjoint of x, and a subalgebra of A is said to be self-adjoint if it contains the adjoint of each of its elements. If A' is also a Banach *-algebra, and if f is an isomorphism of A onto A', then f is called a *-isomorphism if it preserves the involution in the sense that $f(x^*) = f(x)^*$

We naturally want the involution in a Banach *-algebra to be linked in some useful way to the norm. The property $\|x^*\| = \|x\|$ clearly implies that the involution is continuous; for if $x_n \to x$, then

$$\|x_n{}^* - x^*\| = \|(x_n - x)^*\| = \|x_n - x\|$$

Shows that $x_n{}^* \to x^*$. A much stronger relation between the involution and the norm is given by the condition

$$\|x^* x\| = \|x\|^2$$

and any Banach *-algebra which satisfies it is called a B*-algebra. It is easy to see that we have $\|x^*\| = \|x\|$ in every B*-algebra; for

$$\|x\|^2 = \|x^* x\| \le \|x^*\|\,\|x\|$$

shows that $\|x\| \le \|x^*\|$ for every x, so $\|x^*\| \le \|x^{**}\| = \|x\|$, and therefore $\|x^*\| = \|x\|$. It follows from this that relations $\|x^* x\| = \|x^*\|\,\|x\|$ is also true.

Several of the Banach algebras described in ask Banach *-algebras with respect to natural involutions. If X is any topological space, then $\mathscr{C}(X)$ is clearly a commutative B*-algebra relative to the involution defined by f*(x) $= \overline{f(x)}$. The disc algebra, however, is not, for it f is the function defined by f(z) = z, then f* (z) = $\bar{z}$ is not analytic at any point . If H is a non-trivial Hilbert space, then $\mathscr{B}(H)$ is a B*-algebra with ;the adjoint operation T → T* taken as the involution. Since C*-algebras are the self-adjoint Banach subalgebras of $\mathscr{B}(H)$'s, they too are B*-algebras. Finally, the group algebra $L_1(G)$ of a finite group G is a Banach*-algebra with respect to the involution defined by $f^*(g_i) = f\left(g_i{}^{-1}\right)$, and it is easy to see that $\|f^*\| = \|f\|$.

It should be reasonably clear that Banach*-algebras (and especially Banach*-algebras) are modeled along lines suggested by $\mathscr{B}(H)$. We have already called the element x* in such an algebra the adjoint of x. By analogy, we say that x is self-adjoint if x = x*, normal if xx* = x*x, and a projection if x = x* and $x^2 = x$.

Theorem:

If x is a normal element in a Banach-algebra, then* $\|x^2\| = \|x\|^2$.

Proof:

It is obvious that $\|x^2\| \le \|x\|^2$. The inequality in the other direction is a consequence of the following computation :

$$\|x^*\|^2\|x\|^2 = (\|x^*\|\,\|x\|)^2 = \|x^*x\|^2 = \|(x^*x)^*x^*x\| = \|x^*xx^*x\|$$
$$= \|x^*x^*xx\| = \|(x^*)^2x^2\| = \|(x^2)^*x^2\| = \|(x^2)^*\|\,\|x^2\|$$
$$= \|(x^2)^*\|\ \|x^2\| \le \|x^*\|^2\ \|x^2\|.$$

This result suggests more strongly than ever that there are close connections between B*-algebras and algebras of operators on Hilbert spaces. We describe the true state of affairs in this matter at the end of the next section.

THE GEFLAND-NEUMARK THEOREM

We are now in a position to give Theorem 3 its final form.

Theorem 1:

If A is a B-algebra, then the Gelfand mapping $x \to \hat{x}$ is an isometric *-isomorphism of A onto the commutative B*-algebra $\mathcal{C}(\mathcal{M})$.*

Proof:

Since A is commutative, each of its elements is normal, and it follows that $\|x^2\| = \|x\|^2$ for every x. By now suffices to show that $\hat{x}^*(M) = \overline{\hat{x}(M)}$ for each x in A and M in $\mathcal{M}$.

Our first step is to prove that if x is self-adjoint, then $\hat{x}(M)$ is real for every M. We assume the contrary, namely, that there exists an M such that $\hat{x}(M) = \sigma + i\beta$ with $\beta \neq 0$. Since x ;is self-adjoint,

$$y = (x - \sigma 1)/\beta$$

is also self-adjoint. We further note that $\hat{y}(M) = i$, so $y - i1$ is in M. It is obvious from the properties of the involution in A that

$$M^* = \{m^* : m \varepsilon M\}$$

is a maximal ideal; and since it contains $(y - i1)^* = y + i1$, we see that $\hat{y}(M^*) = -i$. If K is any positive number, then

$$(y - iK1)\,(M^*) = -i(1 + K)$$

and $(y + iK1)\,(M) = -i(1 + K)$. It follows from this that $1 + K \le \|y - iK1\| \le \|y - iK1\|$ and, similarly, that $1 + K \le \|y + iK1\|$. On multiplying these two inequalities, we obtain

$$(1 + K)^2 \le \|y - iK1\|\ \|y + iK1\| = \|(y + iK1)^*\|\,\|y + iK1\|$$

$$= \left\|(y+iK1)^*(y+iK1)\right\| = \left\|(y-iK1)(y+iK1)\right\| = \| \ \|\|y+iK1\|$$

$$= \left\|y^2 + k^2 1\right\| \le \left\|y^2\right\| + K^2$$

so $1 + 2K \le \left\|y^2\right\|$. Since K is arbitrary, this is impossible, and this portion of our proof is complete.

We now conclude the proof by showing that if x is any element of A, then $\hat{x}^*(M) = \overline{\hat{x}(M)}$ for every M. It is clear that y = (x + x*)/2 and z = (x – x*)/(2i) are self-adjoint, and that x = y + iz; and therefore, by the result of the above paragraph, we have

$$\hat{x}^*(M) = (y - iz)(M) = \hat{y}(M) - i\hat{z}(M) = \overline{\hat{y}(M)} - \overline{i\hat{z}(M)}$$

$$= \overline{\hat{y}(M) + i\hat{z}(M)} = \overline{\hat{x}(M)}.$$

We already know that if X is any compact Hausdorff space, then $\mathscr{C}$(X) is a commutative B*-algebra. The theorem just proved—it is called the *Gelfand-Neumark representation theorem*—tells us that commutative B*-algebras are simply abstract $\mathscr{C}$(X)'s, in the sense that every such algebra is abstractly identical with $\mathscr{C}$(X) for a suitable compact Hausdorff space X.

There is another Gelfand-Neumark theorem of great interest, which applies to arbitrary B*-algebras. We observed in the previous section that every C*-algebra. The converse of this is also true, for if A is a B*-algebra, then exists a Hilbert space H with the property that A is isometrically *-isomorphic to a C*-algebra of operators on H. General B*-algebras are therefore abstract C*-algebras. The proof of this theorem evidently requires that a suitable Hilbert space be constructed out of the given structure of A. The details of this construction are beyond the scope of this book, and we content ourselves with merely stating the facts.

SOME SPECIAL COMMUTATIVE BANACH ALGEBRAS

Our discussion foreshadowed a generalized form of the spectral theorem, and the principal purpose of the present chapter is to formulate and prove this result. We begin with some additional material relating to Banach algebras of continuous functions. In particular, we keep the promise made in showing that the Stone-Cech compactifcation of a completely regular space is essentially unique.

IDEALS IN $\mathscr{C}$(X) AND THE BANACH-STONE THEOREM

Let X be a compact Hausdorff space, and consider the commutative B*-algebra $\mathscr{C}$(X). If $\mathscr{M}$ is the space of maximal ideals in $\mathscr{C}$(X), then the developments of the previous chapter lead us to expect that $\mathscr{M}$ can be identified with X and that the Gelfand mapping is the identity mapping of $\mathcal{C}(X)$ onto itself.

In order to substantiate this conjecture, we begin by observing that to each point x in X there corresponds a proper ideal M_x in $\mathcal{C}(X)$, defined by

$$M_x = \{f : f \varepsilon \mathcal{C}(X) \text{ and } f(x) = 0\}.$$

M_x is easily seen to be maximal and is thus an element of $\mathcal{M}$, for it is the null space of the multiplicative functional f_{Mx} defined by $f_{Mx}(f) = f(x)$. which assigns to each function in $\mathcal{C}(X)$ its value at x. Since X is compact Hausdorff, and therefore normal, Urysohn's lemma tells us that for each point $y \neq x$ there exists a function f in $\mathcal{C}(X)$ such that $f(x) = 0$ and $f(y) \neq 0$. This shows that $x \to M_x$ is a one-to-one mapping of X into $\mathcal{M}$. Our next step is to prove that this mapping is onto, and for this it clearly suffices to show that if M is any maximal ideal in $\mathcal{C}(X)$, then there exists a point in X at which every function in M vanishes. We assume the contrary, namely, that for each point x in X there exists a function f in M such that $f(x) \neq 0$. Since f is continuous, x has a neighbourhood at on point of which f vanishes. We now vary x to obtain an open cover for X, and we used compactness to infer that this open cover has a finite subcover. Let $f_1, f_2, ..., f_n$ be the corresponding functions in M. M is an ideal, so the function $g = \Sigma_{i=1}^{n} f_i\bar{f}_i = \Sigma_{i=1}^{n} |f_i|^2$ is also in M; and by the manner of its construction, it clearly has the property that $g(x) > 0$ for every x. It follows that g is a regular element of $\mathcal{C}(X)$, and this contradicts the fact that it lies in ;the proper ideal M. We therefore conclude that $x \to M_x$ is a one-to-one mapping of X onto $\mathcal{M}$.

These considerations enable us to identify the set $\mathcal{M}$ with the set X, and in terms of this identification, we regard X and $\mathcal{M}$ as two possibly different compact Hausdorff spaces built on the same underlying set of points. By our work in the previous chapter, we know that the Gelfand mapping $f \to \hat{f}$ is a one-to-one mapping of $\mathcal{C}(\mathcal{M})$. If we use the notation established there, then we find that

$$\hat{f}(M_x) = f(M_x) = fM_x(f) = f(x),$$

so $\hat{f} = f$ and $\mathcal{C}(\mathcal{M}) = \mathcal{C}(X)$. We now recall that any compacted Hausdorff topology on a non-empty set is uniquely determined as the weak topology generated by the set of all its continuous complex functions. If follows from this that X and $\mathcal{M}$ are equal as topological spaces.

We summarize the results of this discussion in

Theorem 1:

Let X be a compact Hausdorff space and $\mathcal{M}$ the space of maximal ideals *in the commutative B*-algebra $\mathcal{C}(X)$. Then to each point x in X there corresponds a maximal ideal M_x defined by*

$$M_x = \{f : f \varepsilon \mathcal{C}(X) \text{ and } f(x) = 0\},$$

and $x \to M_x$ is a one-to-one mapping of X onto $\mathcal{M}$ If this mapping is used to identify $\mathcal{M}$ with X, then $\mathcal{M}$ and X are equal as topological spaces, $\mathcal{C}(X)$ equals $\mathcal{C}(X)$, and the Gelfand mapping $f \to \hat{f}$ is the identify mapping of $\mathcal{C}(X)$ onto itself.

The main idea of this theorem is that the maximal ideals in @(X) correspond in a natural way to the points of X. Our next step is to extend this idea and to obtain a similar characterization of the proper closed ideals in $\mathcal{C}(X)$.

We again consider a compact Hausdorff space X, and we begin our discussion with the observation that to each non-empty closed subset F of X there corresponds a proper closed ideal I(F) in $\mathcal{C}(X)$, defined by

$$I(F) = \{f : f \varepsilon \mathcal{C}(X) \text{ and } f(F) = 0\}.$$

If x is any point not in F, then it follows from the complete regularity of X that there exists a function f in $\mathcal{C}(X)$ such that $f(x) \neq 0$ and $f(F) = 0$. This shows that $F \to I(F)$ is a one-to-one mapping of the class of all non-empty closed subsets of X into the set of all proper closed ideals in $\mathcal{C}(X)$. We shall prove that this mapping is onto, that is, that every proper closed ideal in $\mathcal{C}(X)$ arises in this way from some F.

Let I be a proper closed ideal in $\mathcal{C}(X)$. WE may assume that I is not the zero ideal, for this ideal clearly arises from the full space X. We define F by

$$F = \{x : f(x) = 0 \text{ for every } f \varepsilon I\}.$$

It is easy to see that F is a proper closed subset of X; and since I is contained in some maximal ideal, it follows that F is non-empty. Our task is to prove that I(F) = I, and since it is obvious that $I \subseteq I(F)$, the real problem is to prove that $I(F) \subseteq I$. If f is any function which vanishes on F, we must show that f lies in I. We may evidently assume that $f \neq 0$, so that $\{x : f(x) = 0\}$ is a proper subset of X.

In the first part of our proof, we assume that f vanishes on some open set G which contains F. Since $f \neq 0$, G' is non-empty and is thus a compact subspace of X. For each point x in G', there exists a function g in I such that $g(x) \neq 0$. The technique used in the proof of Theorem A can now be applied again, to yield a finite number of functions $g_1, g_2, ..., g_n$ in I with the property that at least one is non-zero at every point of G'. We next define a function g_0 by $g_0^* = \Sigma_{i=1}^n g_i\bar{g}_i = \Sigma_{i=1}^n |g_i|^2$, and we observe that g_0 is in I and that $g_0(x) > 0$ for every x in G'. By the Tietze extension theorem, the function whose values on G' are given by $1/g_0(x)$ can be extended to a function h in $\mathcal{C}(X)$. It is easily seen that g_0h is in I, that it equals 1 on G', and that $f = fg_0h$, so f is in I.

We now turn to the general case. For each $\in > 0$, the sets K and L defined by $K = \{x : |f(x)| \leq \in/2\}$ and $L = \{x : |f(x)| \geq \in\}$ are disjoint closed subsets of X. K is clearly non-empty, and since $f \neq 0$, L is also non-empty for every sufficiently small $\in$. We assume that $\in$ has been chosen at least this small, so that K and L constitute a disjoint pair of closed subspaces of X. By Urysohn's lemma, there exists a function h in $\mathcal{C}(X)$ such that $g(K) = 0$, $g(L) = 1$, $0 \leq g(x) \leq 1$ for every x. We now define a function h in $\mathcal{C}(X)$ by $h = fg$, and we note that

$$\|f - h\| = \|f(1-g)\| \leq \in.$$

It is evident that h vanishes on the set $G = \{x : |f(x)| < \in/3\}$; and since G is an open set which contains F, it follows from the preceding paragraph that h is in I. This shows that for every sufficiently small positive number $\in$ there exists a function h in I such that $\|f - h\| \leq \in$, and since I is closed, we conclude that f is in I.

We give the following formal statement of our result.

Theorem 2:

Let X be a compact Hausdorff space. Then to each non-empty closed set F in X there corresponds a proper closed ideal I(F) in $\mathcal{C}(X)$, defined by $I(F) = \{f : f \in \mathcal{C}(X) \text{ and } f(F) = 0\}$; and further, $F \to I(F)$ is a one-to-one mapping of the class of all non-empty closed subsets of X onto the set of all proper closed ideals in $\mathcal{C}(X)$.

As an easy consequence of this, we have

Theorem 3:

If X is a compact Hausdorff space, then every closed ideal in $\mathcal{C}(X)$ is the intersection of the maximal ideals which contain it.

Proof:

Since the intersection of the empty set of maximal ideals is $\mathcal{C}(X)$ itself, we may confine our attention to a proper closed ideal I. By Theorem B, I = I(F) for some non-empty closed set F. It is clear that the maximal ideals which contain I are precisely those associated with the points of F. It therefore suffices to observe that a function in $\mathcal{C}(X)$ vanishes on F $\Leftrightarrow$ it vanishes at each point of F.

We have seen in Theorem A that the points and the topology of a compact Hausdorff space X can be recovered from the maximal ideals in $\mathcal{C}(X)$. Since the maximal ideals in $\mathcal{C}(X)$ are objects of a purely algebraic nature, it follows that the compact Hausdorff space X is fully determined, both as a set and as a topological space, by the algebraic structure of $\mathcal{C}(X)$. These observations lead us directly to

Theorem 4:

(the Banach-stone Theorem). Two compact Hausdorff spaces X and Y are homeomorphic $\Leftrightarrow$ their corresponding function algebras $\mathcal{C}(X)$ and $\mathcal{C}(Y)$ are isomorphic.

THE STONE-CECH COMPACTIFICATION (CONTINUED)

It is natural to wonder what can be said along the lines in the case of a topological space X which is not necessarily compact Hausdorff. Regardless of the properties of X, we know from our previous work that $\mathcal{C}(X)$ is a commutative B*-algebra, that its maximal ideal space $\mathcal{M}$ is a compact Hausdorff space, and that $x \to M_x$ is a mapping of X into $\mathcal{M}$. Our difficulty is that without restrictions of some kind on X, we know practically nothing about the properties of the mapping $x \to M_x$. If it happens that this mapping is one-to-one and is also a homeomorphism of X onto a subspace of $\mathcal{M}$, then we observe that X is necessarily completely regular. It is therefore reasonable to assume at the outset that X is completely regular, and we shall see that several interesting conclusions follow this hypothesis.

Theorem 1:

Let X be a completely regular space and $\mathcal{M}$ the space of maximal ideals in the commutative B-algebra $\mathcal{C}(X)$. Then the mapping $x \to M_x$ is a homeomorphism of X onto a subspace of $\mathcal{M}$. Furthermore, if this mapping is used to identify X with its image in $\mathcal{M}$, then (1) X is a dense subspace of $\mathcal{M}$; (2) each function in $\mathcal{C}(X)$ has a unique extension to a function in $\mathcal{C}(\mathcal{M})$; and (3) if Y is a compact Hausdorff space with the properties of $\mathcal{M}$ stated in (1) and (2), then there exists a homeomorphism of $\mathcal{M}$ onto Y which leaves the points of X fixed.*

Proof:

The fact that $x \to M_x$ is one-to-one is immediate from the complete regularity of X, so we may identify X as a set with its image in $\mathcal{M}$. The subset X of $\mathcal{M}$ has two topologies : its own, and its relative topology as a subspace of $\mathcal{M}$. The following arguments show that these topologies are equal. We know that the Gelfand mapping $f \to \hat{f}$ is an isomorphism of $\mathcal{M}(X)$

onto $\mathcal{C}(\mathcal{M})$. Also, just as in the proof we have $\hat{f}(x) = f(x)$ for each f in $\mathcal{C}(X)$ and each x in X. These observations imply that $\mathcal{C}(X)$ is precisely the set of all restrictions to X of functions in $\mathcal{C}(\mathcal{M})$; and since both topologies are completely regular, it follows the weak topology generated by $\mathcal{C}(X)$, so they are equal and X can be regarded as a subspace of $\mathcal{M}$. These observations also show that X is dense in $\mathcal{M}$—for if $\hat{f}$ vanishes on X, then f = 0, $\hat{f} = 0$, and $\hat{f}$ also vanishes on $\mathcal{M}$—and that each function f in $\mathcal{C}(X)$ has a unique extension $\hat{f}$ in $\mathcal{C}(\mathcal{M})$. All that remains is to prove (3). We know that $\hat{f} \rightarrow f$ is an isomorphism of $\mathcal{C}(\mathcal{M})$ onto $\mathcal{C}(X)$; and by the assumptions about Ym, the mapping $f \rightarrow f'$, which assigns to each f in $\mathcal{C}(X)$ its extension f' in $\mathcal{C}(Y)$, is an isomorphism of $\mathcal{C}(X)$ onto $\mathcal{C}(Y)$. Thus $\hat{f} \rightarrow f \rightarrow f'$ is an isomorphism of $\mathcal{C}(\mathcal{M})$ onto $\mathcal{C}(Y)$. If x is a point of X, then this isomorphism clearly carries the maximal ideal in $\mathcal{C}(\mathcal{M})$ corresponding to x over to the maximal ideal in $\mathcal{C}(Y)$ corresponding to x ; so by the Banach-Stone theorem, it induces a homeomorphism of $\mathcal{M}$ onto Y which leaves the points of X fixed.

On comparing this result with $\mathcal{M}$ is homeomorphic, in the manner described, to the Stone-Cech compactification β(X). In this sense, therefore, $\mathcal{M}$ and β(X) can be considered equal to one another, and also to any other compact Hausdodrff space which contains X axs a dense subspace and has the required extension property. In effect, we have shown that the Stone-Cech compactification of a completely regular space X is unique and can equally well be regarded as the maximal ideal of $\mathcal{C}(X)$.

COMMUTATIVE C*-ALGEBRAS

In this final section, we apply the result of the preceding two chapters to the theory of operators on a non-trivial Hilbert space H. We know that $\mathcal{B}(H)$ and all its self-adjoint Banach subalgebras (that is, all C*-algebras of operators on H) are B*-algebras. As a special case of the Gelfand-Neumark theorem we therefore have

Theorem 1:

Let A be a commutative C-algebra of operators on H, and @@ it space of maximal ideals. Then the Gelfand mapping $T \rightarrow \hat{T}$ is an isometric *-isomorphism of A onto $\mathcal{C}(\mathcal{M})$.*

If $\{T_i\}$ is a non-empty set of operators on H, then the smallest Banach subalgebra of $\mathcal{B}(H)$ which contains every T_i is called the Banach subalgebra of $\mathcal{B}(H)$ generated by the T_i's. It is easy to see that this Banach subalgebra of $\mathcal{B}(H)$ is the closure of the set of all polynomials in the T_i's. If N is a normal operator on H, then the Banach subalgebra of $\mathcal{B}(H)$ generated by N and N* is clearly a commutative C*-algebra, and is called the *commutative C*-algebra generated by N.* We now specialize.

Theorem B:

Let N be a normal operator on H, and A the commutative C-algebra generated by N. If $\mathcal{M}$ is the space of maximal ideals in A, then the Gelfand mapping $T \to \hat{T}$ is an isometric *-isomorphism of A onto $\mathcal{C}(\mathcal{M})$.*

As it stands, this result is only a beginning. In order to exploit it effectively, our first task is to show that the spectrum of an operator in A– which is understood to be its spectrum as an element of $\mathcal{C}$(H)—equals its spectrum as an element of A. In proving this, we shall need the following preliminary fact.

Lemma:

Let X be a compact Hausdorff space and A a Banach subalgebra of $\mathcal{C}(X)$. If f is a real function in A which is regular in $\mathcal{C}(X)$, then it is also regular in A.

Proof:

The range of f is clearly a compact subspace of the real line which does not contain 0. If $\epsilon > 0$ is given, then by the Weierstrass approximation theorem (see Problem 35-3) there exists a polynomial p such that $|p(t) - 1/t| < \epsilon$ for every t in f(X). It follows from this that $|p(f(x)) - 1/f(x)| < \epsilon$ for every x, so $\|p(f) - 1/f\| < \epsilon$. Since p(f) is in A and A is closed, we conclude that 1/f is in A.

Theorem 3:

Let A be a commutative C-algebra of operators on H. If an operator T in A is regular in $\mathcal{B}(H)$, then it is also regular in A, and therefore the spectrum of T as an operator on H equals its spectrum as an element of A.*

Proof:

We begin by considering the special case in which T is assumed to be self-adjoint. Let B be the Banach subalgebra of $\mathcal{B}$(H) generated by T and T^{-1}. Since T and T^{-1} are self-adjoint and commute with one another, it is evident that B is commutative C*-algebra; and if $\mathcal{M}$ is its space of maximal ideals then B is isometrically *-isomorphic to $\mathcal{C}(\mathcal{M})$ and T is represented by a real function in $\mathcal{C}(\mathcal{M})$. $C = A \cap B$ is a Banach subalgebra of B; and is therefore isomorphic to a Banach subalgebra of $\mathcal{C}(\mathcal{M})$. Sine T is in C and is regular in B, our lemma shows that T^{-1} is also in C and therefore lies in A.

We now turn to the general case, in which T is not assumed to be self-adjoint. It is clear that $U = TT^*$ is a self-adjoint operator in A, and since it has an inverse $U^{-1} = (TT^*)^{-1} = (T^*)^{-1} T^{-1} = (T^{-1})^* T^{-1}$ in $\mathcal{B}(H)$, we know from the preceding paragraph that U^{-1} is in A. We now make use of the commutativity of A to write the relation $UU^{-1} = I$ in the form $T(T^*U^{-1})T = I$. This shows that $T^{-1} = T^*U^{-1}$, so T^{-1} lies in A and the proof is complete.

This result tells us, in particular, that if N is a normal operator on H, then its spectrum $\sigma(N)$ equals its spectrum as an element of the commutative C*-algebra generated by N. Our next step is to provide a concrete representation for the space of maximal ideals in this algebra.

Theorem 4:

Let N be a normal operator on H, A the commutative C-algebra generated by N, and $\mathcal{M}$ the space of maximal ideals in A. Then the function $\hat{N}$ in $\mathcal{C}(\mathcal{M})$ which corresponds to N under the Gelfand mapping is a homeomorphism of $\mathcal{M}$ onto $\sigma(N)$.*

Proof:

It follows from Theorem C and part (4) that $\sigma(N)$ is precisely the range of the continuous function $\hat{N}$ defined on $\mathcal{M}$. Since both $\mathcal{M}$ and $\sigma(N)$ are compact Hausdorff spaces, it suffices by Theorem 5 to show that $\hat{N}$ is one-to-one. Let M_1 and M_2 be points of $\mathcal{M}$ such that $\hat{N}(M_1) = \hat{N}(M_2)$. Then we also have

$$\hat{N}^*(M_1) = \overline{\hat{N}(M_1)} = \overline{\hat{N}(M_2)} = \hat{N}^*(M_2),$$

so each of the function $\hat{N}$ and $\hat{N}^*$ takes equal values at M_1 and M_2. Since A is the closure of the set of all polynomials in N and N*, every function in $\mathcal{C}(\mathcal{M})$ is a uniform limit of polynomials in $\hat{N}$ and $\hat{N}^*$, and therefore every function in $\mathcal{C}(\mathcal{M})$ takes equal values at M_1 and M_2. We conclude the proof by observing that since $\mathcal{C}(\mathcal{M})$ separates the points of $\mathcal{M}$, it follows that $M_1 = M_2$.

In accordance with this result, we may identify $\mathcal{M}$ with the compact subspace $\sigma(N)$ of the complex plane; and when this identification is carried out, it is easy to see that $\hat{N}(z) = z$ for every z in $\mathcal{M}$. We summarize our conclusions in

Theorem 5:

Let N be a normal operator on H with spectrum $\sigma(N)$, and let A be the commutative C*-algebra generated by N. Then the space $\mathcal{M}$ of maximal ideals in A equals $\sigma(N)$, and the Gelfand mapping $T \to \hat{T}$ of A onto $\mathscr{C}(\mathcal{M})$ is an isometric *-isomorphism which carries N into the function whose values are given by $\hat{N}(z) = z$ for every z in $\mathcal{M}$.

This theorem has a number of simple consequences, of which the following are only a few:

(1) $N = 0 \Leftrightarrow \sigma(N) = \{0\}$;

(2) N is singular $\Leftrightarrow \sigma(N)$ contains 0;

(3) $\sigma(N^*) = \overline{\sigma(N)}$

(4) N is self-adjoint $\Leftrightarrow \sigma(N)$ lies on the real line;

(5) N is unitary $\Leftrightarrow \sigma(N) \subseteq \{z : |z| = 1\}$;

(6) N is a projection $\Leftrightarrow \sigma(N) \subseteq \{0, 1\}$. If it happens that H is finite-dimensional, so that $\sigma(N)$ consists of a finite number of distinct complex numbers $\lambda_1, \lambda_2, \ldots, \lambda_m$, then we can write

$$\hat{N} = \sum_{i=1}^{m} \lambda_i \hat{P}_i,$$

where $\hat{p}_i$ is the function in $\mathscr{C}(\mathcal{M})$ defined by $\hat{p}_i(\lambda_j) = \delta_{ij}$. It is evident from this that

$$N = \sum_{i=1}^{m} \lambda_i P_i,$$

where the P_i's are non-zero pairwise orthogonal projections in A such that $\sum_{i=1}^{m} P_i = I$. This is precisely the spectral resolution of N treated in actually contains the finite-dimensional spectral theorem. We therefore have solid grounds for regarding Theorem E as the generalized form of the spectral theorem discussed in the last paragraph of our promises are fulfilled.

FIXED POINT THEOREMS AND SOME APPLICATIONS TO ANALYSIS

Let f be a continuous mapping of the closed interval $[-1, 1]$ into itself. Fig. 39 suggests that the graph of f must touch or cross the indicated diagonal, or more precisely, that there must exist a point x_0 in $[-1, 1]$ with the property that $f(x_0) = x_0$. The proof is easy.

We consider the continuous function F defined on $[-1,1]$ by $F(x) = f(x) - x$, and we observe that $F(-1) \geq 0$ and that $F(1) \leq 0$. If now follows from the Weierstrass intermediate value theorem that exists a point x_0 in $[-1, 1]$ such that $F(x_0) = 0$ or $f(x_0) = x_0$.

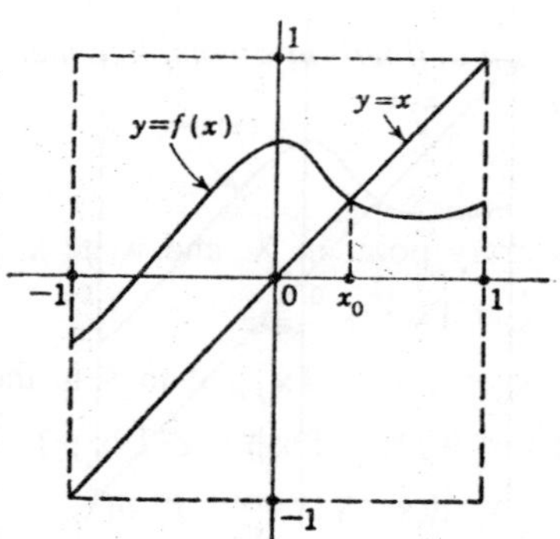

Fig. 4.1

It is convenient to describe this phenomenon by means of the following terminology. A topology space X is called a *fixed point* space if every continuous mapping f of X into itself has a fixed point, in the sense that $f(x_0) = x_0$ for some x_0 in X. The remarks in the above paragraph show that [– 1, 1] is a fixed point space.

Furthermore, the closed disc $\{(xy) : x_2 + y_2 \leq 1\}$ in the Euclidean plane R^2 is also a fixed point space (for a lucid elementary proof of this. Both of these facts are special cases of

Brouwer's Fixed Point Theorem: The closed unit sphere $S = \{x : \|x\| \leq 1\}$ in R^n is a fixed point space.

There are several proofs of this classic result, but since they all depend on the methods of algebraic topology, we refer the reader to Bers. Brouwer's theorem itself is a special case of

Schauder's Fixed Point Theorem: *Every convex compact subspace of a Banach space is a fixed point space.*

For a proof, together with a discussion of other related results. Schauder's theorem was foreshadowed by the work of Birkh off and Kellogg on existence theorems in analysis. We illustrate the relevance of these ideas to such problems by given a full treatment of Picard's theorem on the existence and uniqueness of solutions of first differential equations.

We begin by considering an arbitrary metric space X with metric d. A mapping T of X into itself is called a contraction if there exists a positive real number r < 1 with the property that $d(Tx, Ty) \leq r\, d(x,y)$ for all points x and y in X. It is obvious that such a mapping is continuous. We shall need the following.

Lemma:

If T is a contraction defined on a complete metric space X, then T has a unique fixed point.

Proof:

Let x_0 be an arbitrary point in X, and write $x_1 = Tx_0$,

$$x_2 = T^2x_0 = Tx_1,$$

and, in general, $x_n = T^nx_0 = Tx_{n-1}$. if m < n, then

$$\begin{aligned} d(x_m, x_n) &= d(T^mx_0, T^nx_0) = d(T^mx_0, T^mT^{n-m}x_0) \\ &\le r_m\, d(x_0, T^{n-m}x_0) = r^m\, d(x_0, x_{n-m}) \\ &\le r^m\, [d(x_0,x_1) + d(x_1, x_2) + \ldots + d(x_{n-m-1},x_{n-m})] \\ &\le r^m\, [d(x_0,x_1)\, [1 + r + \ldots + r^{n-m-1}] \\ &< \frac{r^m d(x_0,x_1)}{1-r}. \end{aligned}$$

Since r < 1, it is evident from this that $\{x_n\}$ is a Cauchy sequence, and by the completeness of X, there exists a point x in X such that $x_n \to x$.

We now use the continuity of T to infer that x is a fixed point :

$$T_x = T(\lim x_n) = \lim Tx_n = \lim x_n + 1 = x.$$

We conclude the proof by showing that x is the only fixed point. If y is also a fixed point, that is, if Ty = y, then d(x,y) = d(Tx, Ty) ≤ r d(x,y); and since r < 1, this implies that d(x,y) = 0 or y = x.

This result is the key to

Picard's Theorem : If f(x,y) and ∂f/∂y are continuous in a closed rectangle $R = \{(x,y) : a_1 \le x \le a_2 \text{ and } b_1 \le y \le b_2\}$, and if (x_0,y_0) is an interior point of R, then the differential equation

$$\frac{dy}{dx} = f(x,y) \qquad \ldots(1)$$

has a *unique solution* y = g(x) which passes through (x_0,y_0).

Proof:

Since f(x,y) and ∂f/∂y are continuous in R, they are bounded, and consequently there exist constants K and M such that

$$|f(x,y)| \le K \qquad \ldots(2)$$

and

$$\left|\frac{\partial}{\partial y} f(x,y)\right| \le M \qquad \ldots(3)$$

for all points (x,y) in R. We next observe that if (x,y_1) and (x,y_2) are in R, then the means value theorem guarantees that

$$\left|f(x,y_1) - f(x,y_2)\right| = \left|y_1 - y_2\right|\left|\frac{\partial}{\partial y} f(x, y_1 + \theta\,(y_2 + y_1))\right| \qquad (4)]$$

for some θ such that 0 < θ 1. It now follows from (3) and (4) that

$$\left|f(x,y_1) - f(x,y_2)\right| \le M\left|y_1 - y_2\right| \qquad ...(5)$$

for all (x,y_1) and (x,y_2) in R.[1]

It is convenient at this replace our problem by an equivalent problem relating to an *integral equation*. If y = g(x) satisfies (1) and has the property that $g(x_0) = y_0$, then integrating (1) from x_0 to x yields

$$g(x) - g(x_0) = \int_{x_0}^{x} f(t, g(t))\, dt$$

or $$g(x) = y_0 + \int_{x_0}^{x} f(t, g(t))\, dt. \qquad (6)$$

Conversely, if y = g(x) satisfies (6), then it clear that $g(x_0) = y_0$, and on differentiating (6) we obtain (1). It therefore suffices to show that the integral equation (6) has a unique solution.

To accomplish this, we choose a positive number ∝ such that M∝ < 1 and the closed rectangle R' determined by $|x - x_0| \le \alpha$ and $|y - y_0| \le K\alpha$ is contained in R. We now let X be the set of all continuous real functions y = g(x) defined on the closed interval $|x - x_0| \le \alpha$ such that Ka. X is clearly a closed subspace of the complete metric space $\mathscr{C}[x_0 - \propto, x_0 + \propto]$ and is therefore itself a complete metric space. Our next step is to consider the mapping T of X into itself defined by Tg = h, where

$$h(x) = y_0 + \int_{x_0}^{x} f(t, g(t))\, dt.$$

The fact that T maps X into itself is evident from (2), for

$$\left|h(x) - y_0\right| = \left|\int_{x_0}^{x} f(t, g(t))\, dt\right| \le K\alpha.$$

Furthermore, it follows from (5) that

$$\left|h_1(x) - h_2(x)\right| = \left|\int_{x_0}^{x}\left[f(t, g_1(t)) - f(t, g_2(t))\right] dt\right|$$

$$\le M\alpha \sup\left|g_1(x) - g_2(x)\right|;$$

and since Ma < 1, this shows that T is a contraction on X. We now appeal to our lemma to conclude that the equation Tg = g has a unique solution.

Since this amounts to saying that the integral equation (6) has a unique solution, our proof is complete.

The ideas in this proof have a much wider scope than might be suspected, and can be applied to establish many other existence theorems in the theory of differential and integral equations.

CONTINUOUS CURVES AND THE HAHN-MAZURKIEWICZ THEOREM

A continuous curve is usually thought of as "the path of a continuously moving point," and this rather vague notion is often felt to carry with it the even vaguer attribute of "thinness," or "one-dimensionality." For the case of plane curves, Hordan (in 1887) gave precise expression to this intuitive geometric concept by means of the following definition : if f is a continuous mapping of the closed unit interval I = [0, 1] into the

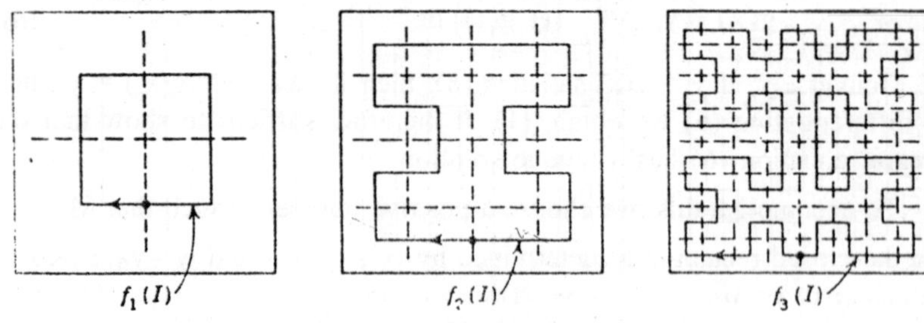

Fig. 4.2

Euclidean plane R^2, then the subset f(I) of R^2 is called a continuous curve. The fame of Jordan's definition rests mainly on Peano's discovery (in 1890) of a continuous curve which passes through every point of a closed square. Curves of this type have come to be called *space-filling curves*.

In Fig. 4.2, we show the first three stages in the construction of a particularly simple example known as Hilbert's space-filling curve. If the square under consideration is S = $\{(x,y) : 0 \leq 1$ and $0 \leq y \leq 1\}$, then the respective curves are the images of I under continuous mappings f_1, f_2, and f_3 of I into S. The process of constructing these curves can be continued in the same way, and it yields a sequence of continuous mapping f_n of I into S. By the manner in which each curve is constructed from its predecessor, it is clear that the sequence $\{f_n\}$ converges pointwise to a mapping f of I into S; and since this convergence is evidently uniform, f is continuous a continuous curve in the sense of Jordan. Furthermore, each point of S lies in f(I), so f(I) is a space-filling curve.

Peano's discovery of space-filling curves was shock to many mathematicians of the time, for it violated all their preconceived ideas of what a continuous curve ought to be. To a few of the others, however, it presented an opportunity. It suggested the very interesting problem of determining what a continuous curve actually is, or in other words, of finding intrinsic topological properties of a subset X of R^2 which are equivalent to the existence of a continuous mapping of I onto X.

Before describing the solution of this problem, we place it in a wider context by extending Jordan's definition. A topological space X is called a *continuous curve* if X is a Hausdorff space and there exists a continuous mapping of I onto X.[1] We know that I is compact and connected any continuous curve is also compact and connected. In the lemmas below, we give two additional properties which every continuous curve must have.

It is convenient to begin by introducing the following concept. A mapping f of one topological space into another is said to be closed if it carries closed sets into closed sets, that is, if f (F) is closed whenever F is closed. We shall use the fact that a continuous mapping of a compact space into a Hausdorff space is automatically closed.

Lemma:

Every continuous curve is second countable.

Proof:

Let f be a continuous mapping of I onto a Hausdorff space X. We must show that X is second countable, that is, that it has a countable open base. I is a separable metric space, so it has a countable open base $\{B_i\}$ and it is easily seen that the class $\{G_i\}$ of all finite unions of the Bi's is also a countable open base of I. Since I is compact and X is Hausdorff, f is closed, and therefore each set $f(G_i')$ is closed,. The class of all sets of the form $f(G_i')$ is thus a countable class of open subsets of X, so it suffices to show that these sets constitute an open base for X.

Let x be a point of X with neighbourhood G. The set $f^{-1}(\{x\})$ is closed and is therefore a compact subspace of I with neighbourhood $f^{-1}(G)$. For each point y in $f^{-1}(\{x\})$, there exists a set in $\{G_i\}$ which contains y and

1. A continuous curve in the sense of this definition is often called a Peano space.

is contained in $f^{-1}(G)$. By the compactness of $f^{-1}(\{x\})$ and the fact that $\{G_i\}$ is closed under the formation of finite unions, there exists a G_i such that

$$f^{-1}(\{x\}) \subseteq G_i \subseteq f^{-1}(G).$$

On taking complements, we obtain

$$f^{-1}(\{x\})' \supseteq G_i' \supseteq f^{-1}(G)' \qquad ...(1)$$

and since the complement of an inverse image equals the inverse image of the complement, we can write (1) in the form

$$f^{-1}(\{x\}') \supseteq G_i' \supseteq f^{-1}(G'). \qquad ...(2)$$

If we now apply f to all members of (2), we get

$$ff^{-1}(\{x\}') \supseteq f(G_i') \supseteq ff^{-1}(G')$$

or $\{x\}' \supseteq f(G_i') \supseteq G'$,

so $\{x\} \subseteq f(G_i')' \subseteq G$,

and the proof is complete.

Lemma:

Every continuous curve is locally connected.

Proof:

Let f be a continuous mapping of I onto a Hausdorff space X. We must show that X is locally connected. Suffices to show; if C is a component of an open subspace G of X, then C is open.

Let A be a component of $f^{-1}(G)$. Then A is connected, and therefore f(A) is connected in G; and since C is a component of G, we see that f(A) is either disjoint from C or contained in C. It follows from this that $f^{-1}(C)$ is a union of components of $f^{-1}(G)$. Since $f^{-1}(G)$ is open and I is locally, tells us that the components of $f^{-1}(G)$ are open, do $f^{-1}(C)$ is open and $f^{-1}(C') = f^{-1}(C')$ is closed. We conclude the proof by observing that since the mapping f is closed, the set $ff^{-1}(C') = C'$ is closed, so C is open.

The above remarks and lemmas establish the easy half of the following famous characterization of continuous curves.

The Hahn-Mazurkiewicz Theorem 1 topological space X is a continuous curve ⇔ X is a compact Hausdorff space which is second countable, connected, and locally connected.

For the remainder of the proof, we refer the reader to Wilder Additional discussions of a descriptive and historical nature can be found.

5

CONVERGENCE (NETS AND FILTERS)

SEQUENCE IN TYPOLOGICAL SPACES

The nation of convergence in fundamental in analysis and topology. In this chapter we shall be mainly concerned with the concept of a generalized sequence called 'net'. The nation of a filter and ultrafilter will also be given in some detail. It is felt that filters are more covenient in general topology, nets in analysis. But before we take up the question of the generalization of the notion of convergence of a sequence, we define some terms concerning sequences,

Recall that a sequence in a set X is a mapping of N into X. If s is a sequence in X, the image s (n) of $n \in N$ is usually denoted by s_n. It is customary to denote the sequences by the classical symbol s_{n-1}, or simply by $< s_n >$. A sequence $< s_n >$ is said to be eventually in a set A if and only if there exists a positive integer m such that $n \geq m \Rightarrow s_n \in A$. A sequence $< s_n >$ in a topological space X is said to converge to point s0 in X if and only if it is eventually in every neighbourhood of s_0 and we then say s_0 is a limit point of the sequence and write $s_n \rightarrow s_0$. A point s_0 in X is said to be a cluster point of a sequence $< s_n >$ in X if and only if $< s_n >$ is frequently in every neighbourhood of s_0, that is every neighbourhood of s_0 contains sn, for finitely many n. A sequence t in a space X is said to be a subsequence of another sequence s in X iff there exists a mapping ϕ of N into N such that $s_o \phi = t$ and $i < j \Rightarrow \phi(i) < \phi(j)$ for all $i, j \in N$.

Definition : *Let $\mathfrak{I}'$ and $\mathfrak{I}$ be two topologies on a set X. Then $\mathfrak{I}$ is said to be sequentially stronger than $\mathfrak{I}'$ iff sn → s0 in $\mathfrak{I} \Rightarrow$ sn → in $\mathfrak{I}$.. Also $\mathfrak{I}$ and $\mathfrak{I}'$ are said to be sequentially equivalent iff each is sequentially stronger than the other.*

Example 1:

Let X = {a, b, c} and $\mathfrak{I}$ = {∅, {a}, {a, b} X}. Find (i) the limit points (ii) the cluster points of the following sequences in X.

(a) < a, b, a, b, a, b,...> (b) < a, a, a,...> *are made on topological spaces, most familiar properties can be retained.*

The above tow examples show that the convergence of a sequence to a point is related to the behaviour of sn as n becomes large as well as to the topology of the range space X. Hence, the notion of order in N is fundamental for convergence. Also the sets whose complements are finite play an important role. The question now arises as to what properties of order in N are essential for the development of a general theory of convergence. Moore and' Smith used the concept of a generalizes sequence later called 'net' by Kelley. Cartan introduced the notion of filter by generalizing some of the essential properties of subsets of N. We shall discuss both these concepts in this chapter.

Theorem 1:

Let $(X, \mathfrak{I})$ be a first axiom space. Then a point x_0 of X in an accumulation point of a subset Y of X if and only there exists a sequence in $Y - \{x_0\}$ which converges to x_0.

Proof :

The 'if' part : Let $s = < s_n : n \in N >$ be a sequence in $Y - \{x_0\}$ which converges to x_0 let V be a $\mathfrak{I}$-nhd of x_0. Since s is eventually in V there exists $n_0 \in N$, such that for every $n \in N$, $n \geq n_0 \Rightarrow s_n \in V$. In particular, $s_{n_0} \in V$. Hence $s_{n_0} \in V \cap (Y - \{x_0\})$. This implies that every nhd of x_0 contains a point of Y distinct from x_0. Hence, x_0 is an accumulation point of Y.

The 'only if' part : Assume that x_0 is an accumulation point of a subset Y of X. Since the space is first axiom, there exists a nested local base at x_0. Let $B(x_0) = \{_{Bn} : n \in N\}$ be this nested local base at x_0 so that $B_n + 1 \subset B_n$ for all $n \in N$. Since x_0 is an accumulation point of Y, every nhd of x_0 must intersect $Y - \{x_0\}$. Hence by the axiom of choice for each $n \in N$, we may choose a point s_n in $B_n \cap (Y - \{x_0\}$, this obtaining a sequence $s = < s_n : n \in N >$ in $Y - \{x_0\}$. We now show that s converges to x_0. Let V be any $\mathfrak{I}$-nhd of x=. Then by the definition of local base $B(x_0)$, there exists $n_0 \in N$, such that $B_{n_0} \subset V$ and since $B_{n+1} \subset B_n$, we have $B_n \subset V \vee n \geq n_0$.

But $s_0 \in B_n$. It follows that $s_n \in V \vee n \geq n_0$. This implies that the sequence s is eventually in V and consequently s converges to x_0.

Theorem 2:

Let $(X, \mathfrak{I})$ be a first axiom space and let $Y \subset X$. Then a point x_0 of X is a adherent point of Y. that is, $s_0 \in Y$ iff there exists a sequence in Y which converges to x_0.

Proof:

(i) This sequence converges to the point b since only $\Im$-nhd of b are {a, b} and X both of which contain all the terms of the sequence. Similarly it converges to c since the only $\Im$-nhd of c which contains all the terms of the sequence is X. But a is not a limit point of this sequence since {a} is a $\Im$-nhd of a such that the sequence are b and c only.

(ii) Since b, c are limit point of this sequence they are certainly cluster points. The point a is also a cluster point of this sequence since the nhds of a are {a}, {a, b} and X each of which contains an infinite number of terms of the sequence. Hence a, b, c are the cluster points of this sequence.

(iii) Since a, b, c are limit point of this sequence, they are certainly cluster points.

Example :

Let X = N and let $\Im$ consists of ∅ and all those subsets of N whose complements are finite. Finite (i) the limit points, (ii) the cluster points of the sequences.

< 1, 3, 5... > *and* < 1, 2, 1, 2, ... > in N.

Solution:

(i) Consider the sequence < 1, 3, 5, ... >. Let V be any $\Im$-nhd of 1. Then by definition of $\Im$, V contains all points of N except a finite number and since all the terms of this sequence are distinct, it follows that the sequence is eventually in V.. Hence 1 is a limit point of this sequence. Similarly 3, 5, 7... are limit points of this sequence.

Now consider the sequence < 1, 2, 1, 2, 1, 2, ... >. The set N – {2} is a $\Im$-open nhd of 1. This sequence is however not eventually in this neighbourhood. Hence 1 is not a limit of this sequence. Similarly 2 is also not a limit point of this sequence.

(ii). Since 1, 3, 5, ... are the limit points of the sequence < 1, 3, 5, ... >, they are certainly cluster points of this sequence. Now consider the sequence < 1, 2, 1, 2, ... >. Let V be any $\Im$-nhd of 1. Then this sequence is frequently in V for it contains an infinite number of terms of this sequence namely, s_1, s_3, s_5, ... Similarly 2 is a cluster point this sequence.

Theorem 3:

Let $\Im$ and $\Im'$ be topologies for a set X with $\Im$ first axiom. Then $\Im \supset \Im'$ if and only if $\Im$ is sequentially stronger than $\Im'$. and $s_n \to s_0$ in $\Im$

Proof:

Suppose $\Im \supset \Im'$ and sn $\to s_0$ in $\Im$. Let V be a $\Im'$-nhd of s0. Then V is a $\Im$-nhd of s_0 since $\Im \supset \Im'$. Since $s_n \to s_0$ in $\Im$, $s_n \in V$ eventually. Hence $s_n \to s_n$ in $\Im'$ and so $\Im$ is sequentially stronger than $\Im'$. Conversely, let $\Im$ be sequentially stronger than $\Im'$. It follows from (1.6) and (1.7) that $CI_{\Im}$ (S) $\subset CI_{\Im}'$ (S) for all $S \subset X$. Let $G \in \Im'$. Then G′. Then G′ is $\Im'$-closed and so $CI_{\Im}'$ (G′) = G. Hence $CI_{\Im}$ (G′) = G′. But $G' \subset CI_{\Im}$ (G′) always. Thus $CI_{\Im}$ (G′) (G′) = G′ and therefore G′ in $\Im$-closed, that is, $G \in \Im$. Hence $\Im' \subset \Im$.

Theorem 4:

Let (X, $\Im$) be a first axiom space and let x_0 be a cluster point of a sequence s = < s_n : n ∈ N > in X. Then there exists a subsequence of s which converses to x_0.

Proof:

Since (X, $\Im$) is a first axiom space, there exists a nested local base B $(x_0) = \{B_n : n \in N\}$ at x_0 so that $B_{n+1} \subset B_n$. Since x_0 is a cluster point of the sequence s, for each B_i in B (x_0) an each n in N, there exists an m in N such that $m \geq n$ and $s_n \in B_i$. We now define sequence ϕ in N as follows: Let ϕ (1) be the least positive integer such that s (ϕ (2)) $\in B_i$; ϕ (2) be the least positive integer greater than ϕ (1) such that s (ϕ (2)) $\in B_2$ and so on. Then s o ϕ = {s (ϕ (n)) : n ∈ N} : n ∈ N} is a subsequence of s. It will now be shown that s o ϕ converges to x_0. Let V be any $\Im$-nhd of x_0. Then by the definition of local base B (x_0), there exists n_0 in N such that $B_{n_0} \subset V$ and since $B_{n+1} \subset B$. we have $B_n \subset V$ for all $n \geq n_0$. But s (ϕ (n)) $\in B_n$. It follows is the $n \geq n_0$. Thus implies that the sequence s o ϕ is eventually in V and consequently s o ϕ converges to x_0.

Theorem 5:

Let (X, $\Im$) be a first axiom space and let $Y \subset X$. Then Y is $\Im$-open iff each sequence converging to a point of Y is eventually in Y.

Proof:

The 'if part': Let every sequence converging to a point of Y be eventually in Y and suppose , if possible, Y is not open. Then there exists a sequence s = < s_0 : n ∈ N > in X – Y which converges to a point $y_0 \in$ Y. Such a sequence surely fails to be eventually in Y. But this is contradiction. Hence Y must be open.

The 'only if part': Let Y be open and let s be a sequence which converges to a point y_0 of Y. Then Y is a $\Im$-open nhd of y_0. Hence by the definition of convergence of sequence, s must be eventually in Y.

Theorem 6:

Let $(X, \Im)$ be a Lindelof space such that every sequence in X has a cluster point. Then X is compact.

Proof:

We have to show that every $\Im$-open cover of X has a finite subcover. Since X is Lindelof, we may assume that the open cover is countable. Let $c = \{G_n : n \in N\}$ be a countable open cover of X. We proceed by induction. Let $H_0 = G_0$ and for each n in N let H_0 be the first member of the sequence of G′ s which is not contained in the union $H_0 \cup H_1 \cup ... \cup H_{n-1}$. If such a choice is impossible at any stage, then the sets already chosen will form a finite subcover of C. In the contrary situation, it is possible to pick a point h_n in H_n for each n in N such that $h_n \notin H_m$ for $m < n$. We thus obtain a sequence $h < n_n : n \in N >$ satisfying the above condition. By hypothesis this sequence must have a cluster point, say x_0. Since $\{H_n : n \in N\}$ covers, $x_0 \in H_n$ for some $n \in N$. and since x_0 is a cluster point of the sequence h, there exists some $i > n$ such that $h_i \in H_n$. But this is a contradiction. Hence c must have a finite subcover and consequently X is compact.

EXERCISES

1. If a sequence $s = < s_n : n \in N >$ in a topological space X has a subsequence $s \circ \phi = \{s (\phi (n) : n \in N\}$ converging to a point x_0 in X.
2. Let X be a T_1-space. Show that $F \subset X$ is closed iff whenever < sn > is contained in F and $s_n \to s_0$., then $s_0 \in F$.

   Let $\Im$ be the countable topology on an uncountable set X. Show that a sequence $s = < s_n >$ in X is convergent iff it is eventually constant, that is, $s_n \to s_0$ iff $s_n = s_0$ eventually.

   [**Hint :** $s_n = s_0$ eventually $\Rightarrow s_n \to s_0$ is clear. Now suppose that s is sequence in X, $s_0 \in X$, and it is false that $s_n = s_0$ eventually. Let $F = \{s_n : s_n \neq s_0\}$. Then $F = \{s_n : s_n \neq s_0\}$. Then F is countable and hence F′ is a nhd of s_0. But it is false that sn $\in$ F′ eventually. Hence $< s_n >$ does not converge to s_0].
3. Give an example of a non-compact space X such that each sequence in X has cluster point.
4. Give an example of a non-Hausdorff space in which every sequence converges to at most one point.
5. Show that each subsequence of a sequence is sequence.

NETS

1. Directed Set
2. Binary Relation Ex.
3. Residual Subset
4. Cofinal Subset.

Directed Sets

Definition : *Let A be a non-empty set. We say that a binary relation $\geq$ on A directs the set A iff*

[D1] : $a \in A \Rightarrow a \geq a$.

[D2] : $a \geq b$ and $b \geq c \Rightarrow a \geq c$ $(a, b, c \in A)$.

[D3] : *Given any two members a and b of A, there exists a member $c \in A$ such that $c \geq a$ and $c \geq b$.*

We say that a follows b in order $\geq$ and that b precedes a iff $a \geq b$.

Example 1:

Let c denote the collection of all finite subsets of a set X. Then c is directed by the relation $\subset$

Solution :

[D1] and [D2] hold in this case also. To prove [D3], we observe that since c consists of finite subsets of X, for any two members A and B of c, $A \cup B$ is also a member of c such that $A \cup B \supset A$ and $A \cup B \supset B$. Hence [D3] is also satisfied.

Example 2:

Let $(X, \mathfrak{I})$ be any topological space and let $x \in X$. Then the collection $N(x)$ of all $\mathfrak{I}$-nhds of x is directed by the inclusion relation $\subset$. For, we have

[D1] : $A \in N(x) \Rightarrow A \subset A$

[D2] : $A \subset B$ *and* $B \subset C \Rightarrow A \subset C$ $(A, B, C \in N(x))$.

In fact [D1] and [D2] are true for any sets whatsoever.

[D3] : *Let A, B be any two members of N (x). Then $A \cap B$ is a also a member of N (x) such that $A \cap B \subset A$ and $A \cap B \subset B$.*

Example 3:

Let sets, N and R are directed by the relation in the usual sense.

Definition : *Let $(A, \geq)$ be a directed set and let $B \subset A$. Then B is said* *to be a residual subset of A iff there exists an element $a_0 \in A$ such that a $\in A$, $a \geq a_0 \Rightarrow a \in B$. If for every $a \in A$, there exists an element $b \in B$ such that $b \geq a$, then B is called a cofinal subset of A.*

It is evident that every residual subset of A is cofinal and that every cofinal subset of A is directed by ≥.

Example 4:

Let N be directed by ≥ in the usual sense. Then M = N – {1, 2] is a residual subset of N. Here 3 is the required element a_0 of the definition. On the other hand the set L = {1, 3, 5, ... } is not a residual subset of N. However L is a confinal subset of N. Note that ≥ also directs L.

Solution:

Let (A, ≥) be a directed system and let f be an arbitrary mapping of A into a set X. Then f is called a not a in X and is denoted by (f, X, A, ≥).

Note that a sequence in X is a particular type of a net in which A = N and N is directed by ≥ in the usual sense.

We shall denote the image f (a) of a ∈ A under f by f_a and a net in X will be often denoted by the symbol ($\{f_a : a \in A\}$, ≥) or simply by

$\{f_a : a \in A\}$.

A net is sometimes called a generalized sequence of Moor. Smith family.

Definition : *Let (f, X, A, ≥) be net and let Y ⊂ X. Then f is said to be in Y iff f [A] ⊂ Y, that is, iff $f_a \in Y \ \forall\ a \in A$, f is said to be eventually in Y iff there exists a residual subset B of A such that f [B] ⊂ Y, that in iff there exists an element $a_0 \in A$ such that for every $a \in A$, $a \geq a_0 \Rightarrow f_a \in Y$, is said to be frequently in Y iff there exists a confinal subset D of A such that f [D] ⊂ Y, that is , iff for each a in A, there exists b ≥ a in A such that fa ∈ Y.*

It is evident that a net f is frequently in Y iff f is not eventually in X – Y.

CONVERGENCE OF A NET IN A TOPOLOGICAL SPACE

Definition : *Let (X, ℑ) be a topological space and let (f, X, ≥) be a net, in X. We say that f converges to a point x0 ∈ X relative to ℑ iff f is eventually in every ℑ-neighbourhood of x0 that is, iff for each $f_a \geq N$ of x_0, there exists an element $a_0 \in A$ such that $a \in A$, $a \geq a_0 \Rightarrow f_a \in N$.*

Example 1:

Let (X, ℑ) be any indiscrete topological space show that every nit in X converges to every point of X

Solution:

Let (f, X, A, ≥) be a net in X and let x be an arbitrary point of X. Then the only nhd of x is X. Since $f_a \in X$ for all a ∈ A, it follows that f converges to x.

This example show that a net can converges to several different points.

Example 2:

Let (X, D) be any discrete topological space. Show that a net (f, X, A, ≥) in X converges to a point $x_0 \in X$ iff is eventually in $\{x_0\}$, that is, iff there exists an $a_0 \in A$ such that $f_a = x_0$ for all $a \geq a_0$ in A.

Solution:

Let f converge to x0. Since $\{x_0\}$ is a D-nhd of x_0 f must be eventually in $\{x_0\}$. Conversely, let f be eventually in $\{x_0\}$. Since every nhd of x_0 contains $\{x_0\}$, it follows that f is eventually in every nhd of x_0. Hence f converges to x_0.

Theorem 1:

Let (X, ℑ) be a topological space and let $Y \subset X$. Then Y is ℑ-open iff no net X – Y can converges to a point in Y.

Proof:

The 'if' part : Assume that no net in X – Y converges to a point in Y and suppose, if possible, Y in not ℑ-open. Then there exists a point x_0. In Y such that every nhd of x_0 intersects X – Y. Let N (x_0) be the collection of all ℑ nhds of x_0. Then N (x_0) is directed by the inclusions relation ⊂. Since $N \cap (X - Y) = \varnothing \ \forall\ N \in N(x_0)$, by the action of choice, for each nhd $N \in N(x_0)$, we may choose a point x (N) in $N \cap (X - Y)$. Now consider the mapping

$$f : N(x_0) \to X - Y.$$

such that $\quad f(N) = x(N) \ \forall\ n \in n(x).$

Then evidently f is a net in X – Y converging to x_0. To see that f converges to x_0., let V be any ℑ-nhd of x_0. Then for each $W \in N(x_0)$ such that $W \geq V$ that is $W \subset V$), we have

$$f(W) = x(W) \in W \cap (X - Y) \subset V.$$

Thus there exists a net in X – Y converging to x0. But this is contradiction. Hence Y must be open.

The 'only if' part : Assume that Y is open, and suppose, if possible, that there exists a net (f, X – Y, A, ≥) in X – Y converging to a point $x_0 \in Y$.

Since Y is a $\Im$-nhd of x_0., by the definition of convergence f must be eventually in Y. This means that f can never be in X – Y. But this is a contradiction. Hence no net in X – Y can converge to a point in Y.

Corollary : 1: *A subset A of a topological space is closed if and only if no net in A converges to a point in X – A.*

Proof:

Let Y = X – A. Then we have to show that X is open iff intent in X – Y converges to a point in Y. Now proceed as in the above theorem.

Corollary : 2. *Let (x, $\Im$) be a first axiom space and let $Y \subset X$. Then Y is $\Im$-open iff no sequence in X – Y converges to a point in Y.*

Proof:

The proof of the 'only if' part is the same as that of the above theorem if we replace the word 'net' by the word 'sequence'. The proof of the 'only if' part follows from first countability of X and the 'if part from the above theorem.

Theorem 2:

Let (X, $\Im$) be topological space and let $Y \subset X$. If x_0 is a point of X, then $x_0 \in \overline{Y}$ if and only if there exists a net in Y converging to x_0.

Proof : The 'if part' : Let (f, Y, A, ≥) be a net in Y converging to $x_0 \in X$, and let N be any neighbourhood of x_0. Since f is eventually in N, there exists $a_0 \in A$ such for every $a \in A$, $a \geq a0 \Rightarrow f_a \in N$. Now $a_0 \geq a0$.

Therefore $f_{a_0} \in N \cap Y$ so that $N \cap Y \neq \varnothing$. Thus every nhd of x_0 contains a point of Y. It follows that x_0 is an adherent point of Y and consequently $x_0 \in Y$.

The 'only if' part : Let $x_0 \in \overline{Y}$. Then every nhd of x_0 must intersect Y. Let N (x_0) be the collection of all $\Im$-nhd of x_0. Then N (x_0) is directed by the inclusion relation $\subset$. Since

$$N \cap Y \neq \phi \;\forall\; N \in N\,(x_0)$$

by the axiom of choice for each N in N (x_0), we may choose a point x (N) in $N \cap Y$. Now consider the mapping

$$f : N\,(x_0) \rightarrow Y : f\,(N) = x\,(N) \;\forall\; N \in N\,(x_0).$$

Then evidently f is a net in Y converging x_0. To see that f converges to x_0, let V be any nhd of x_0. Then for each W in N (x_0) such that $W \geq V$ (that is, $W \subset V$), we have

$$f(\mathring{W}) = x(W) \in W \cap Y \subset V.$$

Theorem 3:

Let $(X, \mathfrak{I})$ be a topological space and let $Y \subset X$. Then a point x_0 of X is an accumulation point of Y if an only if there exists a net in $Y - \{x_0\}$ converging to x_0.

Proof:

The 'if' part : Let $(f, Y - \{x_0\}, A, \geq)$ be a net in $Y - \{x_0\}$ converging to x_0 and let N be a $\mathfrak{I}$-nhd of x_0. Since f is eventually in N, there exists $a_0 \geq A$ such that $a \in A, a \geq a_0 \Rightarrow f_a \in N$. Now $a_0 \geq a_0$. Therefore

$$f_{a_0} N \cap (Y - \{x_0\}).$$

This implies that every nhd of x_0 contains a point of Y distinct from x_0. Hence x_0 is an accumulation point of Y.

The 'only if' part : Let x_0 be an accumulation point of Y. Then every nhd of x_0 must intersect $Y - \{x_0\}$. Let $N\{x_0\}$ be collection of all $\mathfrak{I}$-nhds of x_0. Then $N(x_0)$ is directed by the inclusion relation $\subset$. Since for each $N \in N(x_0)$,

$$N \cap (Y - \{x_0\}) \neq \varnothing$$

we may choose a point x (N) in $N \cap (Y - \{x_0\})$. Now consider the mapping,

$$f : N(x_0) \rightarrow Y - \{x_0\} : f(N) = x(N) \; \forall \; N \in N(x_0).$$

Then f is a net in $Y - \{x_0\}$ converging to x_0 (why ?)

Theorem 4:

Let $(X, \mathfrak{I})$ be a topological space and let $Y \subset X$. Then Y is $\mathfrak{I}$-closed if and only if no net in Y converges to a point $X - Y$.

Proof:

The 'if part' : Assume that no net in Y converges to a point in X–Y and suppose, if possible, Y is not $\mathfrak{I}$-closed. Then there exists an accumulation point x_n of Y such that $x_0 \notin Y$, that is, $x_0 \in X - Y$. Then by the preceding to $x_0 \in X - Y$. But this is a contradiction.

The 'only if' part : Let Y be $\mathfrak{I}$-closed and suppose, if possible, there exists a net f in Y converging to a point $x_0 \in X - Y$. Since $x_0 \notin Y$, f is also a net in $Y - \{x_0\}$ converging to x_0. Hence by the preceding theorem, x_0 is an accumulation point of Y, since $x_0 \notin Y$, Y is not $\mathfrak{I}$-closed. But this is a contradiction. Hence no net in Y can converge to a point in X – Y.

The 'if' part : Assume that every net in X can converges at most one point and suppose, if possible, that the space X is not Hausdorff. Then there exists two distinct points x and Y in X such that every nhd of x intersects every nhd of y. Let N (x) and N (y) be the collection of all $\Im$-nhds of x and y respectively. Then both N (y) and N (y) are directed by the inclusion relation $\subset$. Consider the Cartesian product P = N (x) × N (y). In this product we define a relation $\geq$ as follows. Let (N_1, M_1) iff $N_1 \subset N_2$ and $M_1 \subset M_2$. Then it is easy to see that P is directed by $\geq$. Since every nhd of x intersects every nhd of y, we have $N \cap M = \varnothing \vee (N, M) \in P$. Hence by the axiom of choice we may choose a point x (N, M) in $N \cap M$ for all $(N, M) \in P$.

Now consider the mapping.

$$f : P \to X : f((N, M)) = x(N, M) \forall (N, M) \in P.$$

Then f is a net in X converging to both x and y. To see this, let U be any nhd of a and let V be any b nhd of y. Then for each (N, M) in P such that $(N, M) \geq (U, V)$ so that $N \subset U$ and $M \subset V$, we have

$$f((N, M)) = x(N, M) \in N \cap M \subset U \cap V.$$

Hence $f(N, M)) \in U$ and $f((N, M) \in V$. It follows that f is eventually both in U and V and consequently f converges so both x and y. But this is a contradiction. Hence X must be a Hausdorff space.

The only 'if part' : Let X be a Hausdorff space and let x, y be two distinct points of X. Then there exists a $\Im$-nhd N of x and a $\Im$-nhd M of y such that $N \cap M = \varnothing$. Since a net cannot be eventually in each of two disjoint sects, it is evident that no net in X can converge to both x and y.

Hence a net in X can now or get to at most one point.

Theorem 5:

Let $(X, \Im)$ be a Hausdorff topological space. Then every convergent net has a unique cluster point and this is the unique limit point of the net.

Proof:

We know that in Hausdorff space, every convergent net has a unique limit point. Let p be the unique limit point of a convergent net f in X. Since every limit point is also a cluster point , see ξ_3]. Suppose, if possible, f has another cluster point q distinct from p. Since X is a T_2-space there exist disjoint nhd N_p and N_q of p and q respectively. Since p is the limit of the net f, f is eventually in N_p and since N_4 is disjoint from N_p f cannot be frequently in N_q. But this contradicts our supposition that q is a cluster point of f. Hence f cannot have two distinct cluster points. Accordingly, every convergent net in X has unique cluster point.

Theorem 6:

Let g be a mapping of space X into a space Y. Then g is continuous at $x_0 \in X$ iff whenever a net $\{a : a \in A\}$ converges to x_0 in X, then the net $\{g(f_a) : a \in A\}$ converges to $g(x_0)$ in X.

Proof:

Suppose g is continuous at x_0 and $f_a \to x_0$. To show that $g(f_0) \to g(x)$. Let M be any nhd of $g(x_0)$. Since g is continuous at x_0, $g^{-1}[M]$ is a nhd of x_0. Since $f_a \to x_0$ there exists $a_0 \in A$ such that $a \geq a_0 \Rightarrow f_a \in g^{-1}[M]$. Hence $g(f_a) \to g(x_0)$.

Conversely, let $fa \to x_0 \Rightarrow g(x_0)$ and suppose, if possible, g is not continuous at x_0. Then for some nhd M of $g(x_0)$, $g(N) \not\subset M$ for any nhd of x_0. Thus, for each nhd N of x_0, we can choose $x_N \in N$ such that $g((x_N) \notin M$.

Then

$$\{x_N : N \in N(x_0)\}$$

is a net in X since the collection N (x), of all nhds of x_0 is a directed set by the inclusion relation $\subset$ [see (2.3)]. It is easy to see that $xN \to x_0$. For if N_0 if any nhd of x_0 then for every $N \geq N_0$ (*i.e.*, $N \subset N_0$) in $N(x_0)$, we have $x_N \in N \subset N_0$. But $\{g(x_N) : N \in N(x_0)$ is a net in Y which does not converge to $g(x_0)$. For M is a nhd of $g(x_0)$ such that

$$g(x_N) \notin M \text{ for every } N \in N(x_0).$$

Theorem 7:

A net $\{f_a : a \in \mathfrak{I}\}$ in a product space

$$X = \times \{X_\lambda : \lambda \in L\}$$

converges to $x \in X$ iff for every $\lambda \in L$, the net $\{\pi_\lambda(f_a) : a \in A\}$ converges to $\pi_\lambda(x)$ in X_λ.

Proof:

If $f_a \to x$ in X, then since π_λ is continuous,

$$\pi_\lambda(f_a) \to \pi_\lambda(x)$$

by the previous theorem, for each $\lambda \in \Lambda$.

Conversely, let $\pi_\lambda(f_a) \to \pi_\lambda(x)$ for every $\lambda \in \Lambda$. Let

$$G = \pi_\lambda^{-1}[G_{\lambda_1}] \cap \dots \cap \pi_{\lambda_n}^{-1}[G_{\lambda_n}]$$

be any basic nhd of $x \in X$. Then G_{λ_i} is nhd of $\pi_{\lambda_\lambda}(x)$ in X_{λ_i} for i = 1, 2, ... n.

Since $\pi_{\lambda_i} i (f_a) \to \pi_{\lambda_i} (x)$, for each i there exists $a_i \in A$ such that $a \geq a_i$ implies $\pi_{\lambda_i} (f_a) \to \pi_{\lambda_i}$. Thus if a_0 be picked greater than all a_i i = 1, 2, ... n, we have

$$\pi_{\lambda_i} (f_a) \in G_{\lambda_i}$$

for all $a \geq a_0$. It follows that for $a \geq a_0$.

$$f_a \in f_a \in \bigcup_{i=1}^{n} \pi_{\lambda_i}^{-1} [G_{\lambda_i}] = G.$$

and therefore $f_a \to x$ in X.

ULTRANET

Definition : *A net $\{f_a : a \in A\}$ in a set X is said to be an ultranet (or universal net) iff for each subset S of X, the net is eventually is S or eventually S'.*

It follows from this definition that if an ultranet is frequently in S, then it is eventually in S. In particular, an ultranet in a topological space must converge to each of its cluster points.

Example:

For any direct set A, the map $f : A \to X$, defined by $f(a) = x$ for all $a \in A$ is an ultranet on X, called the trivial ultranet.

Theorem:

Let $\{f_a : a \in A\}$ be an ultranet in X and g a mapping of X into a set Y. Then $\{g(f_a) : a \in A\}$ is an ultranet in Y.

Proof:

Let S be any subset of Y. Then it is easy to see that $g^{-1}[S] = X - g^{-1}[Y - S]$. Since $\{f_a\}$ is an ultranet in X and $g^{-1}[S]$ is a subset of X with $g^{-1}[Y - S]$ as is complement, it follows that {fa} is eventually in either $f^{1}[S]$ or $f^{-1}[Y - S]$, from which it follows that $\{g(f_a)\}$ is eventually in either S or Y – S. Hence $\{g(f_a)\}$ is an ultranet in Y.

EXERCISES

1. Show that a mapping g of a space X into a space Y is continuous at $x_0 \in X$ iff f or every net $\{f, X, A, \neq\}$ converging to x_0, the composition map $g \circ f : A \to Y$ converges to g (x).
2. Let $\{f_a : a \in A\}$ be a net in X. Let $S_1, S_2, \ldots S_n$ be linitely many subset of X and suppose that for each i = 1, 2, ... n. $f_a \in S$, eventually.

Then $f_a \in \bigcup_{i=1}^{n} S$ eventually.

[**Hint** : For each i, choose $a_i \in A$ such that $a \geq a_i$ implies $f_a \in S_i$. Since A is a directed set, there exists $a_0 \in A$ satisfying $a_0 \geq a_i$ for each i. Then $a \geq a_0$ implies $a \geq a_i$ and therefore $f_a \in S_i$ each i]

Definition : *Let (f, X, A, ≥) and (g, X, B, ≥) be two nets. Then g is said to be a subnet of f iff there exists a mapping*

$$\phi : B \to A \text{ such that}$$

(i) g = f o f and

(ii) for each a in A, there exists an element b in B such that f (x) ≥ a for every x ≥ b in B.*

ISOTONE MAP

Definition : *Let (A ≥) and (B, ≥*) be two directed sets. Then a mapping y : A → B is said to be isotone iff*

$$x \geq y \Rightarrow \psi(x) \geq^* \psi(y)\ (x, y \in A).$$

Theorem 1:

Let y be an isotone map of directed sets (B, ≥) into a directed set (A, ≥) such that ψ [a] is confinal in A. Let (f, X, A, ≥) be a net. Then f 0 ψ is a subset of f.*

Proof:

Let g = f 0 ψ. Since ψ is a mapping of B into A and f is a mapping of A into X, it if follows that f 0 ψ = g is a mapping of B into X. Hence, g is also a net in X satisfying the condition (i). Again since ψ is an isotone map of B into A, we have

$$x \geq^* y \Rightarrow \psi(x) \geq \psi(y),\ (x, y \in B).$$

Further since ψ [B] is confinal in A, for every a ∈ A, there exists an element b ∈ B such that

$$\psi(b) \geq a$$

From (1) and (2), we see that for every a in A, there exists an element b in B such that ψ (x) ≥ a for every x ≥* b in B. Hence condition (ii) is also satisfied. It follows that f 0 ψ is a subnet of f.

Corollary : *Let (A, ≥) be a directed set and B be a confinal subset of A so that B is also directed by ≥. Let (f, X, A, ≥) be a net. Then the restriction map of f to B is a subnet of f.*

Proof:

Here the identity map

$$I : B \rightarrow A : I(x) = x$$

is an isotone mapping such that [B] = B is confinal in A. Hence by the above theorem f o I is a subnet of f.

Since I is a mapping of B into A and j is a mapping of A into X it follows that f 0 I is a mapping of B into X such that (f 0 I) (x) = f (i (x)) = f (x) for all x ∈ B. Hence f p I is the restriction map of f to B. It follows that the restriction map of f to B is a subnet of f.

Theorem 2:

Let (X, ℑ) be a topological space and let (f, X, A, ≥) be a net in X. Let c be the collection of subsets of X satisfying the following two conditions.

(i) f is frequently in each member of c.

(ii) It S, T are any two members of c., then there exists a member U of c such that

$$U \subset S \cap T.$$

Then there exists a subset of f which is eventually in each member of c.

Proof:

It is evident from (iii) that c is directed by the inclusion relation ⊂. Now consider the subset B of the cartesian product A × C defined by.

$$B = \{(a, U), a \in A, U \in c \text{ and } f(a) \in U\}.$$

We define a binary relation >> in B as follows ;

Let (a, S) and (b, T) be any tow members of B. Then

$$(a, S) >> (a. T) \text{ iff } a \geq b \text{ and } S \subset T.$$

We now show that B is directed by >>. Hence [D1] and [D2] are obvious. To prove [D3], let (a, S) and (b, T) be two members of B. Then by (iii) there exists U in c such that U ⊂ S ∩ T so that U ⊂ S and U ⊂ T. Again by (i) and by the definition of ≥, there exists an element c in A, such that c ≥ ≥ a and ≥ b and f (x) ∈ U. Then (c, U) ∈ B and such that (c, U) >>

(a, S) and (c, U) >> (B, T). Define a mapping.

$$y : B \rightarrow A : y((a, S)) = S \vee (a, S) \in B).$$

Since (a, S) >> (b, T) ⇒ a ≥ b, it follows that

Hence ψ is an isotone mapping. Further ψ [B] is cofinal in A. [For let a be any element of A. Then a ≥ a. But ψ ((a, S)) = a for all S ∈ c. Hence y ((A, S)) ≥ a. Thus of every element a ∈ A, there exists an element ψ ((a, S)) of y [B] such that ϕ ((a, S)) ≥ a]. It follows by the preceding theorem that the mapping

$$\phi = f\, 0\, \psi : B \rightarrow X$$

is a subnet of f.

We now show that this subnet in eventually in each member of c. Let U_0 be an arbitrary member of c. By (i), there exists a member a_0 of A, such that f (a_0) ∈ U_0. Hence by definition of B, b_0 = (c_0, U_0) is a member of B. Now let b = (a, U) be an arbitrary element of B such that b >> b_0. Then a ≥ a_0 and U ⊂ U_0. We then have

$$\begin{aligned} \phi(b) &= (f\, 0\, \psi)(b) = f(\psi(b)) \\ &= f(\psi(a, U)) = f(a) \in U \subset U_0. \end{aligned}$$

Thus, there exists an element b_0 ∈ B, such that for every b ∈ B. b >> b_0 ⇒ f (b) ∈ U_0. Hence f is eventually in U0.

CLUSTER POINTS OF A NET

Definition : *Let (X, ℑ) be a topological space and let f be a net in X. Then a point x_0 ∈ X is called a cluster point of the net f iff f is frequently in ever ℑ-nhd of x_0.*

Example 1:

Consider the sequence f : N → R defined by setting

$$f(n) = -1 \text{ when } n = 1, 3, 5, 7, \ldots$$

and $$f(n) = n\backslash 2 \text{ when } n = 2, 4, 6, 8\ldots$$

that is the sequence {– 1, 1, – 1, 2, – 1, 3, – 1, ... }.

Then –1 is a cluster point of f since f is evidently frequently in every nhd of –1. Also it is evident that no other real number can be a cluster point of f. Hence –1 is the unique cluster point of this sequence. But this sequence fails to converge to –1 since] –2, 1 [is a U-nhd of – 1 in which does not eventually lie.

Example 2:

Let N be directed by ≥ in the usual sence. Consider the mapping,

$$f : N \rightarrow R$$

defined in such a way that f (N) = Q. Then f is a sequence (and hence a net) in R..

Example 3:

Let N Be the set of all natural members directed by ≥ in the usual sense. Then the mapping.

$$f : N - R : f(n) = n \quad \forall n \in N,$$

is a net in (R, U). This net t has no U cluster points in R. For, let x be any point of R. Then]x – 1, x + 1 [is a U-nhd of x. But evidently f is not frequently in this neighbourhood. Hence x is not a cluster point.

Example 4:

Show that the set of all cluster points of a net in a topological space need not be closed.

Solution:

Consider the cofinite topology $\Im$ on N. Then the mapping $f : N \to N = f(n) = 2n - 1$ is a net in N the set of whose cluster points in {1, 3, 5, ... }. But this set is not $\Im$=closed.

Theorem 1:

Let (X, $\Im$) be a topological space. A point x_0 in X is a cluster point of a net (f, X, A, ≥) iff there exists a subset (G, X, B, ≥) which converges to x_0.

Proof:

The 'if' part : Assume that f has a subset g which converges to x_0. To prove that x_0 is a cluster point of f. Let N be a $\Im$-nhd of x_0 of x_0 and let a_0 by any element of A. Since g is a subnet of f, there exists a mapping f : B → A such that

(i) g = o f φ and

(ii) for each a in A, there exists an element b in B such that f (x) ≥ a for every x ≥* be in B.

Hence be (ii). corresponding to $a_0 \in A$, there exists an element $b_0 \in B$ such that

$$\varphi(x) \geq a_0 \text{ for every } x \geq^* \text{by}.$$

Proof:

The 'if' part : We suppose that for every real valued continuous mapping f of a closed subset F of X into [a, b], there exists a continuous extension of f over X. To show that (X, $\Im$) in normal. Let F_1, F_2 be two closed subsets of X such that $F_1 \cap F_2 = \varnothing$ and let [a. b] be any closed interval. We define a mapping.

$$f : F_1 \cup F_2 \rightarrow [a, b].$$

such that $f(x) = a$ if $x \in F_1$ and $f(x) = b$ if $\in F_2$. This mapping is certainly continuous over the subspace $F_1 \cup F_2$. For if H be any closed subset of [a,b]

$$\text{then, } f^{-1}(H) = \begin{cases} F_1 \text{ if } a \in H \text{ and } b \notin H \\ F_2 \text{ if } b \in H \text{ and } a \notin H \\ F_1 \cup F_2 \text{ if } a \in H \text{ and } b \in H \\ \varnothing \text{ if } a \notin H \text{ and } b \notin H. \end{cases}$$

Now $(F_1 \cup F_2) \cap F_1 = F_1$ and $(F_1 \cup F_2) \cap F_2 = F_2$ so that F_1 and F_2 are closed if $F_1 \cup F_2$. Of course $F_1 \cup F_2$ and $\varnothing$ are closed in $F_1 \cup F_2$. It follows that $f^{-1}[H]$ in closed in $F_1 \cup F_2$. Hence f is a continuous map over the subspace $F_1 \cup F_2$. Therefore ,y hypothesis f can by extended to a continuous map g over X. This means that there exists a continuous map.

$$g = X \rightarrow [a, b]$$

such that $g(x) = a$ if $x \in F_1$ and $g(x) = b$ if $x \in F$. The mapping g now satisfies the condition sated in Urysohn's Lemma and hence $(X, \mathfrak{I})$ is normal.

The 'only if' part. Let $(X, \mathfrak{I})$ be a normal space and let f be a real; valued continuous map of the closed subset of the closed subset F into closed interval [a, b] which for numerical convenience, we take to be [–1, 1]. To show that there exists a continuous extension of f over X. We begin by defining a map

$$f_0 : F \rightarrow [-1, 1],$$

by setting $f_0(x) = f(x) \vee x \in F$.

Let $\quad G_0 = f_0^{-1}\ G_0 = f_0^{-1}\left[\left[-1, -\frac{1}{3}\right]\right], H_0 = f_0^{-1}\left[\left[\frac{1}{3}, 1\right]\right].$

Since $\left[-1, -\frac{1}{3}\right]$ and $\left[\frac{1}{3}, 1\right]$ are closed in [– 1, 1] and f_0 in continuous, it follows that G_0 and H_0 are closed in F and so also closed in X. Furthermore

$$G_0 \cap H_0 = \left(f_0^{-1}\left[\left[-1, -\frac{1}{3}\right]\right]\right) \cap \left(f_0^{-1}\left[\left[\frac{1}{3}, 1\right]\right]\right)$$

$$= \left[\left(f_0^{-1}\left[\left[-1, -\frac{1}{2}\right]\right]\right) \cap \left[\frac{1}{3}, 1\right]\right] = f_0^{-1}[\varnothing] = \varnothing.$$

Since g converges to x_0, there exists an element $p \geq b_0$ in B such that $g(p) \in N$. Now let φ (r). Then $q \in A$ and $q \geq a_0$. Also $f(q) = f(\varphi(p)) = (f(\varphi)(p) \in N$. Thus, we have shown that for each element a_0 in A, there exists an element $q \geq a_0$ in A such that $f(q) \in N$. Hence f is frequently in N. It follows that x_0 is a cluster point of f.

The 'only if' part. Let x_0 be a cluster of a net (f, X, A, ≥) and let N (x_0) be the collection of all $\mathfrak{J}$-nhds of x_0. If L, M are any two members of (N (x_0), then $L \cap M$ is also a member of N (x_0). Also since x_0 is a cluster point of f. f is frequently in each member of N (x_0). Hence by the preceding theorem there exists a subnet of g of f which is eventually in each member of N (x_0). This implies that g converges to x.

Theorem 2:

Let $(X, \mathfrak{J})$ be a topological space and let Y be a $\mathfrak{J}$-closed subset of X. If x_0 is a cluster point of a net (f, X, A, ≥) which is eventually in Y, then $x_0 \in Y$.

Proof:

By the previous theorem, there exists a subnet g of f which converges to x_0. Since f is eventually in Y. g may be considered as a net in Y which converges to x_0. Hence, $x_0 \in \overline{Y}$. Since Y closed, we have $\overline{Y} = Y$. Hence $x_0 \in Y$.

Theorem 3:

Let $(X., \mathfrak{J})$ be topological space and let (f, X, A, ≥) be a net n X. For each a in A, let $M_a = \{f(x) : x \geq a \text{ in } A\}$. Then a point P of X is a cluster point of f iff $p \in \overline{M_a}$ for all $a \in A$..

Proof:

The 'if' part : Let $P \in \overline{M_a}$ $\forall a \in A$ and suppose , if possible, p is not a cluster point of f. Then there exists a nhd N of p and an element a in A such that $f(x) \notin N$ for every $x \geq a$ in A. This implies that $N \cap M_a = \varnothing$. It follows that $p \notin \overline{M_a}$ for this a. But this is a contradiction. Hence p is a cluster point of f.

The 'only if' part : Let p be a cluster point of f and let N be an arbitrary nhd of p. Then f is frequently in N, that is, for each $a \in A$., there exists $x \geq a$ in A such that $f(x) \in N$. Hence $M_a \cap N \neq \varnothing$ for every $a \in A$. Thus each nhd of p intersects M_a for all $a \in A$. It follows that $p \in \overline{M_a}$ for every a in A.

Theorem 4:

A topological space $(X, \mathfrak{J})$ is compact if and only if each net in X has a cluster point.

Proof:

The 'if' part : Let every net in X have a cluster point and let F be a collection of $\mathfrak{I}$-closed subsets of X with FIP. Let C = {D : D is the intersection of a finite subcollection of {F}. Since the intersection of every two members of C is a member of C, it is evident that C is directed by the inclusion relation $\subset$. Since each D is non-empty, by the axiom of choice we may choose a point x (D) in D. Now consider the mapping.

$$f : C \to X : f(D) = x(D) \forall C.$$

Then f is net in X . By hypothesis, f must have a cluster point, say x_0. Let E be an arbitrary member of C. Then for every D ≥ E (*i.e.* D ⊂ E) in c. We have f (D) = x (D) ∈ D ⊂ E. Hence f is eventually in the closed set. E. It follows that is, $x_0 \in E$. Since E was chosen arbitrary,

$$x_0 \in \cap C \subset \cap F \text{ since } F \subset C.$$

Hence $\cap F \neq \varnothing$ and consequently X is compact.

The 'only if' part : Let (X, $\mathfrak{I}$) be compact and let (f, Y, A, ≥) be a net in X. For each a in A, let M_a = {f (x) : x ≥ a in A}. Since A is directed by ≥, the collection {M_a : a ∈ A} has the FIP. Hence { $\overline{M_a}$: a ∈ A} = ≠ ∅. Let p ∈ ∩ { $\overline{M_a}$: a ∈ A}. Then p ∈ { $\overline{M_a}$ ∀ a ∈ A. Hence by the proceeding theorem p is cluster point of f.

Corollary : *A topological space (X, $\mathfrak{I}$) is compact iff each net in X has a subnet which converges to some point in X.*

Proof:

It is an immediate consequence of the above theorem.

EXERCISES

1. Exhibit a sequence < x_n > in a set X and a subset of < x_n > which is not a sequence.
2. Show that every subnet of an ultranet is an ultranet.
3. Show that every net has a subnet which is an ultranet.
4. If every subnet of a net {f_a : a ∈ Λ} in a space X has a subnet covering in x_0 ∈ X (fa) converges to x_0.

FILTERS

Definition : *Let X be any non-empty set. A filter on X is a non-empty family F of subsets of X satisfying the following axioms.*

[F1] : ∅ ∈ F.

[F2] : *If F ∈ F and H ⊃ F, then H ∈ F.*

[F3] : *If F ∈ F and H ∈ F, then F ∩ H = F.*

Remarks:

(i) The axiom [F1] states that no member of F is empty :

(ii) Since F is non-empty, it contains at least one subset, say F, of X and since X ⊃ X, It follows from [F2] that X ∈ F.

(iii) The axiom [F3] implies that if F_1, F_2, ... F_n are members of F, then $F_1 \cap F_2 \cap ... \cap F_n$ is also a member of F. Then by $[F_1]$, $F_1 \cap F_2$ ... $\cap F_n \neq \varnothing$. It follows that F has FIP.

(iv) Since X is the intersection of the empty subfamily of F, it follows from (ii) and (iii) that F is closed for finite intersections.

(v) We observe from [F1] that the power set } (X) is not a filter on X. Thus any filter on X must be a proper subset of P (X).

(vi) If F ∈ F then X – F ∉ F. For if both F and X – F belong to F, then by [F3], F ⊃ (X, – F) = ∅ ∈ F which contradicts [F1].

Definition : *A filter F on X is said to be free iff ∩ {F : F ∈ F} = ∅ and fixed iff ∩ {F : F ∈ F} ≠ ∅.*

Example 1:

Let X be any non-empty set and let x_0 ∈ X. Show that the family F = {F : x_0 ∈ F} is a filter on X called the discrete filter.

Solution:

F is non-empty since {x} ∈ F.

[F1] : Since x_0 ∈ F ∀ F ∈ F. It follows that no member of F is empty and so ∅ ∈ F.

[F2] : Let F ∈ F and H ⊃ F. Now F ∈ F implies that x_0 ∈ F. Since G ⊃ F., we have x_0 ∈ We have x_0 ∈ H. Hence H ∈ F.

[F3] : Let F ∈ F and H ∈ F. Then x_0 ∈ F and x_0 ∈ H. Hence x_0 ∈ F ∩ H. It follows that F ∩ H ∈ F.

Example 2:

Let X be any non-empty set. Then {X} is always a filter on X called the indiscrete filter.

Example 3:

Let X be any non-empty set and let F_0 be a non-empty set and let F_0 be a non-empty subset of X. Then the family F = {F : F ⊃ F_0} is a filter on X.

Solution:

Since $F_0 \in F$, F is non-empty. Further

[F1] : Since F_0 is non-empty and each member of F contains F_0 it follows that each member of F is non-empty. Hence $\varnothing \notin F$.

[F2] : Let $F \in F$ and $H \supset F$. Now $F \in F$. Now $F \in F$. Then $F \supset F_0$ and $H \supset F$, it follows that $H \supset F_0$. Hence $H \in F$.

[F3] : Let $F \in F$ and $H \in F$. Then $F \supset F_0$ and $H \supset F_0$. Hence $F \cap H \supset F_0$. It follows that $F \cap H \in F$.

Example 4:

Let (X, $\mathfrak{J}$) be a topological space and let N (x) be the collection of all $\mathfrak{J}$-nhd of a point $x \in X$ Then N (x) is a filter on X called the neighbourhood filter of X.

Solution:

Since $X \in N(x)$, it follows that N (x) is non-empty.

[F1] : Since each member of N (x) contains x, it follows that no member of N (x) is empty. Hence $\varnothing \notin N(x)$.

[F2] : Let $F \in N(x)$ and $H \supset F$. Then F is a $\mathfrak{J}$-nhd of x and since $H \supset F$. H is also a $\mathfrak{J}$-nhd of x. It follows that

$$H \in N(x)$$

[F3] : Let $F \in N(x)$ and $H \in N(x)$. Then, we have $F \cap H \in N(x)$.

Example. 5:

Let X be any infinite set. Then $F = \{F : X - F \text{ is finite}\}$ is a filter on X called the cofinite filter.

Solution:

Since $X - X = \varnothing$ is finite, $X \in F$ and hence F is non-empty. Further.

[F1] : Since X is infinite and X – F infinite, it follows that F is an infinite set and so no member of F os empty. Hence $\varnothing \in F$.

[F2] : Let $F \in F$ and $H \supset F$. Since X – F is finite, then surely X – H is finite. Hence $H \in F$.

[F3] : Let $F \in F$ and $H \in F$. Then X – F and X – H are both finite. Now

$$X - (F \cap H) = (X - F) \cup (X - H) \quad \text{[De-Morgan law]} \quad ...(1)$$

Since X – F and X – H are finite, it follows from (1) that $X - (F \cap H)$ is finite. Hence $F \cap H \in F$. Thus F is a filter on X, In particular if X = N, then co-finite filter on N is called the Erchet filter.

Example 6:

Show that there exists no filter on the empty set ∅.

Solution:

Suppose, if possible, there exists a filter F on ∅. Now by (ii) of the above remark ∅ ∈ F. But by [F1]. ∅ ∈ F. But this is a contradiction. Hence there exists no filter on ∅.

Example 7:

Let X = {A, b, c} and let

F_1 = {X}, F_2 = {{a, b}, X}, F_3 = {{b, c}, X},

F_4 = {{c,), X}, F_5 = {{a}, {a, b}, {a, c}, X},

F_6 = {{b,}, {a, b}, {b, c}, X}, F_7 = {{c,}, {c, d}, {b, c}, X,}

F_8 = {{a, b}, {b, c}, X}, F_9 = {{a,}, {b}, {a, b}, X},

F_{10} = {{∅, {a}, {a, b}, X}.

Solution:

Then F_1, F_2, F_3, F_4, F_5, F_6, F_7 are filters on X as the reader can easily verify, F_8 is not a filter because it does not satisfy [F3], F_9 is not a filter because it does not satisfy [F2] and [F3], F_{10} is not filter since it does not satisfy [F1] and [F2].

COMPARISON OF FILTERS

Definition : *Let F and F ' be two filters on the same set X, Then F is said to be finer than F' or F' is coarser that F iff F' ⊂ F. If in addition, F ≠ F', then F is said to be strictly finer than F' or F' strictly coarser than F.*

Two filters F and F′ are said to be comparable iff F is finer than F′ or F′ is finer than F. The set of all filters on X is directed by relation ≤ defined by setting F ≤ F′ iff F is coarser than F′.

Example 1:

Let X be an infinite set and let F be a filer on X such that ∩ {F : F ∈ F} = ∅. Show that F is finer than the cofinite filter on X.

Solution:

Let c be the cofinite filter on X. To show that F ∈ c. Suppose, if possible. S c. Then there exists C ∈ c such that C ∈ F. Then X– C is finite. Let X – C = {x_1, X_2, ... x_n}. Now ∩ {F :∈ F ∈ F} = ∅. Therefore, there exists F_1 ∈ F {i = 1, 2, ..., n} such that x_i ∉ F. Since F is a, G = ∩ Fl : 1 ≤

$i \le n\} \in F$. This G will not contain and element of $\{x_1, x_2,... x_n\}$. Hence $X - G \supset X - C$ and consequently $G \in C$. Since F is filter $C \supset G$ implies $C \in F$. But this is a contradiction . Hence F must be finer than the cofinite filter.

Example 2:

Consider the filters defined as in example (4.5). Here F_1 is coarser than all the filters on X. F_5 is finer than F_1, F_2 and F_4 ; F_6 is finer than F_1, F_2 and F_3 ; F_7 is finer than F_2 and F_4, F_6 and F_7 are not comparable etc.

Theorem 1:

Let $\{F_\lambda : \lambda \in \Lambda$ be any non-empty family of filters on a non-empty set X. Then the set $F = \cap \{F_\lambda : \lambda \in \Lambda\}$ is also a filter on X.

Proof:

Since X is a member of each F_λ, X is a member of F. Hence F is non-empty. Further.

[F1] : Since $\varnothing \in F_\lambda \ \forall\ l \in L$, it follows that $\varnothing \notin F$.

[F2] : Let $F \in F$ and $H \supset F$. Now $F \in F \Rightarrow F \in F_\lambda \ \forall\ \lambda \in \Lambda$. Since each F_λ is a filter and $F \in F_\lambda$ it follows that $H \in F_\lambda \ \forall\ \lambda \in L$. Hence $H \in F$.

[F3] : Let $F \in F$ and $H \in F$. Then $F \in F_1$ and $H \in F_\lambda \ \forall\ \lambda \in \Lambda$. Since F_λ is a filter on X for all $\lambda \in L$, it follows that $E \cap H \in F_\lambda \ \forall\ \lambda \in L$. Hence, $F \cap H \in F$.

Corollary : *The filter F obtained above is the coarsest filter which contains A, that is, F is the smallest element of the ordered family of all filters on X containing A.*

Proof:

For, if F′ is any filter containing A, then F′ must contain all finite intersection of members of A and their supersets and hence $F' \supset F$.

Definition : *The filter F defined in the above theorem is said to be generated by A and A is said to be a sub-base of F.*

Observe that for A to be a sub-base it is necessary and sufficient that A has the FIP.

Theorem 2:

Let F be a filter on a non-empty set X and let $A \subset X$. Then there exists a filter F′ finer than F such that $A \in F'$ if and only if $A \cap F \neq \varnothing$ for every $F \in F$.

Proof:

The 'if' part : Let $A \cap F \neq F \neq \emptyset \forall F \in F$. ...(1)

Let $A = \{A \cap F : F \in F\}$. Then A has then FIP as shown below :

Let $\{A \cap Fi : 1 \leq : \leq n\}$ be any finite sub-family of A and let $F = \cap \{F_i : 1 \leq i \leq n\}$. Since each F_i is a member of F, F is also a member of F. Hence.

$$\cap \{A \cap F_i : 1 \leq i \leq n\}$$
$$= A \cap [\cap \{F_i : i \leq i \leq n\}]$$
$$= A \cap F \neq \emptyset \text{ by (1)}$$

Then by the preceding theorem, there exists a filter F′ on X containing A. We now show that $F \subset F'$ contains A, it follows that $A \cap F$ is also member of F′. Since $F \supset A \cap F$ and F′ is a filter, it follows that $F \subset F'$. Thus we have shown that

$$F \in F \Rightarrow F \in F'$$

Hence $F \subset F'$, that is, F′ is finer than F.

The 'only if' part : Let F′ be a filter on X such that $A \in F'$ and $F \subset F'$. To prove that $A \cap F \neq \emptyset$ for every $F \in F'$. Let F be an arbitrary member of F. Since $F \subset F'$, we have $F \in F'$. Also $A \in F'$. Since F′ is a filter, we must have $A \cap F \neq \emptyset$.

Example 1:

Let F be an arbitrary member of F and suppose, if possible, $(X, A) \cap F = \emptyset$. But then $F \subset A$ and so by [F2], $A \in F$ which is a contradiction. Hence $(X - A) \cap F \neq \emptyset$, that is $X - A$ meets every member of F. Then by the preceding theorem there exists a filter F' finer than F such that $X - A \in F'$, that is $A \notin F'$.

Solution:

Let M be any non-empty class of filters an a non-empty set X. Then a filter F' on X is said to be supermum (infimum) of M iff F is finer (coarser) than each member of M and if F' is any filter on X which is finer (coarser) than each member of M then F is coarser (finer) than F'.

Thus a supermum of M is the coarsest of all filters on X which are finer than every member of M. Similarly an infimum of M is the finest of all filters on X which are coarser than each member of M.

Example 2:

Let $M_1 = \{F_2, F_3, F_5\}$

$M_2 = \{F_1, F_2, F_3, F_5\}$

$M_3 = \{F_3, F_2, F_5\}$

and $M_4 = \{F_2, F_5, F_5\}$,

where F_1, F_2, F_3, F_4, F_5 and F_6 are as defined in example (4.5). Find superma and infima of M_1, M_2, M_3 and M_4.

Solution:

Here F_0 is supermum of M_1 since it is the coarser of all filters on X which are liner than each member of M_1 in fact, in this case there is no other filter on X which is filter than each member of M_1. Also F1 is an infimum of M_1, because it is the only filter on X which is coarser than every member of M_1. Observe that in this case, supermum is a member of M_1 whereas infimum is not. M_1 has no supermum since there is no filter on X which is finer than each member of M_2. Also F_1 is an infimum of M_2. Similarly M_3 has no supermum and F_1 is an infimum of M_3. An infimum of M_4 is F_2 since it is coarser than every member of M_4 and finer than F_1 which is the only other filter on X coarser than each member of M_4, M_4, however, has no supermum since there is no filter on X which is finer than every member of M_4 is coarser than every member of M. Supermum of M may or may no exist.

Theorem 3:

Let $M = \{F_\lambda : \lambda \in \Lambda\}$ be a non-empty class of filters on a non-empty set X. Then M has a supermum if and only if for all finite sub-families $\{F_\lambda : \lambda \leq i \leq n\}$ of elements of M and all $F_1 \in F_i$ $(1 \leq i \leq n\}$ the intersection

$$\cap \{Fi : 1 \leq i \leq n\} \text{ is non-empty.}$$

Proof:

The 'if' part : Let the conditions hold and let

$$F = \cup \{F_\lambda : \lambda \in L\}.$$

Then by the conditions of the theorem, it is evident that F, has the FIP. Hence there exists the coarsest filter F′ on X containing F, that is, F′ is the coarsest filter on X which is finer than every member of M. Hence F′ is a supermum of M.

The 'only if' part : Suppose that M has a supermum, say F. The F is the F is the coarsest of all filters on X which are finer then every member of M, that is, F is the coarsest of all filters on X such that $F \cup \{F_\lambda : \lambda \in L\}$. Hence, $\cup \{F_\lambda : \lambda \in L\}$ must have FIP and so the conditions of the theorem hold.

EXERCISES

1. Show that if x is an isolated point of V, then
   $\{D : D = N - (x) \text{ and } N \in N(x)\}$ does not generate a filter on X.
2. Let X be a space and let x be a non-isolated point of X. Show that $\{N : N = V - \{x\}\}$ and $V \in N(x)$, generates a filter on X which contains the filter N (x) of all nhds of x in X.

FILTER BASE

Definition : *Let X be any non-empty set. A filter base on X is a non-empty family B of subsets of X satisfying the following axioms.*

[F*1] : $\varnothing \notin B$

[F*2] : If $F \in B$ and $H \in B$, then there exists a $G \in B$ such that $G \subset F \cap H$.

Example. 1:

Let $(X, \mathfrak{I})$ be a topological space and let $x \in X$. Then the local base B (x) at x is a filter base on X.

Solution:

Since $x \in B \ \forall \ B \in B(x)$, it follows that no member of B (x) is empty. Hence $\varnothing \notin B(x)$. Thus [F*1] is satisfied. Again let $A \in B(x)$ and $B \in B(x)$. Then there exists a $C \in B(x)$ such that $C \subset A \cap B$. Hence [F*2] is also satisfied.

Example. 2:

Let $X = \{a, b. c\}$.

Then $\{\{a\}, \{a, b\}, \{a, c\}\}$ is a filter base in X.

[To obtain filter base on a non-empty set X take any collection A of subsets of X. The collection B of all finite intersection of A is easily seen to be filter base on X].

Example 3:

The family $B = \{A\}$ where A is any non-void subset of a n-void set X is a filter base on X.

Solution:

Here B consists of a single member A. Since A is non-void, $\varnothing \notin B$ and hence [F*1] is satisfied. Again since $A \subset A \cap A$. {F*2] is also satisfied. Hence B is a filter base on X.

Theorem 1:

Let B be a family of subsets of a set X. Then B is filter base on X if and only if the family F consisting of all those subsets of X which contain a member of B is a filter on X.

Proof:

First observe that B is non-empty iff F is non-empty.

The 'if' part : Let F be a filter on X. Then $\varnothing \in$ F. By definition of F, we have B $\in$ F. Hence $\varnothing \notin$ B. Thus [F*1] is satisfied. Again let F $\in$ B and H $\in$ B. Then surely F $\in$ F and H $\in$ F. Hence by [F3], F $\cap$ H $\in$ F and so by definition F, there exists a G $\in$ B such G $\subset$ F $\cap$ H. Thus we have shown that if F, H are members of B, then there exists a member G of B such that G $\subset$ F $\cap$ H and consequently [F*2] is also satisfied. Hence B is a filter base on X.

The 'only if' part : Let B be a filter base on X and let F be family of all those subsets of X which contain a member of B. Since is a filter base, $\varnothing \notin$ B and hence also $\varnothing \notin$ F. Thus [F1] is satisfied. Now let F $\in$ F and H $\supset$ F. Then by definition of F, F contains a member of B, say G so that G $\subset$ F and hence G $\subset$ H. It follows that H $\in$ F.

Hence [F2] is satisfied. Finally let F $\in$ F and H $\in$ F. Then there exists G and K belonging to B such that G $\subset$ F and K $\subset$ H so that G $\subset$ K $\subset$ F $\cap$ H. Since B is a filter base by [F*2], there exists a L $\in$ B such that L $\subset$ G $\cap$ K, and hence L $\subset$ F $\cap$ H.

Hence by definition of F, F $\cap$ H $\cap \subset$ F. This [F3] is also satisfied. Hence F is a filter on X.

Theorem 2:

Let A be any non-void family of subsets of a set X. Then there exists a filter base on X containing A iff A has the FIP.

Proof:

The 'if' part : Let A have the FIP and let B = {B : B is the intersection of a finite subfamily of A}. We shall show that B is a filter base on X. Evidently B $\supset$ A. Since A is non-void, B is so also.

[F*1] : Since A has the FIP, no member of B can be empty. Hence $\varnothing \notin$ B.

[F*2] : Let F $\in$ B and H $\in$ B. Since F and H are finite intersections of members of A, F $\cap$ H must also be a finite intersection of members of A. Hence F $\cap$ H $\in$ B. Since F $\cap$ H $\subset$ F $\cap$ H. [F*2] is satisfied.

The 'only if' part : Let B be a filter base on X containing A. The by [F*1] no member of B is empty. The intersection of every finite subfamily of B contains a member of B. In particular, the intersection of every finite subfamily of A contains a member of B. Hence A has the FIP.

The filter base B defined in the above theorem is said to be generated by A.

Definition : *Let B be a filter base on a set X. Then the filter consisting of all those subsets of X which contain a member of B is called a filter generated by B.*

[Note that such a filter always exists in view]

To filter bases B_1 and B_2 on a set X are said to be equivalent iff they generate the same filter on X.

Definition : *Let F be a filter on a set X. Then a sub-family B of F is called a base of F iff every member of F contains a member of B.*

Remark : *A collection B of subsets of a set X is a filter base on X iff B is a base of some filter on X. Or equivalently, a collection B of subsets of X is a base of some filter on X iff [F*1] and [F*2] are satisfied.*

If A is a sub-base of a filter F. Then the family B of all finite intersections of members of A is a base for F.

Example. 1:

Let X be a non-empty directed set directed by the relation R. For each $a \in$ let

$$s\ (a) = \{a \in X : x\ R\ a\}$$

Then the collection

$$B = \{s\ (a) : a \in X\},$$

is a filter base on X.

Solution:

Since the relation R directs the set X, R is reflexive transitive and is such that if a, b are any two members of X, then there exists a member c ∈ X such that x R a and c R b.

We now verify [F*1] and [F*2].

[F*1] : Since X is non-empty, there is at least one x ∈ X and so there is at least one member s (x) in B. Hence B is non-empty. Further since R is reflective, x ∈ s (x) so that s (x) is non-empty for each x ∈ X. Hence ∅ ∉ B.

[F*2] : Let s (a) ∈ B and s (b) ∈ B. Since a, b ∈ X, there exists an element ∈ X such that x R a and c R b. We now show that s (x) ⊂ s (a) ∩ s (b). Let x ∈ s (c). Then x R c. Since c R a and c R b, by transitivity of R, we have x R a and x R b. It follows that x ∈ s (a) and x ∈ s (b). Thus we have shown that

$$x \in s(c) \Rightarrow x \in s(a)$$

and

$$x \in s(c) \Rightarrow x \in s(b).$$

Hence s (c) ⊂ s (a) and s (c) ⊂ s (b).

Therefore s (c) ⊂ s (a) ∩ s (b).

It follows that B is a filter base on X.

Example 2:

Let (X, ℑ) be a topological space and let N (x) be the neighbourhood filter at x. Then a local bases B (x) at x is a base of the filter N (x).

Solution :

By definition N (x) is the collection of all those subsets of X which contain a member of B (x). Also B (x) is a sub-family of N (x). Hence by the above definition B (x) is a base of the filter N (x).

Theorem 3:

On a set X, a filter F, with base B' is finer that a filter F with base B if and only if every member of B contains a member of B'.

Proof:

The 'if' part : Let every member of B contain a member of B′ and let F be any member of F. Then F contains a member of B, say B′. By hypothesis there exists a member B′ of B′ such that B ⊂ B. Hence, F contains B′ and consequently, F ∈ F′. Thus, we have shown that F ∈ F ⇒ F ∈ F′. Hence F ⊂ F, that is, F′ is finer that F.

The 'only if' part : Assume that F′ is finer that F and let B be any member of B. Then B ∈ F. Since F′ is finer than F we have B′ ∈ F′. Since B′ is a base of F′, there exists a B′ ∈ B′ such that B′ ⊂ B. Thus every member of B contains a member of B′.

Theorem 4:

Two filter basis B and B' on a set X are equivalent if and only if every member of B contains a member of B' and every member of B' contains a member of B.

Proof:

The theorem is an immediate corollary of the preceding theorem. However we give an independent proof below.

Assume that every member of B contains a member of B′ and every member of B′ contains a member of B. Let F and F′ be the filters generated by B and B′ respectively. Let F ∈ F. Then there exists a B ∈ B such that B ∈ F. By hypothesis, there exists a B′ ∈ B′ such that B′ ⊂ B. It follows that B′ ⊂ F. Hence by definition (6.10) F ∈ F′. Thus we have shown that

$$F \in F \Rightarrow F \in F' \text{ and hence } F \in F'.$$

Similarly we can show that F′ ⊂ F. Therefore F = F′. Thus B and B′ generate the same filters and so they are equivalent.

Conversely, assume that B and B′ are equivalent so that they generate the same filter, say F. Let B be an arbitrary member of B. Then B ∈ F. Since B. is also a base of F, there exists a B′ ∈ B′ such that B′ ⊂ B. Thus every member of B contains a member of B′. Similarly it can be shown that every member of B′ contains a member of B.

Theorem 5:

Consider the usual topology U on R^2. Let B (x) be the collection of all circular nhds and B' (x) the collection of all square nhds of a point x ∈ R^2. Then B and B' are equivalent since they generate the same filter on R^2, namely, the neighbourhood filter at x.

Proof:

Do yourse

Theorem 6:

Let f be a mapping of a set X into another set Y.

(i) If B is a filter base on X, then the family C = {f [B] : B ∈ B} is a filter base on Y.

(ii) If C is a filter base on Y such that every member of C intersects f [X], then the family D = {f^{-1} [C] : C ∈ c} is a filter base on X.

Proof:

(i) Let f [B_1] ∈ c and f [B_2] ∈ c where B_1, B_2 ∈ B. Then by [F*2], there exists a member B_2 of B such that $B_3 \subset B_1 \cap B_2$. Therefore f [B_3] ⊂ f [$B_1 \cap B_2$]. But f [$B_1 \cap B_2$] ⊂ f [B_1] ∩ f [B_2]. Hence f[B_3] ⊂ f [B_1] ∩ f [B_2]. Thus C satisfies [F*2]. Further B ≠ ∅ ⇒ f [B] ≠ ∅ ∀ B ⊂ B. Hence [F*1] is also satisfied. It follows that C is a filter base on Y.

(ii) Let f^{-1} [C] be any member of D where C ∈ c. Since C is a filter base, C ≠ ∅. Also by definition of c, C ∩ f [X] ≠ ∅. We claim that f^{-1} [C] ≠ ∅. For if f^{-1} [C] = ∅, then either C = ∅ which is a contradiction or no member of C belong to the range set f [X] so that C ∩ f [X] = ∅ which is again a contradiction. Hence f [C] ≠ ∅. Thus [F*1]] is satisfied.

Now let f–1 [C] and f–1 [D] be any tow members of D where C, D are members of c. Since c is a filter base, there exists a member E of c such that E ⊂ C ∩ D.

But E ⊂ C ∩ D ⇒ f^{-1} [E] ⊂ f^{-1} [C ∩ D]

⇒ f^{-1} [E] ⊂ f^{-1} [C] ∩ f^{-1} [D].

Also E ∈ c ⇒ f^{-1} [E] ∈ D. Thus {F*2] is also satisfied.

Hence D is a filter base on X.

EXERCISES

1. Let f : X → Y, and let B be a filter base on X. Show that f [B] is a filter base on Y, and if F is a filter on X, then f [F] is a filter on Y if and only in f is onto.
2. If B is a filter base, than $\{\overline{B} : B \in B\}$ is also a filter base, and generates a smaller filter.
3. Show that the collection of all discs containing the origin in their interiors is a filter base on R^2.
4. Prove that a filter in a Hausdorff space can converge to at most one point.
5. Let Y be a proper subset of a set X and let F be a filter on Y. Prove that F is not a filter on X {∵ X ∉ F]. But F be a filter base on X,. Also prove that if B is a filter base on Y, then B is also a filter base on X.
6. Show by means of an example that a base for a topology is not a filter base, but a local base is always filter base.

ULTRAFILTERS

Definition : *A filter F on a set X is said to be an ultrafilter (or a maximal filter) on X iff F is not properly contained in any other filter on X that is iff there exists no filter on X strictly finer that F.*

Thus F is an ultrafilter on X iff for every filter F′ on X, F′ ⊃ F implies F′ = F.

It follows that an ultrafilter on X is a maximal element of the collection of all filters on X partially ordered by the inclusion relation ⊂.

Definition : *A filter base on a set X is called an ultrafilter base iff it is a base of an ultrafilter.*

Example:

The filters F_5, F_6 and F_7 defined in example (4.5) are ultrafilters on {a, b, c} since there are no filters strictly finer that these filters.

EXISTENCE OF ULTRAFILTERS

Theorem 1:

Every filter on a set X is contained in an ultrafilter on X.

Proof:

Let F be any filter on X and let M be the class of all filters on X containing F. Then m is non-empty since F ∈ M. Also is partially ordered by the inclusion relation ⊂. Now let H be a linearly ordered subset of M. Then by definition of linear ordering for and tow members F_1, F_2 of H, we have either $F_1 \subset F_2$ or $F_2 \subset F_1$. Let $G = \cup \{F_\lambda : F_\lambda \in H\}$. We claim that G is a filter on X. G is evidently non-empty Also.

[F1] : Since each F_λ is a filter, we have $\varnothing \notin F_\lambda \forall F_\lambda \in H$ and so $\varnothing \notin G$.

[F2] : Let A ∈ G and B ⊃ A. Then $A \in F_\lambda$ for some $F_\lambda \in H$ and since F_λ is a filter. $B \in F_\lambda$. It follows that B ∈ G.

[F3] : Let A ∈ G and B ⊃ G. Then $A \in F_\lambda$ and $B \in F_\lambda$ for some F_λ $F_\mu \subset F_\lambda$, F_μ. ∈ H, Since H is linearly ordered, we have tither $F_\lambda \subset F_\mu$ or $F_\mu \subset F_\lambda$. Hence both A and B belong either to F_λ or to F_μ and so A ∩ B belongs either to F_λ or to F_μ. It follows that A ∩ B ⊂ G.

Accordingly G is a filter on X. Further G is finer than every member of H and so G is an upper bound of H. Thus we have shown that M is a non-empty partially ordered set in which every linearly ordered subset has an upper bound. Hence by Zorn's lemma M contains a maximal on X containing F.

A more general form of the above theorem is the following.

Theorem 2:

Evey filter base on a set X is contained in an ultrafilter on X.

Proof:

The proof of this theorem is the same as that of the above theorem except that M is new the class of all filters, on X containing a filter base instead of a filter. A still more general theorem is the following.

Theorem 3:

Let A be a non-empty collection of subsets of a set X such the A has the FIP. Then there exists an ultrafilter F containing A.

Proof:

Let M = {C : C is a collection of subsets of X with the FIP such that C ⊃ A}. Then M is non-empty since A ∈ M. Also M is partially ordered by the inclusion relation ⊂. Further it is easy to see that every linearly ordered subset of M has an upper bound. Therefore by Zorn's lemma there exists a maximal element F of M. We now show that F is filter on X. Since F ⊃ A and A is non-empty, it follows that F is also non-empty. Further.

[F1] : Since F has the FIP, no member of F can be empty. Hence ∅ ∅ F.

[F2] : Let t A ∈ F and B ⊃ A. Since F has the FIP, it follows that {B} ∪ F also has FIP. Also {B} ∪ F ⊃ F. Hence {B} ∪ F ∈ M. But since F maximal. B ∈ F.

Let A ∈ F and B ∈ F. Since F has thew FIP, it follows the {A ∩ B} ∩ F also has FIP. Also {B} ∪ F ⊃ F. Hence {A ∩ B} ∪ F ∈ M. But since F is maximal, A ∩ B ∈ F. Hence F is the desired ultrafilter.

CHARACTERIZATION OF ULTRAFILTERS

Theorem 1:

A collection F of non-empty subsets of a set X is an ultrafilter on X if the following conditions are satisfied.

(i) F has the FIP

(ii) For every subset A of X, we have A ∈ F or X – A ∈ F.

Proof:

Let F′ be any filter containing F. Let F′ be any member of F′. Let F′ be any member of F′. Then X – F′ ∉ F′ therefore X – R′ ∉ F. Hence by (ii), F′ ∈ F. Then we have shown that F′ F′ ⇒ F′ ∈ F. Hence F′ ⊂ F. But F ⊂ F′. Therefore F′ = F. Accordingly F is an ultrafilter on X.

Theorem 2:

A filter F on a set X is an ultrafilter iff F contains all those subsets of X which intersect every member of F.

Proof:

The 'if' part : Let F be a filter on X satisfying the condition and let F′ be a filter finer than F, so that F′ so that F′ so that F′ ⊃ F. Let F′ ∈ F′. Then by [F1] and [F3], F′ intersects every member of F′ and since F ⊂ F′, it follows that F′ also intersects every member of F. Hence by definition of F, F′ ∈ F. Therefore F′ ⊂ F. Hence F′ = F. Accordingly F is an ultrafilter on X.

The 'only if' part : Let F be an ultrafilter on X and let A be a subset of X intersecting every member of F. Now consider the collection

$$F' = \{F' : F' \supset A \cap F \text{ for some } F \in F\}.$$

We claim that F′ is a filter on X containing F. Surely F′ contains F since F ⊃ A ∩ F ∀ F ∈ F. Moreover,

[F1] : Since A intersects every member of F, we have

$$A \cap F \neq \varnothing \ \forall\ F \in F.$$

Again since F ⊃ A ∩ F F, it follows that

$$F' \neq \varnothing \vee F' \in F'$$

[F2] : Let F′ ∈ F′ and G′ ∈ F ∈. Then F′ ⊃ A ∩ F for some F ∈ F. Hence G′ ⊃ A ∩. It follows that G′ ∈ F′.

[F3] : Let F′ ∈ F′ and G′ ∈ F′. Then F′ ⊃ A ∩ F and G′ ⊃ A ∩ G where F, G ∈ F and so F′ ∩ G′ ⊃ (A ∩ F) ∩ (A ∩ G), that is, F′ ∩ G′ ⊃ A ∩ (F ∩ G). Since F ∩ G ∈ ∈ F, we have F′ ∩ G′ ∈ F′.

Thus it is shown that F′ is a filter containing F. Since F is an ultrafilter, we have F = F′. Now since X ∈ F and A ∩ X = A so that A ⊃ A ∩ X, it follows that A ∈ F and so also A ∈ F.

Corollary : *A filter F on an a set X is an ultrafilter on X if A or X belongs to F for all subsets A of X.*

Proof:

Since a filter has the FIP, the result follows from the above theorem.

Another statement of the above theorem.

A subset A of a filter on set X is an ultrafilter on X iff for every subset A of X we have either X ∈ A or X – A ∈ A.

Theorem 3:

A filter F on a set X is a ultrafilter if and only if for any two subsets A, B, of X such that A ∪ B ∈ F, we have either A ∈ F or B ∈ F.

Proof:

Let F be an ultrafilter on X and let $A \cup B \subset F_0$. To show that either $A \in F$ or $B \in F$. Suppose, if possible, that $A \notin F$. There by [F1], either A or B is non-empty. Assume that $B \neq \varnothing$. Now consider the collection

$$G = \{G : G \cup A \in F\}.$$

Then G is non-void since $B \in G$. We now claim that G is filter on X strictly finer that F.

[F1] : Since $\varnothing \cup A = A \notin F$, we have $\varnothing \notin G$.

[F2] : Let $G \in G$ and $H \supset G$. Then $G \cup A \in F$ and since F is a filter $H \cup A \supset G \cup A$ implies $H \cup A \in F$. Hence by definition of G, $H \in G$.

[F3] : L Remark (4.2) (ii). But $X = (X - A) \cup A$. Hence by hypothesis ether $A \in A \in F$ or $X - A \in F$. Hence by the preceding theorem F is an ultrafilter on X.

Corollary 1 : *If F is an ultrafilter, then for every subset A of X, we have either $A \in F$ or $X - A \in F$.*

Proof:

Since $X \in F$ and $A \cup (X - A) = X$, by the above theorem, we have either $A \in F$ or $A - A \in F$.

Corollary 2 : *If F is an ultrafilter and $A_1 \cup A_2 \cup ... \cup A_n \in F$, then at least one $A \in F$.*

Proof:

Corollary 3 : *If F is an ultrafilter on X and $A_1, A_2, ... A_n$ are subsets of X such that $\cup (A_i : 1 < i < n)$ covers X, then some $A_i \in F$.*

Proof:

Since $\cup \{A_i : 1 < i < n\} = X \in F$, the result follows from the above theorem.

Theorem 4:

A filter F on X is an ultrafilter on X iff A or X – A belongs to F for all subsets A of X.

Proof:

Corollary as above.

Theorem 5:

If F is an ultrafilter on a set X, then F is an ultrafilter base on every superset Y of X.

Proof:

Let F′ be the family of all those subsets A of Y such that A contains a member of F. Then F′ is a filter on Y. Now let B be any subset of Y. Since Y is a superset of X, we have $Y \cap X = X$ and hence

$$(X \cap B) \cup [X \cap (Y - B)] = X \cap [B \cup (Y - B)]$$
$$= X \cap Y = X \in F.$$

since F is an ultrafilter on X, we have either.

$$X \cap B \in F \text{ or } X \cap (Y - B) \in F'.$$

If $X \cap B \in F$, then $B \in F'$, then $Y - B \in F$. Thu~ we have shown that if B is any subset of Y, then either $B \in F'$ or $Y - B \in F$. Hence F′ is an ultrafilter on Y. Hence F is an ultrafilter base on Y.

Theorem 6:

Every filter F on a set X is the intersection of all the ultrafilters finer than F.

Proof:

Let $H = \cap \{ : G$ is an ultrafilter containing $F\}$. To show that $H = F$. Since $F \subset G \vee G.$, we have $F \subset H$. Conversely let $A \in H$. Then $A \in G$ $\forall$ G.

If possible, let $A \notin F$. Then A contains no member of F by [F2]. Hence every $A \in F$ intersects $X - A$. Then exists a filter F′ finer than F and containing $X - A$. There exists an ultrafilter, say G_0, finer than F. Now $X - A \in F' \Rightarrow A \notin F'$ and since $G_0 \supset F'$, we have $A \notin G_0$. But this contradicts (1). Hence $A \in F$. Thus we have shown that $A \in H \Rightarrow A \in F$. Therefore $H \subset F$. It follows that $H = F$.

Theorem 7:

Let A be a subset of X which has non-empty intersection with every member of an ultrafilter F, then A must belong to F.

Proof:

Let $A = \{A\} \cup F$. Evidently, the intersection of any finite number of members of A is non-empty. Hence A generates a filter G containing all l the sets in A, and so, in particular A. But since F is a maximal filter $F = G$, and hence $A \in F$.

Example 1:

Show that in a discrete topological space, every neighbourhood filter is an ultrafilter.

Solution:

In a discrete space X, the neighbourhood filter of a point $x \in X$ is the collection of all subsets of X which contain x. Hence by the preceding example, every nhd filter on X is an ultrafilter on X. Note that in this space, no other filter can be an ultrafilter.

Example 2:

Show that the collection F (p) all subsets of a set X which contains a given element $p \in X$ is an ultrafilter on X. Show filter that {{p}} is an ultrafilter base on X.

Solution:

Here F (p) = $\{F : F \subset X$ and $p \in F\}$. Let A be any subset of X. Then $p \in A$ or $p \in X - A$. Hence either $A \in F$ (p) or X – A (p). Also evidently F (p) has the FIP. Hence F (p) is an ultrafilter on X. Further since {{p}} is a filter base why ?) on X which generates the ultrafilter F (p), it follows that {{p}} is an ultrafilter base on X.

Example 3:

Let F be an ultrafilter on a set X. Show that $\cap$ {R : F $\in$ F} is either empty of a singleton subset of X.

Solution:

Let $G = \cap \{F : F \in F\}$ and let $G \neq \emptyset$. Then there is at least one element x in X such that $x \in G$. Suppose, if possible there is another element y different from x such that $y \in G$. Since F is an ultrafilter, either $\{x\} \in F$ or $X - \{X\} \in F$. If $\{x\} \in F$, then, since $y \neq x$, $y \notin \{x\}$ and hence $y \notin G$, which is contradiction. Similarly if $X - \{x\} \in F$, then $x \notin G$ which is again contradiction. These contradictions arise by supposing $x \neq y$. Hence $x = y$. It follows that $G = \{x\}$. That is, either G is empty or a singleton set.

Theorem 8:

Let f be mapping of set X onto a set Y. If F is an ultrafilter on X, then $G = \{f[F] : F \in F\}$ is an ultrafilter on Y.

We have G is a filter base on Y. Let B be any subset of Y, Then we have

$$X = f^{-1}[Y] = f^{-1}[B \cup (Y - B)] = f^{-1}[B] \cup f^{-1}[Y - B].$$

Proof:

We know that Also , $X \in F$. It follows that $f^{-1}[B] \cup f^{-1}[Y - B] \in F$. Hence $f^{-1}[B] \in F$ or $f^{-1}[Y - B] \in F$, that is, either $f^{-1}[B] = F$ or $f^{-1}[Y - B] = H$ where F, H are members of F. Hence either $f[F]\ f[f^{-1}[B]] = B$ or $f[H] = f[f^{-1}[Y - B]] = Y - B$.

Note that since f is onto, we have $f[f^{-1}[Y - B]] = Y - B$ etc. It follows that either $B \in C$ or $Y - B \in C$. Then C is an ultrafilter on Y

Definition : *Let $\{f_a : a \in A\}$ be a net in X. Then the filter generated by the filter base B consisting of the series*

$$B_{a_0} = \{f_a : a \geq a_0, a_0 \in A\}$$

is called the filter generated by $\{f_a\}$.

Definition : *Let F the filter on X, and let*

$$\Lambda_r = \{(x, F) : x \in F \in F\}.$$

Then Λ_r is directed by the relation $(x_1, F_1) \leq (x_2, F_2)$ iff $F_2 \subset F_1$ and so the map $f : \Lambda_r \Rightarrow X$ defined by $f(x, F) = x$ is an net in X, It is called the net based on F.

Theorem 9:

A net in a set X is ultranet if and only if the filter it generates an ultrafilter.

Proof:

Let $\{F_a : a \in A\}$ be an ultranet in X. Let $S \subset X$ and let be as in. Since $\{f_a\}$ is an ultranet, there is $a_0 \in A$ such that $B_{a_0} \subset S$ or $B_{a_0} \subset X - S$.

[$\because$ $\{f_a\}$ is eventually in either S or X – S]

Therefore either $S \in F$ or $X - S \in F$ where F is the filter generated by B.

Conversely. Let the filter F generated by the net $\{f_a : a \in A\}$ be an ultrafilter. Let $S \subset X$. Then since F is an ultrafilter, either $S \in F$ or $X - F \in F$. By definition of F, there exists $a_0 \in A$ such that $B_{a_0} \subset S$ or $B_{a_0} \subset X - S$. In either case, $\{f_a\}$ is eventually in either S or X – S. Hence $\{f_a\}$ is an ultranet in X.

EXERCISES

1. Show that a filter F on a set X is an ultrafilter only if every net base on F is an ultranet.
2. Show that a necessary and sufficient condition that a filter F on a set X be an ultrafilter is that given any subset A of X, either $A \in F$ or $X - A \in F$.

CONVERGENCE OF FILTERS

Definition : *Let F be a filter on a set X and let A – X. Then F is said to be eventually in A iff $A \in F$.*

Definition : *A filter on a set X is said to be frequently in a subset A of X iff A intersects every member of F, that is, $A \cap F \neq \varnothing \forall S \in F$.*

Remarks:

(i) F is frequently in A if F is eventually in A but non-conversely. For $A \in F \Rightarrow A \cup F = \varnothing \forall F \in F$.

For the converse, consider the filter F = {{b, c}, X} on X = {a, b, c} and the set A = {b}, The filter F is frequently in A since A intersects every member of F but it is not eventually in A since $A \notin F$.

(ii) F is frequently in A iff there exists a filter finer than F and containing A.

Definition : *Let $(X, \mathfrak{J})$ be a topological space and let F be a filter on X. Then F is said to $\mathfrak{J}$-converge or simply converge (or to be convergent) to point $x \in X$ iff F is eventually in each neighbourhood of x, that is, iff every nhd of x is a member of F and we say that x is a limit point (or simply a limit) of F and writer $F \to x$.*

LIMIT POINTS OF A FILTER BASE

Definition : *A filter base B on X is said to converge to a point $x \in X$ iff the filter whose base is B (that is, the filter generated by B) converges to x and we say that x is a limit point of B.*

The set of all limit points of a filter F shall be denoted by lim (F). Likewise lim (B) will denote the set of all limit points of a filter base B.

Theorem 1:

Let $(X, \mathfrak{J})$ be topological space and let F be a filter on X. Then the following statements are equivalent.

(a) : For converges to a point $x \in X$.

(b) : Each neighbourhood of x belongs to F.

(c) : F is finer than the neighbourhood filter N (x) on X.

(d) : For every nhd N of x, there is $F \in F$ such that $F \subset N$.

Proof:

(a) ⇔ (b) follows immediately. (b) ⇔ (c) is an immediate consequence the definition N (x) and , (b) ⇔ (d) follows from [F*2].

Example 1:

Show that in an indiscrete space X, every filter on X converges to every point of X.

Solution:

Let F be any filter on X and let to be any point of X. The neighbourhood filter of x in this case is {X}. Since X ∈ F, it follows that F converges to x. Since x was an arbitrary point of X. We see that F converges to every point of X.

Example 2:

Show that in a discrete space X, the only convergent filters are the neighbourhood filters.

Solution:

We have the neighbourhood filters are the only ultrafilters on X and so by definition these are the only convergent filters.

Example 3:

Show that if a filter F in a topological space X converges to a point x ∈ X, then every filter F finer that F also converges to x.

Solution:

Since F converges to x, F contains the nhd filter N (x) of x, and since F′ is finer than F, F′ also contains N (x) and consequently F′ also converges to x.

Example:

The neighbourhood filter N (x) of a point x in a topological space X converges to x.

Solution:

Do yourself.

Example 4:

Consider the topology $\mathfrak{I}$ = {∅, {a}, {a, b}, X} on X = {a, b, c} and the filters F_1 = {{a,}, {a, b}, {a, c}, X}

F_2 = {{b}, {a, b} {b, c}, X},, F_3 = {{c}, {b, c}, {a, c}, X}

and find the limit points of these filters.

Solution:

We have N (a) = {{a},, {a, b}, (a, c), X}, N (b) = {{a, b}, X} and N {c} = {X}. Since F_1 contains N (a), N (b) and N (c), it follows that a, b, c are the limit points of F_1. Again F_2 contains N (b) and N (c) and does not contain N (a), and hence b, c are the limit points of F. Finally F_2 contains N (c) only and so c is the limit point of F_2. shows.

Theorem 2:

Let $\mathfrak{I}$ and $\mathfrak{I}'$ be topological on a set X. Then $\mathfrak{I}$ is finer than $\mathfrak{I}'$ if and only if F $\to$ x in $\mathfrak{I}$ implies F $\to$ x in $\mathfrak{I}'$ for all filters F in X. Hence $\mathfrak{I} = \mathfrak{I}'$ iff $\mathfrak{I}$ and $\mathfrak{I}'$ have the same convergent filters and corresponding limits.

Proof:

Let $\mathfrak{I}$ be finer than $\mathfrak{I}'$ and let F $\to$ x in $\mathfrak{I}$. Then

$$N'(x) \subset N(x) \subset F.$$

where N′ (x) N (x), denote the nhd filters in $\mathfrak{I}'$ and $\mathfrak{I}$ respectively. Hence F $\to$ x in $\mathfrak{I}'$.

Conversely, suppose that F $\to$ x in $\mathfrak{I}$ implies $\mathfrak{I} \to$ x in $\mathfrak{I}'$ and let $S \subset X$. If $x \in Cl_{\mathfrak{I}}(S)$, then there is a filter base B in S such that B $\to$ x in $\mathfrak{I}$. Then B $\to$ x in $\mathfrak{I}'$ by hypothesis and so $x \in Cl_{\mathfrak{I}'}(S)$. It follows that $Cl_{\mathfrak{I}}(S) \subset Cl_{\mathfrak{I}'}(S)$. It will now be shown that $\mathfrak{I}' \subset \mathfrak{I}$. Let $G \in \mathfrak{I}'$. Then X – G is $\mathfrak{I}$-closed and hence $Cl_{\mathfrak{I}}(X - G) \subset Cl_{\mathfrak{I}'}(X - G) = X - G$ so that X – G that is $\mathfrak{I}$-closed, that is, $G \in \mathfrak{I}$. Hence $\mathfrak{I}' \subset \mathfrak{I}$ *i.e.* $\mathfrak{I}$ is finer than $\mathfrak{I}'$.

Example 1:

Let M be the collection of all those filters on a topological space X which converge to the same point $x \in X$. Then the intersection F of all the filters in M also converges to x. Show further that F = N (x).

Solution:

First note that F is actually a filter on X by virtue of theorem. Since all the filters in M converge to x, the neighbourhood filter N (x) is coarser than each filter in M and consequently coarser than F. Hence F also converges to x. We now show that F = N (x). We have shown above that

$$N(x) \subset F. \quad ...(1)$$

Also sine N (x) converges to x, it is a member of M,

we have

$$F \subset N(x). \quad ...(2)$$

From (1) and (2), we have

$F = N(x)$.

Example 2:

Let $B = \{H_n : n \in N\}$ where $H_n =]\, n, \infty\, [$ and let $F = \{F : F \supset H$ for some $n \in N\}$.

Show that F is a filter on R with base B. Further show that F converges to no point of R relative to the usual topology of R.

Solution :

B evidently satisfies [F1] and [F2] but not [F3] but F satisfies [F3] also as the reader can easily see. Hence F is filter with base

B. [Note that [F2] ⇒ [F2]*. But F does not converge to any point $x \in R$. For] x – 1, x + 1 [is a U-nhd of x which does not belong to F.

Theorem 3:

A filter F on a topological space X converges to a point $x \in X$ every ultrafilter containing F converges to x.

Proof:

If F converges to x, then evidently every ultrafilter containing also converges to x.

Conversely, let every ultrafilter containing x converge to x. Now F is intersection of all ultrafilters' on X finer than F and so F converges to x by above example.

Theorem 4:

Let (X, ℑ) be a topological space and let B be a filter base on X. Then B converges to a point $x \in X$ iff every member of the local base B (x) at x contains a member of B.*

Proof:

Let B converge to x. The filter F generated by B also converges to x, every member of B* (x), belongs to F and consequently as shown above, every member of B* (x) must contain a member of B.

Conversely's let every member of B* (x) contain a member of B and let F be the filter generated by B. Then by definition F consists of all those subsets of X which contain a member of B. Hence by [F2] every member of B* (x) belongs to F. Now let N be any nhd of x. Then by the definition of local base, there exists a member of B of B* (x) such $B \subset N$. Hence,

$N \in F$ by [F2]. It follows that every nhd of x belongs to F and consequently F converges to x. F converges to x.

Corollary : *B converges to x iff every nhd of x contains a member of B.*

Proof:

Let B converges to x and let N be nhd of x. Then there exists $B \in B$ (x) such that $B \subset N$. Hence by the above theorem N must contain a member of B.

Conversely, if every nhd of x contains a member of B. Then surely every member of B* (x) must contain a member of B. Hence, by the above theorem, B converges to x.

Theorem 5:

Let $(X, \mathfrak{I})$ be a topological space and let $A \subset X$. Then a point x in X belongs to the closure A of A iff there exists a filter base on A converging to x.

Proof:

The 'if part': Let B be a filter base on A which converges to x. Then by the preceding theorem for each neighbourhood N of x, there exists a member B of B such that $B \subset N$. Since B is non-empty subset of A, it follows that N contains at least one point of A. Hence x is an adherent point of A and so $x \in \overline{A}$.

The 'only if' part : Let $x \in A$. Then every member of the nhd filter N (x) must intersect A. Now consider the collection $B = \{A \cap N : N \in N (x)]$. Then B satisfies [F*1] since by hypothesis

$$A \cap N \neq \emptyset \vee N \in N (x).$$

It also satisfies [F*2] since if $A \cap N$ and $A \cap M$ are any two members of B, then $(A \cap N) \cap (A \cap M) = A \cap (N \cap M) \in N$. Hence B is a filter base On A. The filter generated by B is evidently the neighbourhood filter N (x). Since N (x) converges to x, the result follows.

Theorem 6:

Let A be a subset of a topological space X and let $x \in X$. Then x is an accumulation point of A iff there exists a filter base on $A - \{x\}$ which converges to x.

Proof:

As above theorem.

Theorem 8:

Let $(X, \Im)$ be a topological space and let $A \subset X$. Then A is $\Im$-open iff A belongs to every filter which converges to a point of A.

Proof:

Let A be a $\Im$-open set and let F be a filter which converges to a point $x \in A$ [Note that there exists at least one such filter, namely, the nhd filter of x]. Then F contains every nhd of x. Since A is $\Im$-open, it is a nhd of x. Hence $A \in B$.

Conversely, let A belong to every filter which converges to a point A. To show that A is open. Let x be an arbitrary point of A. Then the nhd filter N (x) converges to x. By hypothesis, $A \in N(x)$. Hence A is a $\Im$-nhd of x. It follows that A is $\Im$-open.

Theorem 9:

A topological space $(X, \Im)$ is Hausdorff iff every convergent filter in X has a unique limit.

Proof:

Let X be Hausdorff and let F be filter on X converging to two distinct points x and y. Then $F \subset N(x)$ and $F \subset N(y)$ where N (x) and N (y) are the nhd filters of x and y respectively. Since the space X is Hausdorff, there exists nhds N of x and M of y such that $N \cap M = \varnothing$. Then $M \in F$ and $N \in F$ and so by [F3], $N \cap M = \varnothing \in F$ and consequently F has a unique limit. [Note that since F is convergent it must have a limit point].

Conversely, let every convergent filter in X have a unique limit and suppose, if possible the space X is not Hausdorff. Then there exists $x, y \in X$ with $x \neq y$ such that

$$N \cap M \neq \varnothing \vee N \in N(x) \text{ and } \vee M \in N(y)$$

Now $N(x) \cap N(y)$ is also a filter on X, and hence it is also a filter base on X. Hence $N(x) \cap N(y)$ generates a filter F which is finer than both N (x) and N (y). Now N (x) converges to x and N (y) converges to y and consequently F converges to both x and y this is a contradiction. Hence $(X, \Im)$ is Hausdorff.

Theorem 10:

Let $(X, \Im)$ and (Y, V) be two topological spaces and let $x \in X$. Let f be a mapping of X into Y. Then the following statements are equivalent.

(a) : f is continuous at x.

(b) : For every filter base B on X converging to x, the filter base f [B] converges to f (x).

(c) : For every ultrafilter base B on X converging to x the ultrafilter base f [by] on Y converges to f (x).

Proof:

First not that f (B) = {f [B] : B ∈ B}:

(a) ⇔ (b) : Let f be continuous at x and let N be any nhd of f (x). Then f^{-1} [N] is a nhd of x. Let B be any filter base on X converging to x. Then there is a member B of B such that B ⊂ f^{-1} [N]. It follows that f [B] ⊂ f [f^{-1} [N]] ⊂ N. Thus we have shown that every nhd of f (x) contains a member of f (B). Hence f (B) converges to f (x).

Conversely, let (b) hold and let N be a nhd of f (x). Now the nhd filter N (x) converges to x and so by hypothesis, f [N (x)] converges to f (x). Hence there exists a member f [M] of f [N (x)] such that f [M] ⊂ N. Now M ∈ N (x) so that M is a nhd of x. Thus, we have shown that to each nhd N of f (x), there exists a nhd M of x such that f [M] ⊂ N. It follows that f is continuous at x.

(a) ⇔ (c) : Since every ultrafilter base is a filter base. (a) ⇒ (c) follows from (a) ⇒ (b) We now prove (c) ⇒ (a). Suppose, if possible, that f such that f^{-1} [N] is not a nhd of x. Then f^{-1} [N] cannot contain any nhd filter N (x) of x.

Since N (x) has the FIP, N (x) ∪ (X – f^{-1} [N]) has also the FIP. Hence, there exists an ultrafilter F containing N (x) ∪ (X – f^{-1} [N]). Since N (x) converges to x, F surely converges to x.

Now consider the member f {X – f^{-1} [N] of f (F). Evidently, we have N ∩ f [X – F N]] = ∅. Thus, there is a member f [X – f^{-1} [N]] of f (F) which does not intersect the nhd N of f (x). Hence f [F] does not converges to f (x). But this contradicts the hypothesis. It follows that f is continuous at x.

Theorem 11:

A filter F on a non-empty product space X = × {X_λ : λ ∈ L) converges to a point x = {x_λ : λ ∈ Λ} of X if and only if the filter π_λ [F] = {π_λ [F] Λ F ∈ F} for every λ ∈ Λ, converges to $x_\lambda = \pi_\lambda$ (x) in X_λ.

Proof:

Let F converge to x ∈ X. Since π_λ is continuous for every λ ∈ Λ, it follows that the filter π_λ [F] converges to π_λ (x) [Note that every filter is a filter base.

Conversely, if F is a filter on X such that filter π_λ [F] converges to a point x_λ in X_λ for every λ, then let x denote the point of X, whose l[th] coordinate is x. Let N be any nhd of x in X. Then there exists a base member G = × {G_λ λ ∈ Λ} where G_λ is open in X_λ and $G_\lambda = X_\lambda$ for all but a finite number of l's, say $G_\lambda = X_\lambda$ for $\lambda \neq \lambda_1, \lambda_2, \ldots \lambda_n$, such that x ∈ G ⊂ N. Now π_{λ_i} [G] = G_{λ_i}, which is a nhd of x_{λ_i}, i = 1, 2, ..., n.. By hypothesis, π_{λ_i} [F] converges to x_{λ_i}, π_{λ_i} [F] must contain G_{λ_i} for i = 1, 2, ..., n. Hence $\pi_{\lambda_i}^{-1}$ [G_{λ_i}] contains a member of F. Then by definition of filter, $\pi_{\lambda_i}^{-1}$ [G_{λ_i}] ∈ F for i = 1, 2, ..., n, and therefore $\bigcap_{i=1}^{\pi} \pi_{\lambda_i}^{-1}[G_{\lambda_i}] \in F$. But $\bigcap_{i=1}^{\pi} n_{\lambda_i}^{-1}[G_{\lambda_i}] = G$ and so G ∈ F. Since N ⊃ G, N ∈ F by definition of filter. Since N was any nhd of l in Y, F converges to x.

Theorem 12:

If $\{F_\lambda : \lambda \in \Lambda\}$ *ισ any collection of closed subsets of a topological space X, then* $\cap \{F_\lambda : \lambda \in \Lambda\}$ *is a closed set.*

Proof:

Fl is closed " l ∈ L

⇒ F′l is open " l ∈ L

⇒ ∪ {F′l : l ∈ L} is open

[By [T3] of the definition of topology]

⇒ [∩ {Gl : l ∈ L}]′ is open [De-Morgan Law]

⇒ ∩ [Fl : l ∈ L} is closed [By definition of closed set]

Theorem 13:

If F1 and F2 be two closed subsets of a topological space X. Then F1 ∪ F2 is a closed set.

Proof:

F1, F2 are closed

⇒ F′1, F′2 are open

⇒ F1′ ∩ F′2 are open

[By **[T2]** of the definition of topology]

⇒ [F1 ∪ G2]′ is open [De Morgan Law]

⇒ F1 ∪ F2 is closed.

Note that if F1, F2,..., Fn be a finite number of closed subsets of X, their union will also be a closed subset of X. This follows at once by the repeated application of the above theorem.

Theorem 14:

A subset of a topological spaces is open if and only if it is a neighbourhood of each of its points.

Proof :

Let a subset G of a topological space be open. Then for every x $\in$ G $\subset$ G and therefore G is a nhd of each of its points. Conversely let G be a nhd of each of its points, If G = $\varnothing$, then it is open.

If G $\neq \varnothing$, then to each x G there exits an open set G_x such that x $\in$ $G_x \subset$ G. It follows that G = $\cup$ {G_x : x $\in$ G}. Hence G is open, being a union of open sets.

Remark : The above theorem shows that the open sets in a topological space are known whenever neighbourhoods of x are known for every x. This means that two topologies on the same set are identical iff they admit the same neighbourhoods.

For suppose that $\Im$ and $\Im'$ are two topologies for X having the same system of nhds of points in X. Then

G is $\Im$-open $\Leftrightarrow$ G contains a $\Im$-nhd of each of its points

$\Leftrightarrow$ G contains a $\Im'$ nhd of each of its points

$\Leftrightarrow$ G is $\Im'$ -open.

Theorem 15:

Let X be a non-empty set, and for each x $\in$ X, let **N** (x) be a non-empty collection of subset of X satisfying the following conditions.

[M1] : $N \in \mathbf{N}(x) \Rightarrow x \in N$

[M2] : $N \in N(x),\ M \in N(x) \Rightarrow N \cap M \in N(x)$.

Let $\Im$ consist of the empty set and all those non-empty subset G of X having the property that x $\in$ G implies that there exists an N $\in$ **N** (x) such that x $\in$ N $\subset$ G. Then $\Im$ is a topology for X.

Proof:

[T1] : $\varnothing \in \Im$ be definition. We now show that X $\in \Im$. Let x be any arbitrary element of X. Since **N** (x) is non-empty, there is an N $\in$ **N** (x) and so x $\in$ N by **[M1]** . Since N is a subset of X, we have x $\in$ N $\subset$ X. Hence X $\in \Im$.

[T2] : Let $G_1 \in \Im$ and $G_2 \in \Im$

If x $\in G_1 \cap G_2$, then x $\in G_1$ and x $\in G_2$. Since $G_1 \in \Im$ and $G_2 \in \Im$, there exists N $\in$ **N** (x) by **[M2]**. Hence $G_1 \cap G_2 \in \Im$.

[T3] : Let $G_\lambda \in \mathfrak{I} \;\forall\; \lambda \in \Lambda$. If $x \in \cup \{G_\lambda : \lambda \in \Lambda\}$, then $x \in G_{\lambda_x}$ for some $\lambda_x \in L$. Since $G_{\lambda_x} \in \mathfrak{I}$, there exists an $N \in \mathbf{N}(x)$, such that $x \in N \subset G_{\lambda_x}$ and consequently

$$x \in N \subset \cap \{G_\lambda : \lambda : \Lambda\}$$

Hence $\cup \{G_\lambda : \lambda : \Lambda\} \in \mathfrak{I}$. It follows that $\mathfrak{I}$ is topology for X.

EXERCISES

1. Show that if a filter in a metric space converges, it must converges to a unique limit.

   [Hint : Since every metric spaces is a Hausdorff space].

2. Show that a filter F on a space X converges to $x \in X$ iff there exists a filter F′ finer than F such that F′ converges to x.

3. Which filters F on a set X will converge to $x \in X$ if (i) X has the descrete topology, (ii) X has the indiscrete topology ?

4. Show that a net $\{f_a : a \in A\}$ in a set X converges to $x \in X$ iff the filter generated by $\{f_a\}$ converges to x.

5. Let X be an infinite set and F the filter on X generated by the filter base consisting of all complements of finite sets. To which points does F converge if X is given the cofinite topology ?

6. Show that a filter F on a set X converges to $x \in X$ iff the net based on F converges to x.

7. Let Y be a topological sub-space of a space X, let $Y \in Y$ and let B be a filter base on Y. Show that $B \to y$ in Y iff $B \to y$ in X.

CLUSTER POINTS OF A FILTER

Definitions : *(i) Let $(X, \mathfrak{I})$ be a topological space and let F be a filter on X. Then a point $x \in X$ is said to be cluster point (or an adherent point) of F iff F is frequently in each neighbourhood of x.*

(ii) *A point x in a topological space X is said to be a cluster point (or an adherent point) of a filter base B on X iff B is frequently in each nhd of x.*

The set of adherent point of F is called the adherence of F and is denote by adh (F).

Theorem 1:

Let $(X, \mathfrak{I})$ be a topological space and let F be a filter on X. Then the following statements are equivalent :

(a) : $x \in X$ is a cluster point of F.

(b) : Every neighbourhood of x intersects every member of F.

(c) : $x \in \overline{F}$ for all $F \in F$ where $\overline{F}$ denotes the $\mathfrak{I}$-closure of F.

(d) : N (x) $\cup$ F has the FIP where N (x) is the nhd filter of x.

(e) : There exists a filter finer than both N (x) and F.

(f) : There exists a filter finer than F and converging to x.

Proof:

(a) $\Leftrightarrow$ (b) : This follows from definition.

(b) $\Leftrightarrow$ (c) : This is an immediate consequence of the properties of closure $\overline{F}$

(b) $\Leftrightarrow$ (d) : This follows from the fact that

(i) both N (x) and F have the FIP.

(ii) finite intersections of F belong to F and

(iii) finite intersections of N (x) belong to N (x).

(d) $\Leftrightarrow$ (e) : By theorem (5.1), there exists a filter on X containing. N (x) $\cup$ F and consequently containing both N (x) and F iff N (x) $\cup$ F has the FIP. Hence (d) $\Leftrightarrow$ (e).

(e) $\Leftrightarrow$ (f) : Let G be the filter which is finer than both F and N (x). Then G is the filter which converges to x since it contains N (x). Conversely let G be the filter which is finer than F and which converges to x. Then G must contain N (x) by definition.

Theorem 2:

Let (X, $\mathfrak{I}$) be a topological space and let B be a filter base on X. Then a point $x \in X$ is a cluster point of B iff every member of the local base B (x) at intersects every member of B.*

Proof:

x is cluster point of B.

$\Leftrightarrow$ every of x intersects every member of B.

$\Leftrightarrow$ every member of B* (x) intersects every member of B.

Theorem 3:

Let F be an ultrafilter on topological space X and let $x \in X$. Then F converges to x iff x is a cluster point of B.

Proof:

Let F converges to x. Then every nhd of x belongs to F and so every nhd of x intersects every member of F. It follows that x is a cluster point of F.

Conversely, let x be a cluster point of F. There exists a filter F′ finer then F and converging to x. Since F is an ultrafilter, we have F, F′. It follows that F converges to x.

Theorem 4:

Let B be a filter base on a topological space X. Then adh (B) = $\cap$ {B : B $\in$ B} and thus adh (B) is closed in X.

Proof:

x $\in$ adh (B) $\Leftrightarrow$ x is a cluster point of B.

$\Leftrightarrow$ each nhd of d intersects each member of B.

$\Leftrightarrow$ x $\in$ B $\forall$ B $\in$ B.

$\Leftrightarrow$ x $\in$ $\cap$ {B : B $\in$ B}

Hence adh (B) = $\cap$ {B : B $\in$ B}. Since the intersection of closed sets is closed, it follows that adh (B) is closed in X.

Theorem 5:

Let (X, $\mathfrak{J}$) be a topological space, Then the following statements are equivalent.

(a) : X is a compact space.

(b) : Every collection of closed subsets of X with the FIP has a non-empty intersection..

(c) : Every filter on X has a cluster point,

(d) : Every ultrafilter converges.

Proof:

(a) $\Rightarrow$ (b) : See above theorem.

(b) $\Rightarrow$ (c) : Assume (b) and let F be any filter on X. Then F has the FIP. It follows that the collection $\{\overline{F} : F \in F\}$ also has the FIP.

Therefore by (b). There exists at least one point p $\in \cap \{\overline{F} : F \in F\}$. This implies that p $\in \overline{F}$ for every F $\in$ F.

Hence p is a cluster point of F.

(c) ⇒ (d) : Assume (c), that is assume that every filter on X has a cluster point. Then surely every ultrafilter on X has a cluster point which is also a limit point of the filter. Hence every ultrafilter on X converges.

(d) ⇒ (a) : Assume that every ultrafilter on X converges. To show that X is compact. Let C be an open cover of X and suppose, if possible, C has finite subcover. The collection A = {X – G : G ∈ C} has the FIP.

For suppose there is a finite subcollection $\{X - F_i : 1 \leq i \leq n\}$ of A such that $\cap \{X - G_i : 1 \leq i \leq n\} = \varnothing$ Then $X - \cup \{Gi : 1 \leq i \leq n\} = \varnothing$ by De-Morgan Law. This implies that $\cup \{G_i : 1 \leq i \leq n\} = X$ which contradicts our supposition that C has no finite subcover. Thus A must have the FIP. Therefore there exists an ultrafilter F on X containing A. By (d), F converges to some point p ∈ X. Then p is also a cluster point of F.

Hence $p \in \overline{F}$ for every F ∈ F. In particular,

$p \in X - G = X - G \ \forall\ G \in C.$ [∵ X – G is ℑ-closed].

This implies $p \in \cap \{X - G : C\}$

$= X - U \{G \in C\}.$

Hence $p \notin \cup \{G : G \in C\}$ But this contradicts the fact that C is a cover of X.

Hence C must have a finite subcover and consequently X is compact.

Theorem 6: (Tychonoff)

A non-empty product space

$X = \times \{X_\lambda\ \lambda \in \Lambda\}$ is compact iff each coordinate space Xl is compact.

Proof :

If X is compact, then since each X_λ is the continuous image of X under the projection π_λ for every $\lambda \in \Lambda$, it follows that each X_λ is compact. Conversely, let X_λ be compact for every $\lambda \in$. Let F be an ultrafilter on X, $\pi_\lambda [F] = \{\pi_\lambda \{F\} : F \in F\}$ is an ultrafilter on X_λ for every $\lambda \in \Lambda$ and hence $\pi_\lambda [F]$ converges to some point of X_λ.

Then F converges.

Hence $X = \times \{X_\lambda : \in \Lambda\}$ is compact.

SOME MISCELLANEOUS EXAMPLES

Example 1:

Consider the topology

$\mathfrak{I} = \{\phi, \{a\}, \{a, b\}, \{a, c\} ; X\}$

On X = ({a, b, c} and the filters

$F_1 = \{\{b, c\}, X\}$,

$F_2 = \{\{c, a\}, X\}$

$F_3 = \{\{a, b\}, X\}$,

and find the cluster points of these filters,

Solution:

We have

$$N(a) = \{\{a,\}, \{a, b\}, \{a, c\}, X\}$$

$$N(b) = \{\{a, b\}, X\} \text{ and}$$

$$N(c) = \{\{a, c\}, X\},$$

Now a is not a cluster point of F_1 since the nhd {a} does not intersect $\{b, c\} \in F_1$, b is a cluster point of F_1 since every member of N (b) intersects every member of F_1. Similarly, c is a cluster point of F_1. Hence adh (F_1) = {b, c}. Similarly the reader can verify that adh (F_2) = {a, b, c} adh (F_3) = {a, b, c}.

Example 2:

Let x be a cluster point of a filter F on topological space X and let F' be a filter on X coarser than F. Then x is also a cluster point of F'.

Solution:

x is a cluster point of F.

$\Leftrightarrow$ $x \in \overline{F} \ \forall F \in F$.

$\Leftrightarrow$ $x \in \overline{G} \ \forall G \in F'$.

[Since every member of F′ is a member of F]

$\Leftrightarrow$ x is a cluster point of F′.

Example 3:

Let $X = \{a, b, c, d, e\}$ *and let*

$\mathfrak{I} = \{\varnothing, \{b\}, \{c, d\}, \{b, c, d\}, \{a, b, c, d\}, X\}$.

Find (i) interior (ii) exterior (iii) frontier of the following subsets of X.

(a) $A = \{c\}$, *(b)* $B = \{a, b\}$, *(c),* $C = \{a, b, d\}$, *(d),* $D = \{b, c, d\}$.

Solution:

(a) (i) Since A is not a nhd of c, $c \notin A^\circ$.

Hence, $A^\circ = \varnothing$

(ii) Now $A' = \{a, b, c, d, e\}$. It is easy to see that b is an interior point of A' since A' is a nhd of b. But a, d, e are not the interior points of A'.

Hence $\quad$ ext (A) = {b}.

(iii) Since $A^\circ = \varnothing$ and ext (A) = {b}, if follows that

$$\text{fr }(A) = \{a, c, d, e\}.$$

(b) (i) Here b is an interior point of B but a is not.

Hence $\quad B^\circ = \{b\}$.

(ii) Now B' = {c, d, e}. Since c, d, ∈ {c, d} ⊂ B', it follows that B' is a nhd of x, d. Hence c, d are interior points of B'.

Hence ext (B) = $(B')^\circ$ = {c, d}.

(iii) Since $B^\circ = \{b\}$ and ext (B) = {c, d}, it follows that

$$\text{Fr }(B) = \{a, e\}$$

(c) Here (i) $C^\circ = \{b\}$ (ii) C' = {c, e}

so that $\quad$ ext C = $(C')^\circ = \varnothing$

and (iii) Fr (C) = {a, c, d, e} as the reader can easily verify.

(d) (i) Here D is an open set so that it is nhd of each of its points and consequently every point of D is its interior point.

Hence $\quad D^\circ$ D = {b, c, d}.

(ii) D' = (a, e). Since there exits no open set G such that a ∈ G ⊂ D', D' is not a nhd of a. Hence a ∉ $(D')^\circ$. Therefore ext (D) = $(D')^\circ = \varnothing$.

(iii) Since D° = {b, c, d} and ext D = ∅, it follows that

$$\text{Fr }(D) = (a, e\}$$

Example 4:

Show that the spaces (R, U) and (R, S) are Hausdorff.

Solution:

Let a, b be any two distinct points of R so that a ≠ b ans let | a – b | = ε. Then

$$]a - \varepsilon/3,\ a + \varepsilon/3\ [\ \text{and}]\ b - \varepsilon/3,\ b + \varepsilon/3\ [$$

are disjoint U-nhds of a and b respectively.

Also [a – ε/3, a + ε/3 [and [b – ε/3, b + ε/3 [are disjoint S-nhds of a and b respectively. Hence (R, U) and (R, S) are both Hausdorff.

Example 5:

Consider the cofinite topology on an infinite set X and show that it is not T_2.

Solution:

For this topology no two open sets can be disjoint. For suppose, if possible, that G, H are two open disjoint subsets of X so that $G \cap H = \varnothing$. Then

$$(G \cap H)' = \varnothing'$$

i.e., $G' \cup H' = X$. [De-morgan Law]

But G′ and H′ are finite sets so their union is also finite which is a contradiction since X is infinite. Hence no two distinct points can be separated by open sets. Accordingly the space is not T_2.

Example 6:

Find the condition under which a closed (open) set in the subspace Y of a topological space X be closed (open) in the larger space X.

Solution:

Let X = {a, b, c, d, e} and let $\mathfrak{I}$ be a topology on X given by

$$\mathfrak{I} = \{\varnothing, \{a\}, \{a, b\}, \{a, c, d\} \{a, b, c, d\}, \{a, b, e\}, X\}.$$

Let Y = {a, c, e}. Then $\mathfrak{I}_Y = \{A \cap Y : A \in \mathfrak{I}\}$ so that the members of $\mathfrak{I}_Y$ are

$$\varnothing \cap Y = \varnothing,\ Y \cap X = Y,\ \{a\} \cap Y = \{a\},$$

$$\{a, b\} \cap Y = \{a\},\ \{a, c, d\} \cap Y = \{a, c\},$$

$$\{a, b, c, d\} \cap Y = \{a, c\},$$

$$\{a, b, e\} \cap Y = \{a, e\}.$$

Thus $\mathfrak{I}_Y = \{\varnothing, \{a\}, \{a, c\}, \{a, e\}, Y\}$

Now consider the subset A = {a, e}, of Y. Then

$$\text{Int}_Y \{A\} = \{a, e\}$$

and $\text{Int}_X \{A\} = \{a\}$.

Hence in this case $\mathfrak{I}$- interior of A is a proper subset of $\mathfrak{I}_Y$-interior of A, so that $\text{Int}_X (A) \subseteq \text{Int}_Y (A)$.

Example 7:

Consider the usual topology U on *R and find (i) interior (ii) exterior and (iii) frontier of each of the following subset of R.*

(a) $A = [0, 1[$; *(b)* $B = [0, 1[$; *(c)* $C = [0, 1]$;
(d) $D = \{1/n : n \in N\}$; *(e)* N; *(f)* Q.

Solution:

(a) (i) Since A is an open set, it is a nhd of each of its points and so every point of A is its interior point.

Hence $A^\circ = A =]0, 1[$.

(ii) Now $A' =]-\infty, 0] \cup [1, \infty[$.

Here A' is nhd of each of its points except 0 and 1.

Hence ext $(A) = (A') =]-\infty, 0[\cup]1, \infty[$.

(iii) Fr $(A) = [A^\circ \cup \text{ext}(A)(A)]' = \{0, 1\}$.

Thus the frontier of A consists of two points 0 and only.

(b) Proceeding as in (a), we have

(i) $B^\circ =]0, 1[$.

(ii) Here $B' =]-\infty, 0[\cup [1, \infty[$.

Then ext $(B) = (B') =]-\infty, 0[\cup]1, \infty[$.

(iii) Fr $(B) = \{0, 1\}$.

(c) (i) $C^\circ =]0, 1[$.

(ii) Here $C' =]-\infty, 0[\cup]1, \infty[$. Hence

$$\text{ext}(C) = (C') =]-\infty, 0[\cup], \infty[.$$

(iii) Fr $(C) = \{0, 1\}$

(d) (i) Here D cannot be nhd of any its points $1/n$, $n = 1, 2, 3, \ldots$ since there exits no $\varepsilon > 0$ such that

$]1/n - \varepsilon, 1/n + \varepsilon[\subset D$.

Hence no point of D can be its interior point so that $D^\circ = \varnothing$.

(ii) It is easy to see that D' is a nhd of each of its points except 0. Hence

$$\text{ext}(D) = (D')^\circ = [D \cup \{0\}]'$$

(iii) Fr $(D) = [D^\circ \cup \text{ext}(D)]' = D \cup \{0\}$.

(e) (i) It is easy to see that N is not a nhd of any of its points and so $N^\circ = \varnothing$.

(ii) Since **N** has no limit points we have $\overline{N} = N$ so that N is a closed set. Its complement N' is therefore an open set. Hence

ext $(N) = (N')^\circ = N'$.

(iii) $\text{Fr}(N) = [N^\circ \cup \text{ext}(N)]' = [\varnothing \cup \text{ext}(N)]'$

$= [\text{ext}(N)]' = (N')' = N.$

Note that N non-dense since $(\overline{N})^\circ = N^\circ = \varnothing$.

(f) (i) Since Q is not a nhd of any of its points, we have

. $Q^\circ = \varnothing$

(ii) Also $Q' = R - Q$ is the set of all irrational points which is also not a nhd of any of its points so that $(Q')^\circ = \varnothing$.

Hence $\text{ext}(Q) = (Q') = \varnothing$

(iii) $\text{Fr}(Q) = [Q^\circ \cup \text{ext}(Q)]'$

$= [\varnothing \cup \varnothing]' = \varnothing' = R.$

Note that Q is every where dense since $\overline{Q} = R$.

Example 8:

Let (X, d) be a discrete topological space and let A be a subset of X. Find (i) A°, (ii) **ext** *(A), (iii)* **Fr** *(A)..*

Solution:

(i) Since in a discrete space, every set is open, it follows that A is a nhd of each of its points. Hence $A^\circ = A$.

(ii) $\text{ext}(A) = (A') = A'$ [$\because$ B' is open (why ?)]

(iii) $\text{ext}(A) = [A^\circ \cup \text{ext}(A)]'$

$= [A \cup A'] = X' = \varnothing.$

Example 9:

Consider the usual topological (R,a U) and find the U-closure of the following subsets of R.

(i) $A = \{1/n : n \in N\}$. (ii) I (iii) Q. (iv)]0, 1 [. (v) [0, 1[.

(vi) [0, 1]. (vii) $B = \left\{2, \frac{3}{2}, \frac{4}{3}, \frac{5}{4}\right\}$.

Solution:

(i) Here $D(A) = \{0\}$ so that

$\overline{A} = \{1/n : n \in N\} \cup \{0\}.$

(ii) Since $D(I) = \varnothing$, we have $\overline{I} = I$.

(iii) Since $D(Q) = R$, we have

$$\bar{Q} = Q \cup D(Q) = Q \cup R = R$$

(iv) $\overline{]0, 1[} = [0, 1]$ (why ?)

(v) $\overline{[0, 1[} = [0, [$. (vi) $\overline{[0, 1]} = [0, 1]$.

(vii) Let $]\, 1 - \varepsilon, 1 + \varepsilon\, [$ be any nhd of 1. Choose a positive integer n such that

$$\frac{1}{n_0} < \varepsilon.$$

Then $1 + \frac{1}{n} \in]\, 1 - \varepsilon\, [\ \forall\ n \geq n_0$.

But $1 + \frac{1}{n} = \frac{n+1}{n} \in B$.

Hence every nhd of 1 contains infinitely many point of B ans so $1 \in D$ (B). It is obvious that no other point can belong to D (B).

Hence D (B) = {1}, $\therefore\ \bar{B} = B \cup \{1\}$.

Example 10:

Let X = {A, b, c} and let

$\mathfrak{J}$ = {∅, {a}, {a, b}, {a, c}, X}

Find all limit points of the sets

(i) A = {b, c}, (ii). B = {a, c}

Solution:

(i) a is not a limit point of A since {a} is a $\mathfrak{J}$-nhd of a which contains no points of A. Again b is not a limit point of A since {a, b} is a $\mathfrak{J}$-nhd of b which contains no point of A other than b. Also c is not a limit point of A since {a, c} is a $\mathfrak{J}$-nhd of c which contains no point of A other than c. Hence A has no limit points. Thus D (A) = ∅.

(iii) Here a is not a limit point of B since {a} is a $\mathfrak{J}$-nhd of a which contains no point of B other than a. But b is a limit point of B since $\mathfrak{J}$-open nhds of b are {a, b} and X, each of which contains a point of B other than b. Note that here b is not a point of B.

Also c is a limit point of B since there are only two $\mathfrak{J}$-open nhds of c, namely {a, c} and X each of which contains a point of B other than c. Since b, c are limit points of B, we have

$$D(B) = \{b, c\}.$$

Example 11:

Show that the spaces (R, U) and (R, S) are Hausdorff.

Solution:

Let a, b be any two distinct points of R so that $a \neq b$ ans let $|a - b| = \varepsilon$. Then

$$]a - \varepsilon/3, a + \varepsilon/3 [\text{ and}] b - \varepsilon/3, b + \varepsilon/3 [$$

are disjoint U-nhds of a and b respectively.

Also $[a - \varepsilon/3, a + \varepsilon/3 [$ and $[b - \varepsilon/3, b + \varepsilon/3 [$ are disjoint S-nhds of a and b respectively. Hence (R, U) and (R, S) are both Hausdorff.

Example 12:

Consider the cofinite topology on an infinite set X and show that it is not T_2.

Solution:

For this topology no two open sets can be disjoint. For suppose, if possible, that G, H are two open disjoint subsets of X so that $G \cap H = \varnothing$. Then

$$(G \cap H)' = \varnothing'$$

i.e., $G' \cup H' = X$. [De-morgan Law]

But G' and H' are finite sets so their union is also finite which is a contradiction since X is infinite. Hence no two distinct points can be separated by open sets. Accordingly the space is not T_2.

Example 13:

Consider the co-countable topology on an uncountable X and show that it is not T_2.

Solution :

The proof is the same as that of the previous example. In this case $G' \cup H'$ is countable whereas X is uncountable.

Example 14:

Let $X = \{a, b, c, d, e\}$ *and let*

$\mathfrak{I} = \{\varnothing, \{b\}, \{d, e\}, \{b, d, e\}, \{a, c, d, e\}, X\}$.

Then $\mathfrak{I}$ is a topology on X. Consider the subset $A = \{b, c, d\}$. Then c is a limit point of A since the $\mathfrak{I}$-open nhds of c are {a, c, d, e} and X each of which contains a point of A other than c. But b is not a limit point of

A since {b} is a $\Im$-open nhd of b which contains no point of A other than b. The reader can easily verify that the other limit points of A are a, c. Hence

$$D(A) = \{a, c, e\}.$$

Isolated points of A are b and d since b, d belong to A but are not limit points of A. The *adherent points* of A are a, b, c, d, e.

Remark : We have seen that every, derived set in a metric space is closed. In particular, every derived set of a subset of R (with usual topology) is closed. This result, however, does not hold for general topological spaces as example (2) shows. Here

$$D(A) = \{a, c, e\}.$$

which is not $\Im$-closed

Example 15:

Let (X, D) be any discrete topological space and let A be any subset of X. Then no point x ∈ X. Then no point x ∈ X can be a limit point of A since {x} is a D-open set which contains no points of A other then (possibly) x. Thus D (A) = ∅.

Solution:

Do yourself.

Example 16:

Consider any indiscrete topological space (X, I). Let A be a subset of X containing two or more points of X. Then every point x ∈ X is a limit point of A since the only open set containing x is X which contains all the points of A and must therefore contains a point of A other than x. (Since A contains more than one point). Hence **D** *(A) = X.*

If A = {p}, consisting of a single point p. Then all the points of X other than p are limit points of A for if x (≠p) ∈ X, then the only open set containing x is X which contains the point p of A which is different from x. Hence **D** *(A) = X − {p} in this case.*

If A = ∅, then evidently no point of X can be a limit point of A ans so **D** *(A) = ∅ in this case.*

Solution:

Do yourself

Example 17:

Find the derived set of each of the following subset of R relative to (i) the U-topology. (ii) the I-topology and (iv) the D-topology.

(a)]0, 1[(b) [0, 1[(c)]0, 1]

(d) N i.e. the set of all natural numbers,

(e) the set Q of all rational numbers.

(f) $A = \{1/n : n \in N\}$.

(g) $B = \{-1/n : n \in N\}$.

(h) $C = \{(2n + 1)/n : n \in N\}$.

(i) $D = \{x : x = 0 \text{ or } x \in Q_+ \text{ and } x^2 < 2\}$.

i.e. the set of all non-negative rationals whose square is less than 2.

Solution:

(a) (i) Here all the points of closed interval [0, 1] are **U**-limit points of]0, 1[. Hence

D (]0, 1 [) [0, 1].

Note that 0 is a limit point because any open nhd of 0 must contain an interval of the form]– ε, ε [, ε > 0 which surely contains points of]0, 1[other than 0. As a matter of fact every nhd of 0 contains uncountably many points of]0, 1[ans so 0 is a point of condensation also. Similarly other points of [0, 1] are points of condensation of the set [0, 1[.

(ii) Any S-nhd of 0 must contain an interval of the form [0, ε [, ε > 0. Since [0, e[is S-open nhd of 0, which contains points of]0, 1 [other than 0, it follows that 0 is a S- limit point of [0, 1[. But is not a S-limit point of]0, 1[since [1, 1[[is a S-nhd of 1 which contains no point of]0. 1[.Also all points of]0, 1[are S limit points of]0, 1[as the reader can easily see. Hence D – ( ] = [0, 1 [for the S-topology for R.

(iii) For the indiscrete topology for R every point of R is a limit point of]0, 1[. Hence D ( ] 0, 1 [ ) = R

(iv) For discrete topology for R no point of R can be a limit point of]0, 1. Hence D- ( ]0, 1 [ ) = ∅.

(b) D ([0, 1 [ ) = 0, 1] for the U-topology.

(i) D ([0, 1 [ ) = 0, 1[for the S-topology.

(iii) D ([0, 1 [ ) = R for the I topology.

(iv) D ([0, 1 [ ) = ∅ for the D topology.

(c) As in (b), we have

(i) D ( ]0, 1]) = [0, 1] for the U-topology.

(ii) D ( ]0, 1]) = [0, 1] for the S-topology.

(iii) D ( ]0, 1]) = R for the I-topology.

(iv) D ( ]0, 1]) = ∅ for the D-topology.

(d) (i) The set N has no U-limit point since for each x ∈ R we can choose e > 0 so small that the nhd] x – ε, x + ε[contains no natural number other than x. Hence D (N) = ∅.

(ii) As in (i), we shall get D (N) = ∅ for the S-topology

(iii) D (N) R for the I-topology (why ?),

(iv) D (N) = R for the D-topology,

(e) (i) Since rational points are every where dense, every nhd of x ∈ R must contain infinitely many rational points. Hence every point x ∈ R is a limit point for the U-topology. Hence

D (Q) = R

(ii) Proceeding as in (i) we shall have D (Q) = R for the S-topology.

(iii) As before

D (Q) = R for the I-topology.

(iv) D (Q) = ∅ for the D-topology.

(f) (i) Here 0 is the only limit point of A for the U-topology for R since for each nhd] – ε, ε [, ε > 0 of zero, we can choose n so large that the interval [0, /n] is contained in] – ε, ε[so that each nhd of 0 contains points of A other than 0. No other point of R can be the limit point of A as the reader can easily see. Hence D (A) = {0}.

(ii) For S -topology also 0 is the only limit point of A.

Hence D (A) = {0}.

(iii) D (A) = R for I-topology.

(iv) D (A) = ∅ for D-topology.

(g) (i) As in (f), we have D (B) = {0} for the U-topology.

(ii) D (B) = ∅ for the S-topology.

Note that here 0 is not a limit point of B for the S-topology. Since [0, 1] is a S-nhd of 0 which contains no point of B.

(iii) D (B) = R for the I-topology

(iv) D (B) = ∅ for the D-topology.

(h) We leave it to the reader to verify that

(i) D (C) = {2} for the U-topology.

(ii) D (C) = {2} for the S-topology.

(iii) D (C) = R for the I-topology.

(iv) D (C) = ∅ for the D-topology.

(1) (i). Here every point of the closed interval [0, √2] is a limit point of D for the U-topology. Hence

D (D) = [0, √2].

Note that √2 is a limit point of D since every nhd of √2 must contain infinitely many rational points whose square is less than 2.

(ii) Here √2 is not a limit point of D for the S-topology since [√2, 2[is a S-nhd or √2 which contains no rational number whose square is less than 2. Also all the points of the interval [0, √2[are the limit points of D for the S-topology. Hence D (D) = [0, √2 in this case.

(iii) Here D (D) = R for the I- topology.

(iv) R (D) = ∅ for the D-topology.

Example 18:

Consider the topology $\mathfrak{J}$ on N which consists of ∅ and all subsets of N of the form

$$A_n = \{n, n + 1, n + 2 ... \}, n \in N$$

Find (i) the derived set of the set A = {3, 12, 28, 35}

(ii) those subset G of N for which D (G) = N.

(iii) D ({1}).

Solution:

(i) Here open sets containing any point m ∈ N are the sets A_n where n ≤ m. If m_0 $\mathfrak{J}$ 34, then every open set containing m_0 also contains 35 ∈ A which is different from m_0.

Hence $m_0 \leq 34$ is a limit point of A. On the other hand, if $m_0 < 34$, the open set

$$A_{m_0} = \{m_0, m_0 + 1, m_0 + 2, ...\}$$

contains no point of A different from m_0. Hence $m_0 > 34$ is not a limit point of A.

D (A) = (1, 2, 3, ..., 33, 34}

(ii) Let G be any infinite subset of N. Then G is not bounded from above. Hence if p be any point of N, then every open set containing p must contain points of G other than p.

Hence **D** (G) = N.

Note that if F is a finite subset of N bounded above by n_0 then the open set $A^*_{n_{0+1}}$ contains no point of F. Hence $n_0 + 1 \in N$ is not a limit point of F and so N (F) ≠ N.

(iii) The only nhd of 1 is N which contains n_0 point of {1} other then 1. hence 1 is not a $\mathfrak{I}$-limit point of {1}.

Also if p is any point greater then, then nhd A_p of p contains no point of {1} and so p > 1, is not a limit point of {1},. Hence no point of N can be a limit point of {1}. Thus

$$D(\{1\}) = \varnothing.$$

Example 19:

Explain how 'neighbourhood' can be used as primitive notion to define a topology on a set.

Solution:

See the above theorem.

Example 20:

Let X be a non-empty set and for each point x ∈ X, let **N** *(x) denote the collection of subset of X containing x. Show that N (x) satisfies all the conditions and determine the induced topology on X.*

Solution:

[N0] Since Y is non-empty, there exists a point x ∈ X. Since X contains x, we have

$$X \in N(x) \text{ and so, } N(x) \neq \varnothing,$$

[N1] : By hypothesis, x belongs to each member of N (x), that is N ∈ N (x) ⇒ x ∈ N.

[N2] : Let N ∈**N** (x) and let M ⊃ N. Then by definition of N (x), x ∈ N. Since M ⊃ N, we have x ∈ M. Hence M ∈ N (x). Thus we have shown that

$$N \in N(x),\ M \supset N$$

$$\Rightarrow \quad M \in N(x)$$

[N3] : Let M, N ∈ N (x). Then x ∈ M and x ∈N and so x ∈ M ∩ N. Hence M ∩ N ∈ N (x).

[N4] : Let N ∈ N (x). By definition of N (x), N has the property that N ∈ N (x) for every x ∈ N and so [N4] is also satisfied.

Now we fine dht topology $\mathfrak{I}$. A subset G of X is opening the induced topology iff G $\in$ N (x) for every x $\in$ G.

Since by def. of N (x) every subset of X has this property, the induced topology of X is the discrete topology.

Example 21:

Let X = {a, b, c} and let

$$N(a) = \{\{a\}, \{a, b\}, \{a, c\}, Y\}$$

$$N(b) = \{\{a, b\}, X\} \; N(c) = \{\{a, c\}, X\}$$

(i) Show that N (a), N (b) and N (c) satisfy all the conditions.

(ii) Let $\mathfrak{I}$ be the topology for X induced in the sense by N (a), N (b) and N (c). Find all $\mathfrak{I}$-open subsets of X.

(iii) Find all the $\mathfrak{I}$-nhds of a, b and c and verify that they coincide with N (a), N (b), N (c) respectively,

Solution:

(ii) It is easy to see that the conditions [N0], [N2] and [N3] are satisfied by N (a), N (b), and N (c). The condition [N4] is also satisfied. Take from example, {a, b} $\in$ N (a). Then {a} $\in$ N (a) is such that {a} $\subset$ {a, b} and {a} $\in$ N (y) $\vee$ y $\in$ {a}. Note that there is only one y $\in$ {a}, namely a. Again take {a, c} $\in$ N (c). Then {a, c} $\subset$ {a, c} such that {a, c} $\in$ N (a) and {a, c} $\in$ N (c) so that {a, c} $\in$ N (y) $\vee$ y $\in$ {a, c}. The reader is advised to verify [N4] for other members of N (a), N (b) and N (c).

(ii) Now we select the members of $\mathfrak{I}$ as follows.

G $\in$ $\mathfrak{I}$ iff G $\in$ N (x) $\vee$ x $\in$ G.

Then $\varnothing \in \mathfrak{I}$ since the statement $\varnothing \in$ N (x) $\vee$ x $\in \varnothing$ is trivially true.

{a $\vee$ x $\in$ {a},} $\in \mathfrak{I}$ since {a} $\in$ N (a) $\vee$ x $\in$ {a}

{b} $\notin \mathfrak{I}$ since {b} $\notin$ N (b).

{c} $\notin \mathfrak{I}$ since {c} $\notin$ N (c.)

{a, b} $\in \mathfrak{I}$ since {a, b} $\in$ N {a} and {a, b} $\in$ N {b}.

{b, c} $\notin \mathfrak{I}$ since {b, c} $\notin$ N (b)

{a, c} $\in \mathfrak{I}$ since {a, c} $\in$ N (a) and {a, c} $\in$ N {c}.

X $\in \mathfrak{I}$ since X $\in$ N (x) $\vee$ x $\in$ X.

Hence $\mathfrak{I}$ = {{$\varnothing$, {a}, {a, b}, {a, c}, X}

(iii) Let N* (a), N* (b), N* (c) denote the $\mathfrak{I}$-nhds of a, b, c respectively. Then

N* (a) = {{a,}, {a, b}, {a, c}, X}, N* (b) = {{a, b}, X},

N* (c) = {{a, c}, X}

Here N* (a) = N* (a), N* (b) = N (b) and N* (c) = N (c).

Another way of generating a topology is given by the following theorem.

Example 22:

Let X = {a, b, c} and let

N (a) = {{a}, {a, c}}, N (b) = {{b}, {,b, c}}, N (c) = {X}

(i) Show that N (a), N (b) and N (c) satisfy the conditions [M1], [M2] of the preceding theorem.

(ii) if ℑ is the topology for X induced (in the sense of the preceding theorem) by N (a), N (b) and N (c), then all ℑ open subsets of X.

(iii) Find all ℑ closed subsets of X.

(iv) Is (Y, ℑ) a door space ?

Solution:

(i) a ∈ {a} and a ∈ {a, c}

Also {a} ∩ {a, c} = {a} ∈ N (a).

Hence N (a) satisfies [M1] and [M2]. Similarly it can be shown that N (b) and N (c) satisfy [M1] and [M2].

(ii) Now we select the members of ℑ as follows :

{a} ∈ ℑ since {a} ∈ B (a) st. a ∈ {a} ⊂ {a}.

{b} ∈ ℑ since {{b} since {b} st. b ∈ {b}.

{c} ∉ ℑ since there is no N ∈ N (c) st. c ∈ N ⊂ {c}.

{a, b} ∈ ℑ since {a} ∈ N (a) st. ⊂ {a, b} and

{b} ∈ N (b) s.t. {b} ⊂ {a, b}

{b, c} ∉ ℑ since c ∈ {b, c} but there is no N ∈ N (c) st. Similarly {c, a}

X ∈ ℑ since ∨ x ∈ X. there is an N ∈ N (x).

s.t. x ∈ N ⊂ Y.

Hence ℑ = {∅, {a}, {b}, {a, b}, X}

(iii) ℑ-closed subset of X are.

X, {b, c}, {c, a} {c}, ∅.

(iv) Since every subset of X is either ℑ-open or ℑ-closed, it follows that (X, ℑ) is a door space.

Example 23:

Which of the following subsets of R are U-neighbourhoods of 1 ?

(i) [0, 2[, (ii) [0, 2], (iii) [1, 2],

(iv)]1, 2],(v) [1, 2[, (vi) [0, 2] $- \left\{1\frac{1}{2}\right\}$

(vii) $[0,2]-\left\{\frac{1}{2}\right\}$ *(viii) R.*

Solution:

(*i*) Since]0, 2[is U-open set containing 1, it follows that]0, 2[is U-nhd of 1.

(ii)]0, 2] is U-nhd of 1 since]0, 2[is U-open set such that 1 ∈]0, 2 [⊂] [0, 2].

(iii) [1, 2] is not a U-nhd of 2 since there is no U-open set contained in [1, 2] and containing 1.

(iv)]1, 2] is not a U-nhd of 1 since]1, 2] does not contain 1.

(v) [1, 2[is not a U-nhd of 1 since there exits no U-open set G such that 1 ∈ G ⊂ [1, 2[.

(vi) [0, 2] $- \left\{1\frac{1}{2}\right\}$ is a U-nhd of 1 since [0, $1\frac{1}{4}$ [is a U-open set such that

$$1 \in -, 1\frac{1}{4} [\subset [0, 2] - \left\{1\frac{1}{2}\right\}.$$

(*vii*) [0, 2] $- \left\{\frac{1}{2}\right\}$ is a U-nhd of 1 since] $- -\frac{3}{4}$, 2 [is a U-open set *i.e,*

$$1 \in \left]\frac{3}{4}, 2\right[\subset [0,2]-\left\{\frac{1}{2}\right\}.$$

Example 24:

Let X be a topological space. If FD is a closed subset of X and x ∈ F', prove that there exists a nhd N of x such that N ∩ F = ∅.

Solution:

Since F is closed, Fℑ is open and Fℑ contains a nhd of each its points. Hence there exits a nhd N of, x such that

$$N \subset F' \text{ i.e., } N \cap F = \varnothing.$$

Properties of neighbourhoods : We now prove several essential properties of neighbourhoods which are usually known *Neighbourhood axioms.*

Example 25:

A topological space $(X, \Im)$ it said to be a door space iff every subset of X is either open or closed. Give an example of a door space.

Solution:

Let $X = \{a, b, c\}$ and let

$\Im = \{\varnothing, \{b\}, \{a, b\}, \{b, c\}, X\}$.

Then closed sets are

$X, \{a, c\}, \{c\}, \{a\}, \varnothing$

Hence all the subsets of X are either closed or open and consequently $(X, \Im)$ is a door space.

Example 26:

If $a \in R$, show that $\{a\}$ is a closed set in the usual topology for R.

Solution:

$\{a\}' =]-\infty, a[\cup], a, \infty[$. But $]-\infty$ a[and]a, ∞[are U-open sets as the reader can easily show. Hence their union is also U-open. It follows that $\{a\}'$ is U-open. Therefore $\{a\}$ is U closed.

Example 27:

Let a, $b \in R$ where $a < b$. Show that the closed interval [a, b] is closed in the usual topology for R.

Solution:

$[a, b]' = \{x \in R : x < a \text{ or } x > b\}$.

$=]-\infty, a [\cup] b, \infty [$

which is U-open, being the union of two U-open sets.

Hence [a, b] is U closed.

Example 28:

Show that the right half-open interval [a, b[is neither U-closed nor U-open. But it is both S-open and S-closed.

Solution:

Do yourself

Solution:

$[a, b['=]-\infty, a [\cup [b, \infty[$. ...(1)

Since there exists no U-open interval containing b and contained in [a, b[', is follows that [a, b[' is not U-open. Hence its complement [a, b[is not U-closed. We also see that [a, b[is not U-open since there exist not U-open interval containing a and contained in [a, b[. Thus [a, b[is neither U-closed nor U-open. Similarly the left half-open interval]a, b] is neither U-closed nor U-open.

[a, b[is S-open by definition. Since [–∞, a [and [b, ∞ [are both S open, we see from (1) that [a, b[' is S-open and so [a, b[is S-closed.

Example 29:

Let (X, ℑ) be a co-finite space. Since ℑ-open sets are those whose complements are finite, it follows that all finite subsets of X together with X are ℑ-closed.

Example 30:

Give two examples of a proper non-empty subset of a topological space such that it is both open and closed and prove your assertion.

Solution:

Consider the topology

$$\mathfrak{I} = \{\varnothing, \{a\}, \{b\}, \{c\}, \{a, b\}, \{b, c\}, \{c, a\}, X\}$$

on $X = \{a, b, c\}$

Here the sets {a}, {b}, {c}, {a, b}, {b, c}, {c, a} are all both open and closed. The set {a} is closed since it is the complement of {b, c} etc. In fact, in every discrete topological space (X, **D**) consisting of more than one element, every subset of X is both open and closed. As another example consider the topology.

$$\mathfrak{I} = \{\varnothing, \{a\}, \{c, d\}, \{a, c, d\}, \{b, c, d, e\}, X\} \text{ on } Y = \{a, b, c, d, e\}.$$

Here {a} is open. Also {a} is closed since it is the complement of the open set {b, c, d, e}.

Example 31:

Give an example to show that the union of an infinite collection of closed sets in a topological space is not necessarily closed.

Solution:

Let (R, U) be the usual topological space and let fn = [1/n, 1], n ∈ N so that Fn is a closed interval of R.

$$\cup \{Fn : n \in N\} = \{1\} \cup \left[\frac{1}{2}, 1\right] \cup \left[\frac{1}{4}, 1\right] \ldots =]0, 1[.$$

Since]0, 1] is not closed, it follows that the union of an finite collection of closed sets is not necessarily closed.

Example 32:

Show that every finite subset of R is a U-closed set.

Solution:

We have shown that every singleton {a} is a U-closed subset of R. Since a finite subset of R is the union of a finite number of singletons, it follows every finite subset of R is U closed.

Example 33:

Give an example of a topological space different from the discrete and indiscrete spaces in which open sets are exactly the same as closed sets.

Solution:

Let X = R (the set of reals) and let $\mathfrak{I}$ consist of the empty set and all those subsets G of R having the property that $x \in G$ implies $-x \in G$. Then (R, $\mathfrak{I}$) is the required space.

We first prove that $\mathfrak{I}$ is actually a topology for R. We have

[T1] : $\varnothing \in \mathfrak{I}$ by definition. Also $R \in \mathfrak{I}$ since

$x \in R \Rightarrow -x \in R$

[T2] : Let $G_1 \in \mathfrak{I}$ and $G_2 \in \mathfrak{I}$ and let $x \in G_1 \cap G_2$. But $x \in G_1 \cap G_2$

$\Rightarrow x \in G_1$ and $x \in G_2$.

$\Rightarrow -x \in G_1$ and $-x \in G_2$ [By def. on $\mathfrak{I}$]

$\Rightarrow -x \in G_1 \cap G_2$

Hence $G_1 \cap G_2 \in \mathfrak{I}$

[T3] : Let $G_1 \in \mathfrak{I}$ for all $\lambda \in L$ let $x \in \cup \{G_\lambda : \lambda \in \Lambda\}$

Now $x \in \cup \{G_\lambda : \lambda \in \Lambda\} \Rightarrow x \in G_\lambda$ for some $\lambda \in \Lambda$

$\Rightarrow -x \in G_\lambda$ [$\because G_1 \in \mathfrak{I}$]

$\Rightarrow -x \in \cup \{G_\lambda : \lambda \in \Lambda\}$.

Hence $\cup \{G_\lambda : \lambda \in \Lambda\} \in \mathfrak{I}$.

It follows that $\mathfrak{I}$ is a topology for R. We now show that every open set is closed. Let G be any open set. We shall show that G is also closed, that is, R – G is $\mathfrak{I}$ open. We have

$$x \in R - G \Rightarrow x \notin G$$
$$\Rightarrow -x \notin G \quad [\because G \in \mathfrak{I}]$$
$$\Rightarrow -x \in -G.$$

It follows that R – G is open and so G is closed.

Similarly it can be shown that every closed set is $\mathfrak{I}$-open.

Example 34:

Let $\mathfrak{I}$ and $\mathfrak{I}'$ be two topological spaces on a set X such that $\mathfrak{I}'$ is coarser then $\mathfrak{I}$ and let x be a $\mathfrak{I}$ cluster point of a filter F on X.

Then show that x is also a $\mathfrak{I}'$ cluster point of F.

Solution:

Let N (x) and N′ (x) be the nhd filters of x relative to $\mathfrak{I}$ and $\mathfrak{I}'$ respectively. Since $\mathfrak{I}' \subset \mathfrak{I}$, it follows that $N'(x) \subset N(x)$. Hence x is a cluster point of F relative to $\mathfrak{I}$.

$\Leftrightarrow$ every member of N (x) intersects every member of F.

$\Leftrightarrow$ every member of N′ (x) intersects every member of F.

$\Leftrightarrow$ x is a cluster point of F relative to $\mathfrak{I}'$.

Example 35(a):

Show that every discrete space is totally disconnected.

Solution:

Let (X, D) be any discrete space and let x, y be any two distinct points of X. Then A = {x} and B = X – {x} are both non-empty open disjoint sets whose union is X such that $x \in A$ and $y \in B$. It follows that (X, D) is totally disconnected.

Example 35(b):

Given two examples of locally connected spaces which are not connected.

Solution:

(i) let X be a discrete space containing more than one point. Then X is disconnected. For it $x \in X$, then {x} is a non-empty proper subset of X which is both open and closed in X. X is locally connected.

(ii) Let Y be the union of two disjoint open intervals on the real line, that is, let Y =]a, b[$\cup$] c, d[where $a < b \leq c < d$. Then Y as a subspace of (R, U) is locally connected but not connected as shown below. Since [a, b[and [c, d[are open in R, they are also open in Y.

Thus Y is the union of two non-empty disjoint sets both open in Y and therefore Y is disconnected. To show that Y is locally connected, let x be any point of Y and G_x any open nhd of x in Y. Then we can always find $\varepsilon > 0$ such that $]x - \varepsilon, x + \varepsilon[\subset G_x$. But $]x - \varepsilon, x + \varepsilon[$ is connected in **R** and hence connected in Y.

Thus every open nhd of x in Y contains an open connected nhd of x in Y and therefore Y is locally connected.

Example 35(c):

Give an example of a space which is connected but not locally connected.

Solution:

Consider the subspace Y of the Euclidean plane **R**2 defined by $Y = A \cup B$ where

$$A = \{(0, y)\} : y \in [-1, 1]\} \text{ and}$$

$$B = \{(x, y) : x \in]0, 1] \text{ and } y = 1/x\}$$

Now B is the image of the interval]0, 1[under the continuous mapping f defined by $f(x) = x \sin(1/x)$. Since]0, 1] is connected, B is also connected. It is easy to see that every nhd of a point $x \in A$ contains an infinite number of points of B. Hence every point of A is a limit point of B. Also no point outside $A \cup B$ can be a limit point of B. Hence $\overline{B} = A \cup B = Y$.

It then follows that $\overline{B}$ i.e., Y is connected. However, X is not locally connected; since it can be easily shown that each point $x \in A$ has nhd which does not contain any connected nhd of x.

Remark : *Examples 3 and 4 show that local connectedness neither implies nor implied by connectedness.*

Example 35(d):

Show that the space (R, S) is totally disconnected.

Solution:

Let x, y be any two distinct real numbers and let x < y. Let $A =]-\infty, y[$ and $B = [y, \infty[$. Then A, being U-open is also S-open since S is finer than U. Also B is S-open. Since x < y, we have $x \in A$. Also $y \in B$. Further $A \cup B = R$ and $A \cap B = \varnothing$. Hence $A \cup B$ is a disconnection of R with $x \in A$ and $y \in B$. It follows that (R, S) is totally disconnected.

Example 36(a):

Show that *the set Q of all rational numbers with the relativized usual topology is totally disconnected.*

Solution:

Let x, y be any two distinct rational numbers and let $x < y$. Then there exists an irrational number p such that $x < p < y$. Let A = Q ∩] –∞, p [and B = Q ∩], ∞[.

Then A ∪ B = (Q ∩) –∞,

p [) ∪ (Q ∩]p, ∞[) = Q

Also A ∩ B = ∅. Hence A ∪ B is a disconnection of Q with x ∈ A and y ∈ B. It follows that Q is totally disconnected.

Example 3(b):

Show that the set of all irrational numbers with the relativized usual topology is totally disconnected.

Solution:

Solution of this problem is on the same lines as that of the preceding example since between any two irrational numbers there is a rational number.

EXERCISES

1. Show that a space X is compact iff every filter on X has a cluster point.
2. Show that a filter F on a space X has x as a cluster point iff there is a filter finer than F which con verges to x.
3. Let X be a non-empty set and let F be a filter on X. Show that the set of all limit points of X is the intersection of the sets of cluster points of those filters on X which are finer than F, that is.

   lim, (F) = ∩ {adh (G) : G ⊃ F}.